Degradation, Stabilization, and Recycling of Packaging Materials

Degradation, Stabilization, and Recycling of Packaging Materials

Muhammad Rabnawaz
Michigan State University
East Lansing, Michigan, United States

Susan E.M. Selke
Michigan State University
East Lansing, Michigan, United States

Ian Wyman
Michigan State University
East Lansing, Michigan, United States

WILEY

Published by John Wiley & Sons, Inc., Hoboken, New Jersey.
Published simultaneously in Canada.

Library of Congress Cataloging-in-Publication Data Applied for:

Hardback ISBN: 9781394294268
Epdf: 9781394294282
Epub: 9781394294275
Obook: 9781394294299

Cover Design: Wiley
Cover Image: © hh5800/Getty Images

Set in 9.5/12.5pt STIXTwoText by Straive, Pondicherry, India

Contents

Preface

This book is a critical resource for researchers, students, and industry professionals in the field of materials science and packaging. While this book is aimed at packaging students/professionals, it will equally benefit students and professionals working in fields such as automotive, construction, furniture, and other related sectors. It offers a detailed analysis of various packaging materials and evaluates efforts to enhance their sustainability.

Organized into two parts, the first section (Chapters 2–6) of this book provides a comprehensive examination of how environmental factors such as heat, shear, light, air, packaged products, and stress affect packaging materials, focusing on the chemistry of their deterioration and stabilization methods. The second section (Chapters 7–11) explores solid waste management, recent developments in recycling, and key principles of eco-friendly packaging design, culminating in an extensive discussion of legal and regulatory aspects. This book presents advanced technical knowledge that demystifies the sustainable packaging landscape, equipping readers with technological and policy advancements, and is written with the aim of providing a valuable asset that will promote a more sustainable future.

Readers will gain the ability to:

- Grasp the complex interactions that exist between packaging materials and various environments, appreciating their practical relevance.
- Comprehend and articulate plastic degradation processes and have a firm understanding of methods to address these degradation processes.
- Acquire insights into the degradation behaviors of wood, paper, glass, and metal, and suggest specific stabilization strategies for each of these materials.
- Evaluate challenges related to solid waste, devise effective disposal strategies, and advocate for recycling and waste minimization in current waste management practices.

- Evaluate the latest developments in plastic recycling, demonstrating an understanding of contemporary issues and their solutions.
- Apply new legislation as a framework for implementing and promoting the use of sustainable packaging.

Book Outline

Chapter 1. Introduction
Chapter 2. Plastics: Stabilization and Degradation
Chapter 3. Wood: Degradation and Stabilization
Chapter 4. Paper/Paperboard: Degradation and Stabilization
Chapter 5. Glass: Degradation and Stabilization
Chapter 6. Metal: Degradation and Stabilization
Chapter 7. Solid Waste Issues
Chapter 8. Recycling of Metal and Glass
Chapter 9. Paper and Paperboard Recycling
Chapter 10. Plastics Recycling
Chapter 11. Legal, Regulatory, EPR, and Green Design

1

Introduction

This book is the only book that offers comprehensive insights into all packaging materials, how packaging materials interact with their surroundings, and how they are affected by their environments, such as packaged products, heat, light, air, stress, and so on. The book also covers strategies used to stabilize these materials against their environment. In addition, the impact of stabilizing strategies on recycling is also discussed. Other topics include solid waste management, recycling, and strategies for reducing waste. The book concludes with a focus on advancements in plastic recycling, packaging-related laws, and principles of eco-friendly packaging design.

In this chapter, we will provide a brief introduction to packaging sustainability issues. Subsequently, Chapters 2–6 will take a deeper look at individual packaging materials, their advantages and disadvantages, as well as the sustainability issues that should be considered for each. In particular, packaging materials that will be discussed in these chapters will include plastics (Chapter 2), wood (Chapter 3), paper (Chapter 4), glass (Chapter 5), and metals (Chapter 6). The degradation pathways and stabilization strategies of some of these materials will also be discussed. Issues relating to packaging's impact on municipal solid waste systems and how this waste is handled, and how issues are addressed will be discussed in Chapter 7. Subsequently, the recycling of packaging materials will be discussed in Chapters 8–10, with the recycling of metals and glass being covered in Chapter 8, paper and paperboard recycling being the focus of Chapter 9, and plastics recycling being discussed in Chapter 10. Last, Chapter 11 will discuss packaging policies, extended producer responsibility (EPR), legal frameworks, and other policies, as well as green design principles and how they are being used to strengthen, incentivize, mandate, and promote packaging sustainability. It is hoped that this book will provide valuable insight into the landscape, challenges, and opportunities in sustainable packaging.

Degradation, Stabilization, and Recycling of Packaging Materials, First Edition.
Muhammad Rabnawaz, Susan E.M. Selke, and Ian Wyman.

1.1 General Introduction

The packaging sector has a market value of 1.1 trillion USD globally as of 2023, which is projected to increase [1]. Therefore, it is evident that the packaging sector has a key influence on the economy and the environment. The vast majority of retail items sold today come in some form of packaging, whether these goods be food, medicines, cosmetics, consumer electronics, or other merchandise. Given the consumable nature of many of these products (especially food), it becomes readily apparent why packaging has such an impact on our everyday life, the economy, and the environment. In this chapter, we will briefly introduce some general concepts relating to packaging, which will be followed by a more detailed discussion in the subsequent chapters.

As we begin to look into packaging, it will be helpful to discuss some general concepts as well as some terms that are relevant to this topic.

1.2 What Are Some Ideal Properties of Packaging?

There are a number of roles that packaging fulfils, such as protecting goods against damage (including impacts in some cases), helping to maintain freshness, preventing contamination and spoilage of the product, conveying information about the product to consumers (this is usually done through labels, which can either be added onto the package or are printed directly on the package), as well as attracting the attention of consumers to help promote sales.

Depending on the application and the goods being packaged, a number of properties can be desirable and should be considered before a new package is selected. One of these properties includes its ability to prevent the passage of water vapor, oxygen, or other gases into the product (these are often referred to as barrier properties). Another consideration to take into account is the water or grease resistance of a packaging material, mainly if the product is susceptible to damage or contamination by water or oils. The weight of the packaging material is another important consideration, especially if the goods are likely to be shipped over long distances. The mechanical properties of a particular format can also be an important consideration, as in some cases, flexibility may be desirable, while rigidity may be preferable in other circumstances. The raw materials required to produce the packaging will also be an important factor in many cases. Ideally, these raw materials should be inexpensive and readily available. Due to environmental concerns, there is a trend toward seeking renewable biobased materials rather than those that are derived from nonrenewable petroleum. Alternatively, it is also highly desirable from a sustainability perspective to utilize recycled materials as feedstocks or raw materials to produce new packaging. Another consideration is the recyclability of the packaging at the end of its service lifetime, as some materials are easier to recycle than others.

A given material will have certain inherent advantages and drawbacks, depending on the intended packaging application. For example, plastics are relatively lightweight and tend to have good barrier properties. On the other hand, plastics are often derived from petroleum-based resources (although biobased plastics are also available), and they tend to break down into microplastics, which can proliferate in the environment. Meanwhile, paper and paperboard packaging is highly desirable from a sustainability perspective but has poor water and grease resistance and offers only very limited barrier protection in the absence of coatings. These are a few examples of packaging materials, but we will discuss all this in greater depth in Chapters 2–5.

1.3 Liquid Resistance and Barrier Properties

As noted above, the barrier properties, as well as the oil and grease resistance of a packaging material, are often an important consideration when one is designing or selecting a new packaging material or format. Although these properties may not seem to directly impact the sustainability of packaging material, they are key criteria that are often required in order for the packaging to effectively fulfill its purpose of protecting the packaged goods. In addition, these properties do have some implications toward sustainability as the packaging with good liquid resistance and barrier properties can help to prevent food spoilage or damage to the packaged goods, thereby helping to minimize waste. Meanwhile, materials that do not meet these criteria will not be suitable candidates for new packaging, even if they are highly sustainable.

Some tests that are performed to evaluate liquid resistance include Cobb tests [2], kit rating tests, as well as contact angle measurements. Cobb tests are employed to measure the amount of water that has been absorbed by paper and paperboard over a certain timespan (typically 60, 180, or 1800 seconds) in units of g/m^2. Meanwhile, kit rating tests are typically employed to evaluate the oil resistance of a given surface and have ratings in the range of 0–12, with a kit rating of 12 corresponding to the highest kit rating on this scale [3]. These tests are performed with a series of liquids with different viscosities and surface tensions (or "aggressiveness"), which have kit ratings in the range of 1–12, with the liquid with the kit number of 1 being the least aggressive (least likely to stain the paper) and the liquid with the kit number of 12 being the most aggressive. The kit rating of a paper substrate is assigned to the kit number of the liquid with the highest kit number that did not leave a stain. There may seem to be a slight discrepancy between the kit ratings of a sample (0–12) and the kit numbers of the test liquids [1–11]. This situation exists because a sample with a kit rating of 0 would have become stained by the least aggressive test liquid with a kit number of 1 and thus would be considered to have "failed" the kit rating test with that liquid, thus resulting in the sample being assigned a kit rating of 0. Meanwhile, a high contact angle can suggest that the liquid has resistance against that particular liquid. The contact angle can be influenced by various properties of an underlying solid, such as its surface energy and roughness. Some ways through which gas barrier properties are measured include the water vapor transmission rate, oxygen transmission rate, as well as carbon dioxide transmission rate (CO_2TR) [4], which are the respective rates at which water vapor, oxygen, and carbon dioxide permeate through a material. Lower transmission rates would correspond to higher gas barrier properties. While high-barrier properties are often desirable, in some cases a certain degree of breathability may be desired, such as with the packaging of produce.

1.4 End-of-Life (EoL) Outcomes

The EoL scenario or outcome refers to the eventual fate of a product. In the context of this book, the EoL outcomes being focused on are packaging materials, or used packaging. Before environmental issues became a focus of concern, there were fewer EoL scenarios for packaging than there are today. In the past, some of the main outcomes for used packaging were disposal at landfills, littering, or in some cases they were simply burned. In situations where a material was reused or recycled in the past, this was often done primarily for economic considerations rather than environmental. The range of EoL scenarios has expanded greatly in recent years, largely due to the push for sustainable packaging. From an environmental perspective, some EoL outcomes are much more desirable than others. For example, landfilling is a rather wasteful EoL outcome, as the packaging materials that are sent to landfills are no longer used as feedstock or source material for subsequent use. In addition, waste in landfills can, in some cases, leach into the nearby ecosystem. While landfilling is still in use, efforts have been underway to minimize reliance on landfilling as an EoL outcome. Recycling efforts have grown significantly in recent years. Efforts have been underway to adopt biodegradable or compostable packaging, which can break down either in the environment or under composting conditions, respectively.

1.5 Life-Cycle Assessment (LCA) and Techno-Economic Analysis (TEA)

LCA is an evaluation or analysis that takes the full life-cycle of an item (such as packaging) into account [5, 6], with a particular emphasis on its environmental impacts throughout its full life-cycle. While EoL scenarios refer to the final outcomes for packaging materials or waste at the end of their service lifetime, the LCA is much broader in scope, and takes into account the full lifetime of a material. This can span from obtaining the raw materials or feedstocks used to manufacture the package, to production (and even including the source of the feedstocks used to produce a given packaging product), use of that product, as well as its EoL outcomes. It also takes factors such as the impacts arising from the transportation of that material into account, as well as energy requirements needed for production as well as processes, such as recycling. Before a new packaging product is developed and manufactured, it is important for manufacturers to conduct a detailed LCA to evaluate the environmental costs that will arise over the full lifespan of that product.

TEA is another evaluation that is primarily directed toward the manufacturing phases of production, but also downstream or upstream phases can be included [7]. While LCA focuses more on the environmental aspects, TEA is often used to evaluate the economic and technical performance of a process of product, such as a form of packaging [7]. Both LCA and TEA can have critical roles in ensuring that a new packaging format is environmentally sustainable as well as viable from economics and technology perspectives.

1.6 Open-Looped Versus Closed-Loop Processes

An open-loop system or process is not fully self-contained, with either new inputs or materials (often through the extraction of virgin materials) [8] being needed for the production process, or some of the materials degrading with repeated use so that they are eventually disposed of and leave the production loop. It is noteworthy that although recycling is a key tool for achieving environmental sustainability, in many cases even recycling is not a fully closed-loop process. For recycling to be a truly closed-loop system, the material should be recycled for an infinite number of cycles. In the case of open-loop recycling, the material is not recycled indefinitely and eventually leaves the cycle. In addition to avoiding the extraction of virgin materials, factors such as waste generation, water usage [9], and energy inputs must be considered in order to achieve a truly closed-loop system.

In addition, the terms open-loop and closed-loop are used to refer to systems in which an item is recycled into either a different item, or alternatively into the same item. In this context, the term open-loop refers to a system in which an item is recycled into a different product, which often is of a lower grade than the original product. In the case of closed-loop recycling the item is recycled into the same type of product (bottle-to-bottle recycling would be one such example).

1.7 Recycling

As mentioned above, achieving a closed-loop system is a key requirement for developing sustainable packaging. One of the most important ways to achieve sustainability is through recycling efforts. While recycling will be discussed in further depth in Chapters 7–9, we will briefly introduce a couple of concepts regarding recycling here. There are various types of recycling, and some materials are inherently more recyclable than others (we note that the recycling of materials such as paper, glass, and metals are discussed in Chapter 8, while plastics recycling is discussed in Chapter 9). Two of the main forms of plastic recycling are mechanical and chemical recycling. In the case of mechanical recycling [10], the used packaging material does not undergo significant chemical changes. In contrast, the material is broken down chemically into other materials, such as feedstock materials, when it is subjected to chemical recycling (in the case of plastic recycling, the polymer may be broken down into its monomers, for example). Some examples of chemical recycling processes [11] can include pyrolysis, chemolysis, fluid catalytic cracking, gasification, as well as hydrogen techniques. While this provides for the recovery of feedstocks, in some cases, the breakdown of material into fuel for use as an energy source is sometimes considered to be a form of chemical recycling.

Plastic recycling can also be divided into four categories known as primary, secondary, tertiary, and quaternary recycling [10]. Both primary and secondary recycling are forms of mechanical recycling. In the case of primary recycling, here postindustrial scrap is reused for the same purpose by grinding and reprocessing (e.g., bottle to bottle). More processing is involved for secondary recycling than for primary recycling, here postconsumer materials are first sorted, the size of the polymer materials is reduced (such as via grinding processes), and then extrusion is performed. Meanwhile, tertiary recycling is a form of chemical recycling, and is used for materials that are no longer suitable for mechanical recycling. In the case of plastics, the polymer can be reverted into its feedstock monomers. Last, quaternary recycling is used for energy recovery purposes, and is considered to be a lower-value recovery process compared to tertiary recycling. In addition, greenhouse gas generation may become a more significant issue in quaternary recycling cases.

Depending on the recycling process involved or its effectiveness, a recycled material may be converted into something of greater, equal, or lesser value. The conversion of a waste material into a value-added material that has greater value is referred to as upcycling, while the conversion of a recycled material into a material with less value is referred to as downcycling.

1.8 Biodegradable and Compostable Packaging

As mentioned earlier, the EoL outcomes for packaging materials are an important consideration. Ideally, the packaging materials should be reused or recycled as much as possible. However, a certain amount of packaging waste will inevitably enter the environment, whether it occurs through spills, littering, or ending up in landfills. In these situations, it is important that the material should degrade in an acceptable manner without proliferating in the environment. One way through which this can be accomplished is by ensuring that the packaging is biodegradable or compostable. There is a deeper discussion regarding the definitions of biodegradable and compostable in Sections 2.10.2 and 7.3.2. The two terms sound quite similar, but there is a subtle difference between them. It should be noted that the term biodegradable is often misused. In a broad sense, the term biodegradable simply means that the bonds that hold material together (such as in plastic-, paper-, and wood-based products) can be broken via microbial activity. In a practical case, this biological degradation should occur completely at a somewhat rapid pace, but the conditions determining whether a material is considered to be biodegradable for practical use are often ill-defined. Meanwhile, a compostable material is meant to biodegrade under specific conditions, where thermophilic conditions are achieved in a certain period. For a product to be labeled as compostable in the United States, it must satisfy the American Society for Testing and Materials (ASTM) standards D6400 and D6868, which stipulate that the material must disintegrate within 90 days without leaving harmful residue [12]. Composting criteria demand proper disintegration during composting and no adverse impacts of the compost on plant growth. Home composting is materials that decompose at natural temperatures with common microorganisms. Currently, there are no home composting ASTM or BPI standards. Only the Australian AS 5810 standard, used for TUV Austria's OK compost HOME certification, demands disintegration in six months and complete composting in a year. Thus, AS 5810 is a key indicator for home compostable products [13]. ASTM's D20 committee on plastics is working on a test method for home composting. The D20 committee is also working with D34 (waste management committee) to develop tests for the degradation scenarios when biodegradable materials are in landfills [14].

1.9 Concluding Remarks

Ideally, a well-chosen packaging material should combine several sustainability criteria, such as recyclability, renewability, cost-effectiveness, and biodegradability/compostability. While achieving environmental sustainability is a key goal, it remains vital that the packaging is still able to fulfill its function in a highly effective manner. The choice of suitable material has a critical influence on whether the packaging will effectively fulfill its function in a sustainable manner. In addition, economics is an important factor, as sustainable packaging will not be developed if its cost is too high, and thus its economic viability is another important consideration. The upcoming chapters will provide a detailed look at the packaging comprised of different materials, the recycling considerations and processes used for these packaging formats, as well as the role that policies fulfill in the development of sustainable packaging materials. Hopefully, this book will allow students, researchers, industry professionals, and decision-makers in both industry and government to successfully navigate the challenging yet very critical field of sustainable packaging.

References

1 Mordor Intelligence. (2023). Packaging market size & share analysis - growth trends & forecasts (2023–2028). https://www.mordorintelligence.com/industry-reports/global-packaging-market (accessed 14 December 2023).

2 Smithers. (2024). Water absorbency. https://www.smithers.com/industries/packaging/manufacturers-and-users/packaging-materials-testing/paper-testing-other-properties/water-absorbency (accessed 19 November 2024).

3 Applied Paper Technology, Inc. Grease resistance – kit test. https://appliedpapertech.com/test-methods/grease-resistance/ (accessed 19 November 2024).

4 Brandenburg, J.S. (2020). Chapter 10 - Packaging design: functions and materials. In: *Controlled and Modified Atmospheres for Fresh and Fresh-Cut Produce* (ed. M.I. Gil and R. Beaudry), 185–210. Academic Press https://doi.org/10.1016/B978-0-12-804599-2.00011-9.

5 Golisano Institute for Sustainability, Rochester Institute of Technology. (2020) What is life cycle assessment (LCA)? https://www.rit.edu/sustainabilityinstitute/blog/what-life-cycle-assessment-lca (accessed 15 December 2023).

6 Quist, Z. (2024) Life cycle assessment (LCA) – complete beginner's guide. Ecochain Technologies. https://ecochain.com/blog/life-cycle-assessment-lca-guide/ (accessed 19 November 2024).

7 Mahmud, R., Moni, S.M., High, K., and Carbajales-Dale, M. (2021). Integration of techno-economic analysis and life cycle assessment for sustainable process design – a review. *J. Clean Prod.* 317: 128247.

8 Tshidimba, D. (2022). Closing the loop on the circular economy. https://www.rolandberger.com/en/Insights/Publications/Closing-the-loop-on-the-circular-economy.html (accessed 17 December 2023).

9 Tshidimba, D. (2022). Closing the loop on the circular economy. https://www.rolandberger.com/en/Insights/Publications/Closing-the-loop-on-the-circular-economy.html (accessed 19 November 2024).

10 Schyns, Z.O.G. and Shaver, M.P. (2021). Mechanical recycling of packaging plastics: a review. *Macromol. Rapid Commun.* 42 (3): e2000415.

11 Ragaert, K., Delva, L., and Van Geem, K. (2017). Mechanical and chemical recycling of solid plastic waste. *Waste Manag.* 69: 24–58.

12 Elevate Packaging. (2023). Biodegradable or compostable: what's the difference? https://elevatepackaging.com/blog/biodegradable-vs-compostable (accessed 17 December 2023).

13 BPC Instruments AB. (2024). AS 5810 Standard. https://bpcinstruments.com/as-5810-standard/ (accessed 19 November 2024).

14 Ervin, J. P. (2023). Biobased Plastic vs. Biodegradable Plastic – What's the Difference? ASTM International. https://sn.astm.org/features/bioplastics-biodegradable-plastics-standards-impact.html (accessed 19 November 2024).

2

Plastics

2.1 Introduction

Plastics have become an indispensable part of our modern lives in numerous ways, ranging from packaging to automobiles. The terms plastics and polymers are often used interchangeably, but there is a difference. In particular, "polymers" refer to large molecules made of repeating units of one or more species of atoms or groups of atoms. A couple of common examples include polyethylene (PE) and rubber. The term plastics refers to those polymers that are solid at room temperature and can be melt-processed repeatedly. Common examples of plastics include PE, polyethylene terephthalate (PET), polystyrene (PS), etc.

On the other hand, there are polymers that cannot be melt-processed owing to their crosslinked structures. Some common examples of these crosslinked materials include rubber, phenolics, and cellulose. These crosslinked structures are held together by **covalent bonds** (as is the case with rubber or phenolics) or via hydrogen bonding interactions (as is the case with cellulose). Note that there are two types of crosslinking. Physical crosslinking involves the use of noncovalent interactions, such as hydrogen bonding or ionic interactions, to form crosslinks between polymer chains. These interactions are reversible under certain conditions. On the other hand, chemical crosslinking involves the use of covalent bonds to link polymer chains together. These are permanent and are not reversible under normal conditions. Vitrimers are an emerging class of materials that belong to chemical crosslinking, but they are still reversible due to the use of dynamic reversible bonds.

Degradation, Stabilization, and Recycling of Packaging Materials, First Edition.
Muhammad Rabnawaz, Susan E.M. Selke, and Ian Wyman.

2.2 How Are Polymers Named?

Let us discuss how polymers are named. Many polymers are named simply by adding the prefix "**poly**" in front of the name of the monomer from which the polymer has been synthesized. For example, PE is a polymer that is prepared from the monomer ethylene. Likewise, polypropylene is a polymer that is prepared from the monomer propylene. As another example, PS is a polymer that is prepared from the monomer styrene.

Some polymers are named after their structures rather than the monomers that they are prepared from. For example, PET is prepared from ethylene glycol and terephthalic acid. However, it is named PET because it has ethylene and terephthalate groups within its structure.

2.2.1 Classification of Polymers

Two well-known classification systems for polymers include:

1) Classification based on the polymer structure (Carothers, 1929) [1]
2) Classification based on the mechanism (Flory, 1953) [2]

2.2.1.1 Classification Based on the Polymer Structure

According to Carothers, there are two classes of polymers based on their structures, namely, condensation and addition polymers. These types of polymers will be briefly described below.

Condensation Polymers: According to Carothers, condensation polymers are formed from polyfunctional monomers via condensation reactions involving the release of small molecules such as water, HCl, and so forth (see **Figure 2.1**). Some polymers, such as cellulose, are also condensation polymers as they can hypothetically be prepared via the condensation of beta-glucose.

Addition Polymers: These polymers are formed from monomers without the loss of small molecules. Some common examples of addition polymers include PS (**Figure 2.2**), PE, etc.

Carothers' polymer classification revolves around functional groups participating in polymerization and has overlooked the polymerization mechanism.

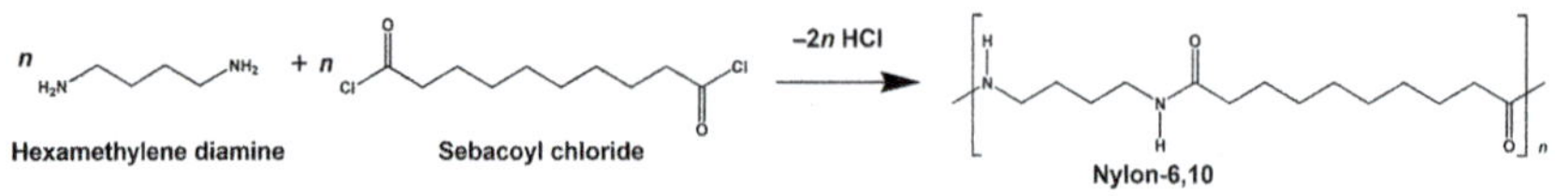

Figure 2.1 A condensation reaction leads to nylon-6,10.

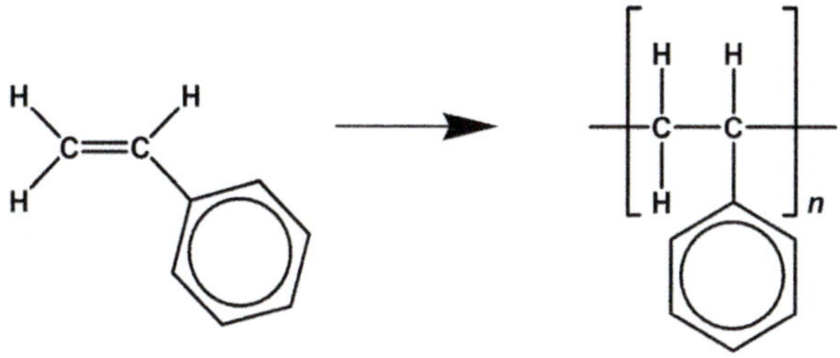

Figure 2.2 An additional polymerization leads to polystyrene, where styrene (left) undergoes free-radical vinyl polymerization to yield polystyrene (right).

2.2.1.2 Classification Based on the Mechanism

Flory classified polymers according to their **polymerization mechanisms**, and this classification system included two categories: Step growth and chain-growth polymers [2].

Step-growth polymerizations are processes that occur through a sequence of reactions, where the monomers react with other monomers to form a dimer, which then combines with another monomer to form a trimer, or with another dimer to form a tetramer. This process continues, progressively building up large polymer chains and reaches maximum polymer size when the yield approaches 100% (refer to **Figure 2.3** for more detail).

Chain-growth polymerizations result in the growth of polymer chains within a matter of minutes, such that there are substantial quantities of high molecular weight polymers and a considerable number of monomers present at any given time. The molecular weights of polymers synthesized via these reactions do not change significantly during the reaction, though more and more polymers are formed as the polymerization proceeds.

2.2.1.3 Classification Based on the Source

In this classification system, polymers can be classified as synthetic, natural, or bio-based polymers depending on their source.

Synthetic polymers are those that are produced in industry or laboratory settings through nonbiological reactions. These polymers, which include widely used plastics such as PE, polyvinyl chloride (PVC), and more, are often preferred due to their cost-effectiveness and customizable performance features. Synthetic polymers can be derived from petrochemical as well as biobased feedstocks. For example, poly(lactic acid) (PLA) is biobased but it is produced in industry/lab and is thus a synthetic polymer.

In contrast, natural polymers such as cellulose and polyhydroxyalkanoates (PHAs), are derived from nature. These polymers are biodegradable, but they often lack the necessary properties to be used in packaging or other applications

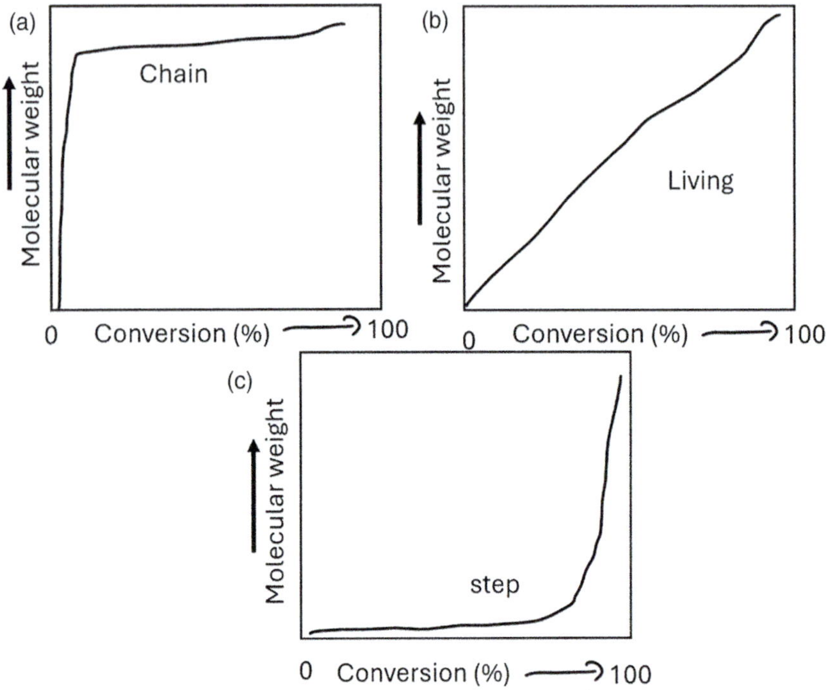

Figure 2.3 Plots of molecular weight versus percent monomer conversion for chain-growth (a), living (b), and step-growth polymerization reactions (c).

and some of them are difficult to melt-process; thus, their use is limited as compared to synthetic polymers. There is also another category of polymers that is known as bio-based polymers and this represents polymers that are either naturally derived or those produced from biobased feedstock. An example of this type of polymer would be PLA. Although PLA is synthesized in laboratories or industrial settings, it is produced from naturally sourced materials such as plants.

2.2.1.4 Classification Based on Cost and Performance

Polymers can also be classified based on their cost and performance. Under this classification, there are three distinct categories: commodity, engineering, and specialty polymers.

Commodity polymers, which are distinguished by their low cost and high-volume use, are commonly utilized for applications below 100 °C. Such polymers can be heated briefly for 30 min or so up to 130 °C or so, but overall, their usage is primarily limited to ambient temperature applications. Examples include polyethylene, polypropylene, PS, PET, etc.

Engineering polymers, on the other hand, are more expensive and often used in lower volumes but have more resilience at high temperatures. Polycarbonates,

Thermoplastics

- Some examples include: Polyethylene (PE), polypropylene (PP), polystyrene (PS), polyethylene terephthalate (PET), polyvinyl alcohol (PVA), ethylene-vinyl acetate (EVA), polyvinylidene chloride (PVDC), nylons, and so forth.
    - Most packaging materials.
    - Thermosets can have linear or branched structures.

Thermosets

- Some examples include: Epoxies, vulcanized rubber, some polyesters, and so forth.
    - Some adhesives and coatings.
    - Thermosets have crosslinked structures.
        - The crosslinks are created via "curing" treatment.

Figure 2.4 Common examples of thermoplastics and thermosets.

Kevlar, and polyetheretherketone (PEEK), are all examples of engineering polymers. PET was once an engineering polymer, but became a commodity plastic because technologies emerged that reduced its cost, thus demonstrating that the category can change as technology progresses.

Specialty polymers are created for specific applications. For example, ethylene vinyl alcohol (EVOH) is used for high gas barrier applications, while polyvinylidene chloride (PVDC) is employed for high water, oil, gas, and water vapor barrier purposes. Specialty polymers tend to be costlier than commodity plastics.

2.2.1.5 Classification Based on Thermal Behavior

Polymers can be classified into two categories based on their thermal behavior: thermosets and thermoplastics. **Thermosets** are a type of polymer that cannot be remelted when they are subjected to high temperatures and shear force, whereas **thermoplastics** can be repeatedly remelted and reshaped when they are exposed to heat and shear force. Notably, the plastics employed in packaging are typically thermoplastics. It is important to note that thermosets constitute approximately 20% of all produced polymers, while thermoplastics account for the remaining 80%. Examples of thermosets and thermoplastics are shown in **Figure 2.4**.

2.3 Molecular Architecture

Polymers can also be categorized based on their molecular architecture, which refers to the arrangement of monomers during polymerization. This arrangement can lead to **(1)** linear, **(2)** branched, or **(3)** crosslinked structures/architecture. It is important to note that both homopolymers and copolymers can exhibit these molecular architectures – that is, they can be linear, branched, or crosslinked in structure.

What's the big deal about molecular architecture? Indeed, a polymer's properties depend on various factors, one of which is its molecular architecture. PE serves as a textbook example of how molecular architecture influences a polymer's properties. **Linear PE, also known as high-density PE,** is crystalline and opaque with good durability, while branched PE, **also known as low-density PE,** is more flexible and transparent. This illustrates that even when the same ethylene monomer is used to create both linear and branched PE, the differences in molecular architecture result in unique properties for each polymer.

Crosslinked polymers such as those used to form rubber offer unique properties such as excellent wear resistance, chemical resistance, and longevity. Crosslinked polymers are less common in packaging, where their use is mostly limited to lids and caps for chemical containers, as well as liners for food cans.

2.3.1 Homopolymers and Copolymers

Polymers can be synthesized from one or more than one type of monomer. When a polymer is created using more than one kind of monomer, it is referred to as a copolymer. Conversely, a polymer formed from a single type of monomer is termed a homopolymer.

The primary reason for copolymerization is to modify and enhance certain properties of polymers. For instance, by incorporating a small percentage of ethylene monomer during the production of polypropylene, the resulting polymer exhibits improved optical clarity and flexibility. Examples of homopolymers and copolymers are shown in **Table 2.1**.

As we have seen in the previous section describing molecular architecture, there are various ways in which polymers are arranged. This is also true with copolymers, and there are various structural classes of copolymers. **The most common type of copolymer is a random copolymer**, in which two (or more) types of monomer units are mixed together in a random manner along the polymer chain. Examples of random copolymers are polyethylene vinyl acetate, polyethylene vinyl alcohol, and so on. In contrast, the different monomers are arranged in distinct "blocks" in the case of a block copolymer. A couple of common examples of block copolymers are PS-*block*-PE oxide and PE-*block*-polypropylene. On the other hand, the different monomers are arranged in an alternating manner in an alternating copolymer such as PS-*alt*-maleic anhydride.

2.3.2 Polymer Molecular Weights

Small molecules like water and carbon dioxide possess absolute molecular weights because these are formed in one or two steps and are kinetically and thermodynamically very stable so that they do not react further under the conditions that they are formed. However, the scenario is different for polymers. The molecular

Table 2.1 Examples of various homopolymers and copolymers and their nomenclature.

Homopolymers			
Monomer nomenclature	**Monomer formula**	**Repeating unit**	**Polymer nomenclature**
Ethylene	$CH_2{=}CH_2$	$-[CH_2CH_2]-$	Polyethylene
Propylene	$CH_2{=}CH_2CH_3$	$-[CH_2CH_2CH_3]-$	Polypropylene
Vinyl chloride	$CH_2{=}CHCl$	$-[CH_2CHCl]-$	Poly(vinyl chloride)
Styrene	$CH_2{=}CH_2C_6H_5$	$-[CH_2CH_2C_6H_5]-$	Polystyrene
Acrylonitrile	$CH_2{=}CHCN$	$-[CH_2CHCN]-$	Polyacrylonitrile
Vinyl acetate	$CH_2{=}CHOCOCH_3$	$-[CH_2CHOCOCH_3]-$	Poly(vinyl acetate)
Tetrafluoroethylene	$CF_2{=}CF_2$	$-[CF_2CF_2]-$	Polytetrafluoroethylene
Copolymers			
Monomer	**Monomer**	**Monomer**	**Copolymer nomenclature**
Acrylonitrile	Butadiene	Styrene	Acrylonitrile butadiene styrene (ABS)
Propylene	Ethylene		Ethylene propylene (E/P)
Vinyl chloride	Vinyl acetate		Vinyl chloride/vinyl acetate
Vinyl chloride	Ethylene		Vinyl chloride/ethylene

weight of a synthetic polymer does not have a fixed value; instead, it spans a range of values. **This situation arises because polymer synthesis involves repeating steps**. Each polymer chain is formed by consuming hundreds and sometimes even thousands of monomers. Alterations in the parameters during the polymerization, such as the concentration, temperature, and chain termination, lead to differences in polymer chain lengths. Therefore, the molecular weight of a polymer is typically expressed in terms of the **average molecular weight.** Number average molecular weight (M_n), weight average molecular weight (M_w), and viscosity average molecular weight (M_v) are common ways to express the average molecular weight of a polymer. Each of these averages provides unique information about the distribution and size of molecules within a polymer sample (Figure 2.5). These types of molecular weights are briefly described below.

$\boldsymbol{M_n}$ (Number Average Molecular Weight): This is the total weight of all the molecules in a sample divided by the number of molecules. It is the simplest form

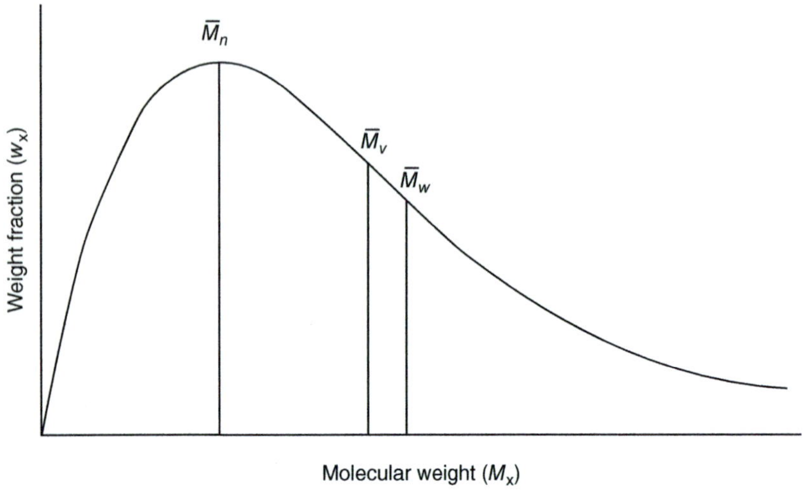

Figure 2.5 Plot of the weight fraction versus the molecular weight (M_n) [3]. This image is Reproduced from Odian, G, 2024/ with permission of John Wiley & Sons.

of average molecular weight, and it is sensitive to the presence of small molecules. The value of M_n is calculated via equation 2.1.

$$M_n = \frac{\Sigma n_i M_i}{\Sigma n_i}, \tag{2.1}$$

where $\boldsymbol{n_i}$ is the number fraction or mol fraction of polymer chains with the molecular weight M_i.

$\boldsymbol{M_w}$ (Weight Average Molecular Weight): This average molecular weight considers both the mass and the number of molecules in a sample. M_w is more sensitive to the presence of larger molecules and is calculated by equation 2.2.

$$M_w = \frac{\Sigma n_i M_i^2}{\Sigma n_i M_i}, \tag{2.2}$$

$\boldsymbol{M_v}$ (Viscosity Average Molecular Weight): This average molecular weight is determined by measuring the viscosity of a polymer solution. This calculation involves comparing the viscosity of the dilute polymer solutions containing the polymer in question with that of the pure solvent. M_v is particularly sensitive to larger molecules in the distribution.

Another important aspect of polymers that is related to their molecular weight is their dispersity ($Đ$), which is sometimes called the polydispersity or polydispersity index (*PDI*). The term dispersity is currently the accepted term employed by IUPAC [4, 5], while the terms polydispersity or polydispersity index were more commonly used in the past, although those terms are still frequently used by many researchers and may still appear in some publications. The dispersity of a polymer refers to its molecular weight distribution or to the spread of its molecular weight. The dispersity **Đ** is calculated by dividing the mean of the M_w by the mean of the M_n as shown in equation 2.3:

$$Đ = \frac{\bar{M}_w}{\bar{M}_n}, \tag{2.3}$$

For the dispersity, a value of **1.0** will indicate that a given polymer sample has a uniform molecular weight distribution. For typical polymer samples, the range of dispersities is typically between 1 and 25, with those having lower dispersities closer to **1** having a narrower molecular weight distribution than those with higher dispersity values. The polymerization method has a strong influence on the dispersity of the resultant polymer, as some techniques tend to offer narrower distributions than others. For example, anionic polymerization will typically yield polymers with low dispersities, with the resultant dispersity becoming larger in the order from anionic, to step growth, to radical, olefin coordination, and to random-branching polymerization. Typical dispersity values of polymers produced via these polymerization methods are shown in Table 2.2.

2.4 Polymer Characterization Techniques

The properties of polymers are determined by their molecular weights, dispersity, and chemical composition. Therefore, it is critical to characterize them to find key information regarding their structure, molecular weight, dispersity, or other

Table 2.2 Various polymerization methods and typical dispersity ranges obtained via each approach.

Polymerization method	Typical dispersity range
Anionic polymerization	1.01–1.02
Step-growth polymerization	1.5–2.0
Radical polymerization	2.0–5.0
Olefin coordination polymerization	5–25
Random branching polymerization	Large

important properties. Various techniques can be used to characterize polymers. **Three approaches** that are widely used for the characterization of polymers include magnetic resonance spectroscopy (NMR), size exclusion chromatography (SEC), and intrinsic viscosity measurements. These techniques will be discussed below, along with a very brief discussion of microscopic characterization techniques.

2.4.1 Nuclear Magnetic Resonance Spectroscopy

NMR spectroscopy can provide useful structural information about a given polymer sample. This type of spectroscopy utilizes the resonance behavior of a given nuclide, with some common examples of nuclides that are suitable for NMR characterization including proton (^{1}H), carbon-13 (^{13}C), phosphorus-31 (^{31}P), fluorine-19 (^{19}F), and nitrogen-15 (^{15}N), among others [6]. However, **^{1}H NMR spectroscopy** is the more commonly used technique compared to other types of NMR spectroscopy, as it offers high sensitivity (partially due to the high natural abundance of ^{1}H compared to other hydrogen isotopes).

NMR spectroscopy offers valuable insight regarding the chemical environment that surrounds a given nuclide, as certain neighboring functional groups will cause the resonance signal exhibited by that particular nuclide to shift to either a higher or lower frequency (which is often referred to as the "chemical shift" and is measured in units of ppm). This is largely because the resonance signal of a nuclide such as a proton is affected by the density of the electron cloud that surrounds it [7, 8]. In general, a nearby atom or functional group that tends to pull electrons away from the given nuclide will cause the resonance signal of that nuclide to shift to a higher frequency. This phenomenon is often referred to as "deshielding" due to the fact that electrons which had "shielded" that nuclide have been drawn away. Conversely, nearby functional groups which may increase the density of electrons that surround that nuclide will cause the resonance frequency or chemical shift of that nuclide to shift to a lower frequency. Consequently, helpful information regarding the functional group that a particular nuclide is part of (or is neighbored by) can be determined based on the resonance frequency or chemical shift of that particular nuclide.

In addition to the resonance frequency, the neighboring nuclei can affect the splitting pattern (this splitting pattern is often called the multiplicity) exhibited by the resonance signal of a particular nuclide, and this is especially applicable for ^{1}H NMR spectroscopy. In general, a multiplet will exhibit $2NI + 1$ peaks, where n represents the number of nuclei that are attached to the neighboring atoms [9], and I is the magnetic spin of the nuclide in question. In the case of a proton, the magnetic spin I is ½, and thus the splitting rule can be simplified to $N + 1$ for 1H NMR spectroscopy. Consequently, if a proton has no adjacent proton neighbors, it will exhibit a single peak (called a singlet). Alternatively, the resonance signal of a proton with one proton neighbor will be split into two peaks, with this multiplet being called a doublet. Likewise, protons with two, three, or four peaks will respectively form multiplets with three peaks (giving a triplet), four peaks (giving a quartet), or five peaks (to give a quintet). It should be noted that the aforementioned splitting patterns are observed provided that the coupling constant (which has a distinct frequency and is denoted by the symbol *J*) between all of the adjacent protons and the proton in question are the same. If that is not the case, the splitting pattern will become more complex.

In addition to the chemical shift and multiplicity, another important piece of information that can be gained from ^{1}H NMR spectroscopy is the signal integration. This provides information regarding the number of protons that are present in a given sample corresponding to a particular signal. In particular, this signal integration corresponds to the relative area corresponding to that signal. Through comparison of these signal integrations, it can be possible to use these signal integrations to help determine the relative amounts of functional groups within a polymer, thereby enabling one to compare the relative amounts of different repeat units that are present within the polymer (this can be especially helpful in cases involving copolymers) or the amount of branching groups that are attached to a polymer's backbone. Provided that the proton signals corresponding to the end groups are clearly visible, it can also become possible to calculate the degree of polymerization or molecular weight of the polymer through comparison of signal integrations for resonance signals corresponding to defined protons of the end groups with those of protons within the repeating units of the polymer. **Prior to a polymerization reaction,** NMR characterization can also be employed to confirm that the monomers and other starting materials are indeed the desired chemicals, thus helping to prevent wasteful unsuccessful reactions that would otherwise result due to the use of incorrect or degraded starting materials.

Other information that can be gained via ^{1}H NMR spectroscopy includes details regarding the **microstructure of the polymer**, such as which isomers are present and the tacticity of the polymer or whether any side-products may have been formed during the synthesis of the polymer. NMR spectroscopy can also be used to evaluate the **kinetics of a polymerization reaction.** If samples are collected from a reaction mixture, it becomes possible to monitor the progression of a polymerization reaction. This can help researchers to determine whether the reaction is proceeding in a well-controlled manner, whether the reaction has produced the desired reaction and can be stopped, or whether the reaction is generating an undesired mixture of products so that corrective action may be needed during the course of the reaction (such as by changing the feed ratio or other conditions).

Although NMR spectroscopy is a very useful analytical tool, one limitation that presents an issue with the characterization of polymers arises due to the slow molecular motion or rotation of polymers. As a consequence, the T_2 spin–spin relaxation (also called transverse relaxation) time or transverse relaxation time is fast in comparison with the polymer's rotation, thus causing the polymer's signal to coalesce as a broad peak [8, 10]. Due to this coalescence, it may not be possible to distinguish between two nearby resonance signals, and information regarding the multiplicity or coupling behavior may not be available.

2.4.2 Size Exclusion Chromatography

SEC, which is also often called gel permeation chromatography (GPC), can provide valuable information about the molecular weight and dispersity of a polymer sample. Through this characterization method, the polymer chains within a polymer sample are separated according to their molecular weight. With SEC characterization, the stationary phase within each column is made up of numerous beads that contain small pores. As a polymer sample is passed through the column via a flowing mobile phase (solvent such as chloroform), the shorter polymer chains of a polymer sample will tend to enter within these pores to a greater extent than the longer chains. Consequently, the shorter polymer chains will move more slowly through the column because they spend more time being "held up" within the pores of the stationary phase, and thus, the shorter polymer chains will have a longer elution time. In contrast, the longer polymer chains are less likely to enter these pores, and thus, they will progress through the column at a faster rate than their smaller counterparts, and consequently, they will have a shorter elution time. In essence, because the longer polymer chains are excluded from the pores of the stationary phase, they have a smaller volume through which they may potentially transverse (the spaces between the beads of the stationary phase), thus leading to their faster elution through the column. On the other hand, shorter polymers have a larger volume available in the pores through which they would potentially transverse as they pass through the column (the spaces between the beads in addition to the pores within the beads), leading to their longer elution time [8]. The width/broadness of an SEC elution signal provides valuable information regarding the dispersity of that sample. In particular, a narrower peak corresponds to a smaller dispersity.

Unfortunately, many polymers used in packaging are difficult to analyze by SEC. For instance, PET is challenging to dissolve in common organic solvents, making it hard to determine its molecular weight by SEC. Similarly, PE and PP require high-temperature SEC analysis, which is not commonly available to lab researchers. This leads us to our next topic, and that is the need for viscosity measurements for plastics.

2.4.3 Viscosity

The viscosity (η) is another important characteristic of a polymer that has an important influence on its applicability. In general, viscosity refers to the resistance of a liquid against the flow or refers to a fluid's internal friction [11]. To measure a polymer's viscosity, the polymer is dissolved in an appropriate solvent to yield a dilute solution. The intrinsic viscosity is related to the viscosity of dilute polymer solution and is typically determined using an Ubbelohde or Ostwald viscometer. The relationship between intrinsic viscosity and the M_v is expressed as equation 2.4:

$$\left[\eta\right] = K^{M_v^a} \tag{2.4}$$

This equation is also known as the Mark–Houwink equation, and it relates the intrinsic viscosity $[\eta]$ of a polymer to its molecular weight M_v, where $[\eta]$ is the intrinsic viscosity, and M_v is the viscosity average molecular weight of the polymer. The values of k and "a" are specific to the polymer–solvent system at a given temperature. For random coil polymers (e.g., many synthetic polymers in good solvents), the value of "a" is typically in the range of 0.5 to 0.8.

In many cases, especially when the molecular weight distribution of the polymer is narrow, the viscosity-average molecular weight M_v is close to the weight-average molecular weight M_w, and thus one can estimate M_w via viscosity measurements.

Entanglement between polymer chains has an important influence on the viscosity, and during viscosity measurements, the critical molecular weight for entanglement (M_C) refers to the molecular weight at which the viscosity begins to increase rapidly due to chain entanglement (Figure 2.6). Meanwhile, the

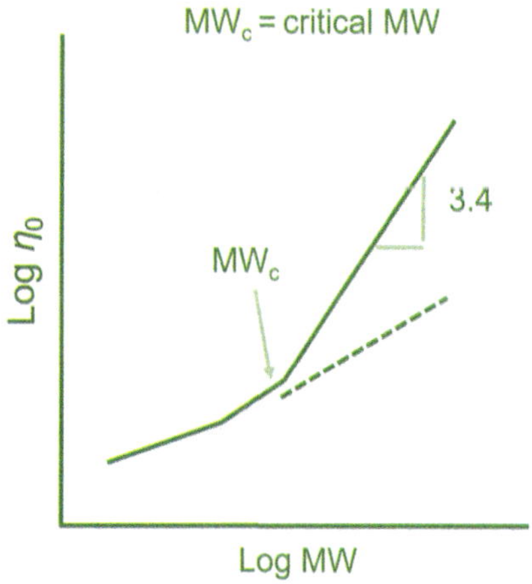

Figure 2.6 Influence of the molecular weight on the viscosity of a polymer.

entanglement molecular weight (*M*e) is the molecular weight of the segment of a polymer chain that exists between two adjacent entanglement points.

2.5 Microscopy Techniques

Characterization techniques such as transmission electron microscopy (TEM), scanning electron microscopy (SEM) [12, 13], and atomic force microscopy (AFM) are also employed in polymer science to characterize their micro- and nano-scale structures. Both TEM and SEM are related to one another in that they rely on a beam of electrons to illuminate the samples being analyzed. Since the wavelength of an electron is much shorter than that of visible light, much higher image resolutions can be obtained via TEM and SEM than is offered via optical microscopy. TEM is generally employed to characterize the interior of a structure, while SEM is employed for analysis of the surface of a structure.

AFM characterization also provides morphological information about a surface. While SEM and TEM provide two-dimensional images of a specimen, AFM can provide three-dimensional images and can thus provide information regarding the roughness of a sample. AFM images are recorded in a very different manner than TEM or SEM images, as in this case, they are recorded with the use of a sharp tip at the end of a cantilever, which essentially serves as a probe that scans the surface of a specimen. Forces between the specimen and the AFM tip cause the cantilever to deflect. A laser is used to track these deflections and an image is generated by means of a photodiode [14, 15].

2.6 Physical State of a Polymer

Polymers are often semicrystalline, implying that they possess both crystalline and amorphous structures. Semicrystalline polymers possess crystalline regions in which the polymer chains are arranged into well-ordered structures, such as lamellae or fringed micelles (Figure 2.7) [16, 17]. These crystalline domains are often called crystallites. On the other hand, amorphous polymers, which are also known as glassy polymers, lack crystallinity, meaning that they do not have organized molecular structures and consequently the polymer chains have a random arrangement and can become intertwined with one another.

Polymers capable of crystallization have a distinct melting temperature, above which they turn into a liquid. These polymers also have a glass transition temperature (T_g), above which they become rubbery but not liquid. In contrast, fully amorphous/glassy polymers only possess a T_g and do not have a specific melting point.

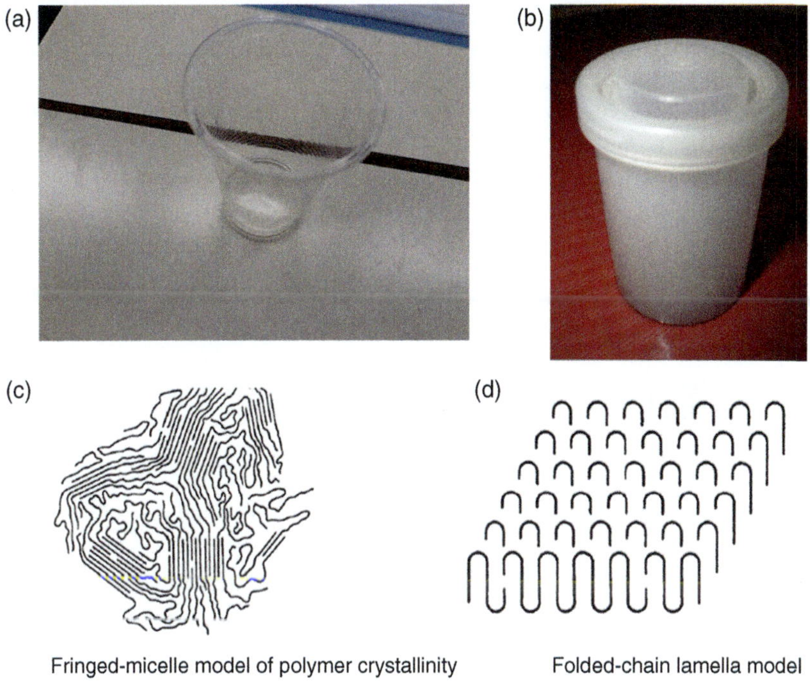

Figure 2.7 Photographs of a transparent cup **(a)** and an opaque plastic photo film container **(b)** cups made of amorphous and crystalline plastic, respectively. Schematic depictions of two crystallization models, including the fringed micelle model **(c)** and the folded-chain lamellar model **(d)** [16]. Image **(a)** Latimeria 2004 / wikimedia Commons / CCBY-SA 4.0 **(b)** KowCFLemont / Wikimedia Commons / CCBY-SA 3.0 Images **(c)** and **(d)** Reproduced from Odian, G, 2004 / with permission of John Wiley & Sons.

The transparency of plastic is determined by the degree of crystallinity and the size of the crystals. Crystals larger than the wavelength of visible light will cause the plastic to appear opaque. Conversely, smaller crystals scatter less light, making the plastic more transparent. One well-known example of this phenomenon is that when films are stretched, their transparency increases because the larger crystals break down into smaller ones due to stress applied to the orientation.

The crystallinity of a polymer depends on various factors such as composition, regularity, and stereochemistry. Generally, polymers with regular structures are more crystalline than those with irregular structures. Techniques such as optical microscopy, SEM, and AFM are used to determine the type and size of crystalline domains in polymers but cannot quantify the degree of crystallinity.

Wide-angle X-ray scattering (WAXS) or X-ray diffraction (XRD) are techniques that directly probe the crystalline structure of a material, making them suitable for determining the degree of crystallinity.

Differential scanning calorimetry (DSC) is also widely used to quantify the degree of crystallinity.

A relatively new concept related to amorphous regions includes flexible and rigid amorphous domains. Rigid amorphous domains have less chain segment mobility and are typically located next to the crystalline domains, while flexible amorphous regions are farther from the crystalline domains and have higher chain segment mobility.

2.7 Thermal Transitions

DSC is employed to determine thermal transitions such as the melting temperature, glass transition temperature, degree of crystallization, cold-crystallization temperature, and crystallization temperatures. As demonstrated in Figure 2.8, the sharp endothermic peak at a higher temperature identifies the melting temperature (T_m).

The glass transition temperature (T_g) is represented by a slight endothermic dip in the DSC plot, appearing at a lower temperature, and it does not return to the previous level. The crystallization temperature (T_c) is denoted by an exothermic peak that may appear between the T_g and the T_m. The glass transition temperature

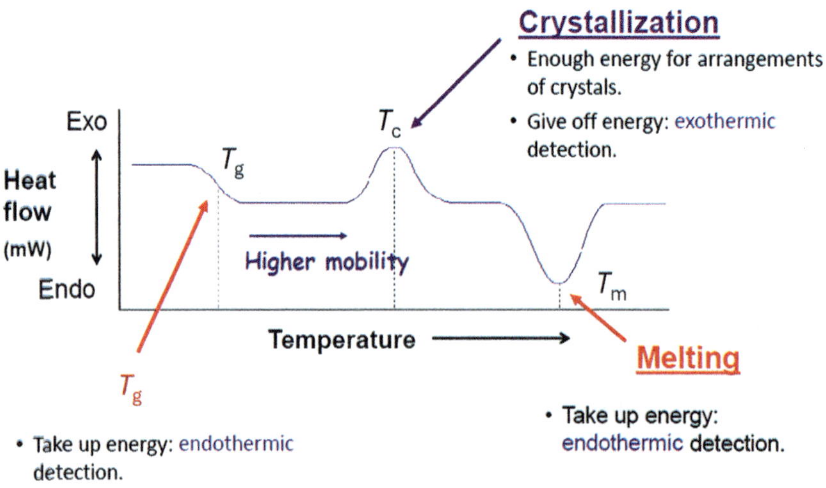

Figure 2.8 Various characteristic transition temperatures in a typical DSC trace of a polymer sample.

(T_g) provides valuable insights into how a polymer tends to behave at lower temperatures. On the other hand, the melting temperature (T_m) indicates the maximum upper temperature at which the plastic can be used effectively while also offering insights into the temperature at which the polymer can be melt processed. Often, polymers are melt-processed at temperatures which are 50 °C above their T_m.

2.8 Mechanical Properties

One of the most important properties of plastic is its mechanical properties. Mechanical properties refer to the response exhibited by materials when they are subjected to various forces (such as tension, compression, etc.). The most commonly investigated mechanical properties of polymers are their tensile and impact strengths. Tensile strengths are determined by analyzing the stress-strain behavior of a material and by obtaining plots of the stress versus the strain. During these experiments, the plastic is stretched at a constant rate (ASTM rates are 10 or 100 mm/min). The force required to achieve this stretching is measured. In particular, the stress refers to the amount of force that is applied to a given cross-sectional area, as shown in equation 2.5:

$$\text{Stress} = \frac{\text{Force}}{\text{Cross}-\text{sectional area}} = \frac{F}{A}, \tag{2.5}$$

where F denotes the force (in pounds or newtons), and A denotes the area (in square inches or square meters), so that the units for stress are lb/in^2 (in the imperial system) and N/m^2 (in the metric system)

Meanwhile, strain refers to the elongation exhibited by a sample with respect to its original length, as expressed by equation 2.6:

$$\text{Strain} = \frac{\text{Elongation}}{\text{Original length}} = \frac{\Delta L}{L_0}, \tag{2.6}$$

where L_0 denotes the original length, and ΔL is the change in length or elongation.

Tensile testing provides valuable information regarding the tensile strength, yield strength, strength at break, strain at yield, strain at break, and the Young's modulus. The Young's modulus is a parameter that describes the stiffness of a material, and it is sometimes called the elastic modulus [18]. In the initial stages of a stress-strain curve (well before the yield point is reached), a stretched plastic will normally exhibit elastic behavior and will follow Hooke's law. In this regime, the stress-strain curve will exhibit a linear plot, and the Young's modulus can be determined from the slope of this plot until the elastic limit is reached, where the plot is no longer linear.

The yield point is defined as the point where you get increased strain without needing increased stress. Meanwhile, yield strength is the stress at the yield point. Ultimate tensile strength (also called tensile strength) refers to the maximum stress for a material on its stress-stress curve.

Meanwhile, the tensile strength at break corresponds to the stress that a plastic is able to endure when it breaks [19]. The strain at break, which is also called the strain at failure, is determined based on the elongation that is observed at the breaking point with respect to the original length and is calculated via equation 2.7:

$$\text{Strain at break} = \frac{L_f - L_0}{L_0}, \tag{2.7}$$

where Lo denotes the original length, and L_f is the length of the plastic sample when it failed or broke [20]. In a similar manner, the strain at yield is the strain exhibited by a plastic sample at its yield point.

In this section, we have provided a general introduction to some of the concepts relating to plastics. In the remainder of this chapter, we will discuss the degradation of plastics (Section 2.2) and subsequently cover the stabilization of plastics (Section 2.3).

2.9 Degradation of Polymers/Plastics

2.9.1 Overview

Organic compounds contain at least one carbon–hydrogen (C—H) or carbon–carbon (C—C) bond. Such bonds are prone to degradation (breakage, oxidation, and reduction), and this leads to changes in the chemical and physical properties of organic compounds.

Polymers used in packaging are organic polymers, which means they are comprised predominantly of C—C and C—H bonds, and like small organic compounds, they are also prone to degradation.

Generally, degradation of a polymer or plastic can occur when the polymer is exposed to one or more of the following environments, e.g.:

- Mechanical degradation occurs due to mechanical stresses.
- Thermal degradation resulting from exposure to heat, often in combination with oxygen and/or moisture.
- Radiation degradation is induced by high-energy radiation.
  - Photodegradation is a type of radiation degradation resulting from exposure to sunlight.

- Chemical degradation occurs upon exposure to chemicals such as oxygen, ozone, atmospheric pollutants, water, acids, and bases.
  - Oxidation results from exposure to oxygen.
- Biological degradation is due to the action of enzymes/microorganisms.

At ambient temperatures and in the absence of sunlight and stress, plastics used in packaging applications are very stable and are not likely to undergo any significant changes.

2.9.2 Impact of Degradation on Polymer Properties

Polymer degradation can affect both its chemistry and molecular weight. Polymer degradation results in both fragmentation due to chain scission and crosslinking due to chain coupling. Fragmentation or chain scission involves the breakdown of polymer chains into smaller segments, which decreases their molecular weight and broadens the molecular weight distribution or dispersity. In contrast, crosslinking involves the coupling of polymer chains, leading to an increase in their molecular weight with a notable change in their flow properties. As the crosslinking increases, it becomes increasingly challenging to melt process the polymer. Furthermore, the chemical degradation of a polymer leads to an alteration in its chemistry. New functional groups appear, resulting in diminished performance, reduced stability, and often discoloration, and so on.

2.10 Wanted Versus Unwanted Degradation in Polymers

2.10.1 Unwanted/Undesirable Degradation

Polymer degradation is most often undesirable both during its synthesis and during processing. Polymers are melt-processed at elevated temperatures (often 150–300 °C) to produce films, sheets, bottles, injection molded parts, and so forth. At these temperatures, polymer degradation can happen. In addition to heat, shear force is also applied during melt-processing, which also results in polymer degradation. The purpose of shear force is to homogenize polymer melts, reduce melt viscosities, and remove any trapped air/vapors from the polymer melt.

Sterilization of packaged products is a common practice in the industry, especially for food and medical packaging. Several methods are used for sterilization, such as heat, radiation, plasma, gas, chemicals, etc. Polymer degradation also happens during the sterilization process.

Sunlight causes polymer degradation because of the UV-C radiation in the sunlight, which can break chemical bonds in polymers. The light used in our homes and offices does not have UV light with the exception of fluorescent light that may have a low level of UV, and thus polymers are highly stable when they are used for indoor applications. For outdoor applications, UV stabilizers are added to delay degradation in polymers.

2.10.2 Wanted/Desirable Degradation

In some cases, it is desirable for plastics to degrade. Some examples of polymers that are designed for applications where polymer degradation is leveraged for benefit include:

Biodegradable Plastics: These plastics are often used in applications such as packaging, agriculture, and disposable goods. The intention is for the plastic to break down over time when it is exposed to the environment, thereby reducing the formation of persistent microplastics and its environmental impacts. Some examples of biodegradable plastics are PHAs, poly(glycolic acid) (PGA), and so forth, and these polymers do not form persistent microplastics.

In the purest sense of the term, biodegradable plastics simply means plastics that have bonds that are susceptible to the action of biological species such as microorganisms. For this to be useful in practice, the degradation must be reasonably fast and complete, and must occur in the environment (with the microbial population) that the item is likely to be exposed to. Unfortunately, these qualifications and conditions are often not specified, leading to considerable misuse of the term.

Compostable Plastics: As is the case with biodegradable plastics, compostable plastics are designed to biodegrade, but in this case, the degradation occurs in an industry composting environment. Note that compostable plastics biodegrade at elevated temperatures in industry compost conditions while biodegradable plastics (as first defined above, implicitly assuming reasonable complete degradation at short time frames) biodegrade under ambient conditions in soil and water. PLA is an example of a compostable plastic, and depends on initial hydrolysis at elevated temperatures to reduce the molecular weight sufficiently to permit the biodegradation to occur at a reasonable rate.

Thus, using the more restrictive definition, all biodegradable plastics are compostable, but the opposite is not true. In contrast, using the more general definition of biodegradable, all compostable plastics are biodegradable, but the opposite is not true.

Oxo-biodegradable Plastics: Oxo-degradable plastics are defined as plastics that can be degraded by the combined actions of oxidation and microorganisms. Commercial oxo-degradable plastics are typically produced by the use of additives in nondegradable plastics. Oxo-degradable additives (pro-oxidant additives) are metal catalysts that are added into nondegradable plastics such as polyolefins during the manufacturing process. These catalysts can speed up the oxidation of nondegradable plastic, thus leading to the breakdown of the polymer chains. Owing to environmental issues, some regions globally have prohibited the use of certain plastics. The EU Packaging Directives and the US Plastics Pact have advised the gradual elimination of these plastics. Major concerns include the time frames required for breakdown and incomplete decomposition leading to the production of persistent microplastics.

Biomedical Applications: Some plastic materials are used in pharmaceutical applications to create controlled-release systems for drugs. Here, the polymer degrades over time within the body, thereby releasing the drug at a controlled rate. Similarly, bioabsorbable polymers are used in medical devices such as sutures, implants, and so forth. These materials gradually degrade in the body, thus eliminating the need for secondary surgery to remove the device. An example of such a material is PGA, which is often used in sutures.

Agricultural Applications: In agriculture, degradable polymers are used to make mulch films, for seedlings, and slow release of pesticides/fertilizers. Some of these materials are biodegradable, while the main degradation mechanism in others is photodegradation or hydrolysis.

2.11 Do all Polymers Degrade at the Same Rate?

The answer is No. Polymer degradation is dependent on its chemical makeup, which determines which polymer will be stable against what kind of stimuli. In addition to the polymer itself, additives and contaminants present in the polymer also affect its stability. The difference between an additive and a contaminant is that the additive was placed there deliberately, while the presence of the contaminant is inadvertent. For example, unreacted/residual monomers in plastics are contaminants, while heat stabilizers added for polymer protection during melt processing are additives. Further, previous degradation of a polymer, such as oxidation during processing, often increases its susceptibility to future degradation.

2.11.1 How Can We Know if a Polymer Is Stable or not Against a Particular Degradation Environment?

The general stability of a polymer can be predicted by examining its chemical structure. However, this is a starting point, as it does not account for contaminants and exposure history.

Polymers Prone to Thermal-oxidative Degradation

1) **Polymers that have tertiary carbons** (carbons that are attached to three other C atoms) are prone to degradation. For example, polypropylene (which has a tertiary C where each methyl group is attached to the backbone) is less thermally

stable than HDPE, which has few tertiary carbons (only where branching occurs). This is because tertiary C—H or tertiary C—C bonds can readily break because the resultant tertiary carbon radicals are more stable than secondary carbon radicals, and thus, this C—H or C—C bond is a weak bond in the system.

2) **Polymers that have allylic carbon bonds** are less stable against heat. The allylic carbon is a carbon that is adjacent to an alkene. Allylic radicals are more stable, and thus, allylic C—C or C—H bonds are easy to break to form stable allylic radicals. For example, polybutadiene rubber is very prone to degradation because of the allylic C—C or C—H bonds that are present.
3) **Polymers that are vulnerable to moisture at high temperatures:** Polyesters such as PET, PLA, and PHAs are degraded by moisture at high temperatures as they have water hydrolyzable ester linkages. In contrast, polymers with C—C backbones such as PS, PE, and polypropylene (PP) are not affected by moisture at high temperatures.
4) **Contaminants in plastic may increase susceptibility to degradation**. Additives, depending on their type, may either promote or inhibit degradation. In fact, as we will see later, they may provide protection against degradation for a time and then later increase susceptibility to degradation. For example, the use of any acidic additive will increase the degradation of polyesters such as PET.
5) **Crystallinity:** Crystallinity reduces rates of oxidation through a reduction in oxygen availability. Amorphous polymers tend to oxidize homogeneously and oxidize significantly more rapidly above their glass transition temperature (T_g) than below their T_g, while semicrystalline polymers tend to oxidize mostly in the amorphous areas. The overall extent of oxidation will be less in crystalline polymers compared to their amorphous counterparts (at the same exposure conditions). However, performance loss due to oxidation can be greater in crystalline polymers since oxidation tends to damage the spherulite (crystallite) boundaries – affecting the "adhesion" between the crystallites. The availability of oxygen tends to have a large influence on rates of oxidation. For example, PS oxidizes readily in solution but is relatively stable in its solid form. PS at normal use temperatures is below its T_g, so diffusion of oxygen through the material is relatively slow – certainly slow compared to its diffusion through the liquid.

2.12 Types of Polymer Degradation

Degradation processes are commonly classified either by the cause of degradation or by the major process involved in degradation. In either case, the real-life situation is often complicated by the fact that frequently, there tend to be multiple causes and processes acting together.

Our approach will be to focus mostly on the cause as an organizing strategy, so we will discuss:

1) Thermal degradation,
2) Oxidative degradation,
3) Photodegradation,
4) Chemical degradation, and
5) Mechanical degradation

2.12.1 Thermal Degradation

Thermal degradation is degradation caused by heat alone or by shear in the presence of heat (in which case it is classified as **thermal-mechanical degradation**). The underlying principle is that at a high enough temperature, all chemical bonds can (and will) be broken. Because bonds in polymers differ in strength, they differ in susceptibility to thermal degradation; the stronger the bond, the more energy is required to break it down.

The basic measure of susceptibility to thermal degradation is the bond dissociation energy.

The basic process involved in pure thermal degradation is bond cleavage, generating two free radicals, as shown in Figure 2.9.

Chain degradation can occur in the main chain (often called the backbone chain) of a polymer, such as the cleavage of C—C bonds in the main chain, as shown above, and can also occur in the side/pendant group. However, the main chain cleavage has a strong detrimental impact because C—C bonds are more susceptible and also, main chain cleavage generally causes more drastic changes in polymer properties than reactions involving side groups, so they concern us more. As can be seen in Figure 2.10, the activation energies for the cleavage of primary, secondary and tertiary carbon bonds differ.

These free radicals (substances with unpaired electrons) are unstable, so they tend to react further, resulting in unsaturation and generation of more free radicals, as we will see later in this subsection. Thus, the process tends to reinforce itself (i.e., it tends to be self-catalyzing).

Thermal degradation can also result in "**unzipping**" – the reverse of polymerization, which is also called depolymerization. This unzipping process is depicted as the backward reaction in Figure 2.11.

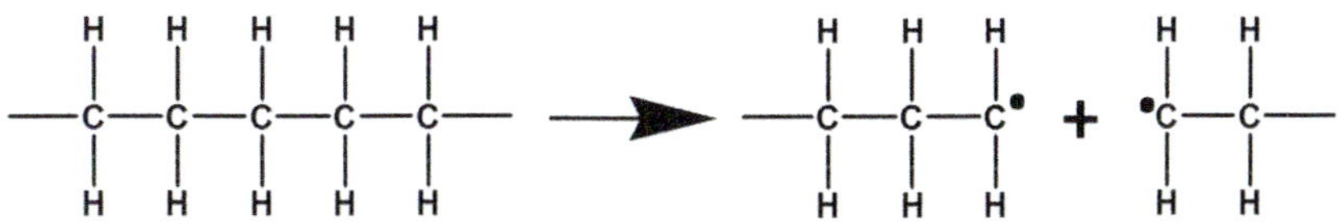

Figure 2.9 Bond cleavage generates two polymeric free radicals.

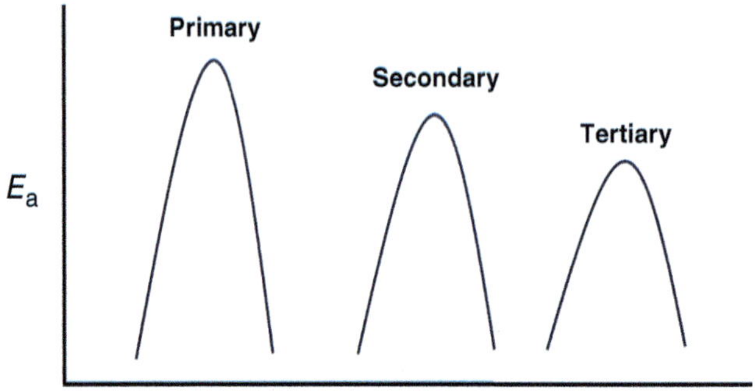

Figure 2.10 Illustration of activation energies for the cleavage of primary, secondary, and tertiary carbon–carbon bonds. (Note that the E_a shown in the figure are not drawn according to scale. Also, the horizontal diagram is not meant to indicate a parameter or numerical value here, but rather, the three activation energies are placed adjacent to one another for visualization purposes.).

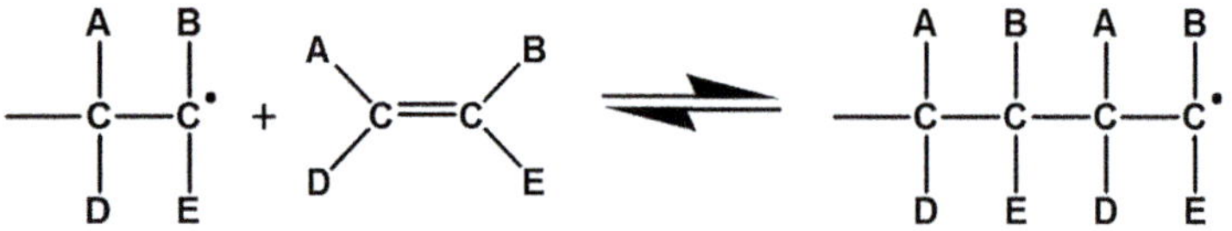

Figure 2.11 Polymerization and depolymerization reactions. The forward reaction (represented by the arrow from left to right) is polymerization, while the backward reaction is unzipping (depolymerization).

The difference between unzipping and random chain scission is that random chain scission, even when carried out to a very great extent, does not result in a high yield of monomers, whereas unzipping produces mostly monomers. Also, random chain scission results in a rapid decrease in molecular weight, whereas unzipping produces rapid weight loss (if the monomers are volatile) but a slower change in the molecular weight of the remaining polymer.

Relatively pure thermal degradation is encountered in pyrolysis, the process of heating a material (in the absence of oxygen) to nearly totally degrade the material, thereby producing non-polymer products that are generally at least partly gaseous. Many, but not all, of the reactions involved in pyrolysis, lead to the formation of free radicals.

Since the bond dissociation energy is the most relevant parameter in determining thermal stability, it is useful to take a look at the dissociation energies of some bonds that are commonly found in polymers, as shown in Table 2.3.

We can see the relationship between bond energy (bond stability) and polymer degradation by examining a graph of the half-weight loss temperature (T_h) versus the bond dissociation energy. The half-weight loss temperature is the temperature

Table 2.3 Bond dissociation energies of various single bonds (kJ/mol) (the bond that is dissociating is shown as a "dash" line that is placed furthest to the right in each structure for structures having multiple dashes, or alternatively as the only dash line for structures drawn with only one dash line) [21].

Bond	Dissociation energies (kJ/mol)	Bond	Dissociation energies (kJ/mol)
CH_3—F	460.2	CH_3—CH_3	377.4
C_2H_5—F	447.4	C_2H_5—CH_3	370.3
CH_3—Cl	350.2	*n*—C_3H_7—CH_3	372.0 ± 2.9
C_2H_5—Cl	352.3 ± 3.3	*n*—C_4H_9—CH_3	371.5 ± 2.9
CH_2=$CHCH_2$—Cl	298.3 ± 5.0	*i*—C_3H_7—CH_3	369.0
C_6H_5—H	433.9	*t*—C_4H_9—CH_3	363.6
C_2H_5—H	420.5	CH_2=CH—CH_3	359.8 ± 5.9
i—C_3H_7—H	410.5	CH_2=$CHCH_2$—CH_3	317.6 ± 3.8
t—C_4H_9—H	400.4 ± 2.9	CH_2=$CHCH(CH_3)$—CH_3	302.5 ± 6.3
$C_6H_5CH_2$—H	375.5	C_6H_5—CH_3	433.9 ± 4.2
CH_2=$CHCH_2$ H	369 ± 3	C_6H_5—CH_2—CH_3	325.1 + 4.2
HO—OH	210.45 ± 0.04	C_6H_5—$CH(CH_3)$—CH_3	318.8 ± 8.4
HO—H	497.32 ± 0.02	C_6H_5—$C(CH_3)_2$—CH_3	303.3 ± 8.4
HOO—H	365.71 ± 0.17	*n*—C_4H_9—H	421.3
t—C_4H_9O—OH	186.2 ± 4.2	*n*—C_3H_7—H	422.2 ± 2.1
i—C_3H_7O—OH	185.8 ± 6.3	C_2H_5O—OH	178.7 ± 6.3

at which half the weight of the polymer has been lost (at a specified rate of temperature increase in an inert atmosphere) due to the formation of volatile products of the reaction.

PE is more stable than PP because secondary bonds are more readily cleaved than primary bonds. This is evident from the thermogravimetric analysis (TGA) thermograms of various waste plastics, including PE, PP, PS, and PET, which were reported by Miandad and coworkers, as shown in Figure 2.12 [22]. It can be seen that PP undergoes weight losses at lower temperatures than PE, which is consistent with a lower stability for PP in comparison with PE.

2.12.1.1 The Chemistry of Thermal Degradation

There are three general mechanisms involved in thermal degradation, although usually more than one mechanism is in operation simultaneously. These mechanisms include random scission, depolymerization, and elimination of

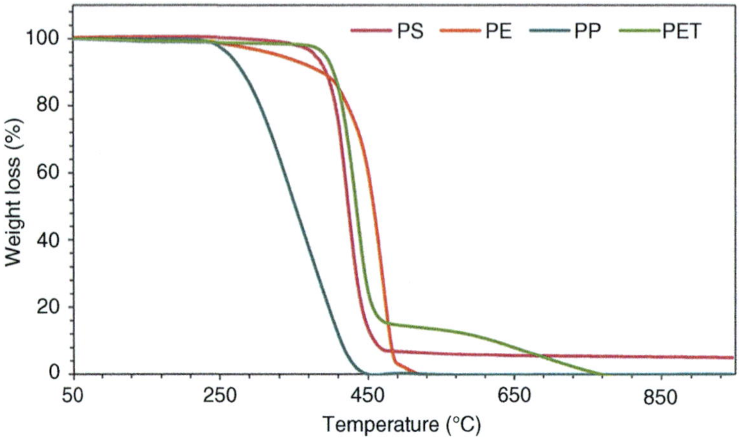

Figure 2.12 Thermogravimetric analysis of various forms of plastic waste, including polystyrene (PS), polyethylene (PE), polypropylene (PP), and poly(ethylene terephthalate) (PET). This plot is taken from: Miandad, R., 2019 / Frontiers Media S.A. / CC BY 4.0.

low molecular weight fragments (other than monomers). We will briefly discuss these mechanisms below.

Random Scission Random scission, as the name implies, involves the random cleavage of bonds in a polymer's backbone chain or side chains. When scission happens in the backbone, it results in a rapid decrease in molecular weight, with almost no monomer formed during the early stages of the reaction. Random chain scission is typically the mechanism that predominates when polyolefins are exposed to elevated temperatures in the absence of oxygen. For example, Table 2.4 shows the volatile products produced by pyrolysis of polypropylene at 438 °C [23].

Depolymerization In depolymerization reactions, monomer units are cleaved from the chain ends, so that this process is essentially the reverse of the polymerization reaction (as shown in the backward reaction in Figure 2.11). This mechanism results in very little change in molecular weight during the early stages of the reaction, in contrast with random chain scission, where molecular weight change is rapid. In the case of addition polymers, the mechanism for depolymerization predominantly involves free-radical reactions and is essentially the same as the **"unzipping"** reaction described above. While costly and environmentally harmful catalysts are often required to achieve depolymerization, Rabnawaz et al. [24] have recently demonstrated that PS could be efficiently

Table 2.4 Volatile products obtained from the pyrolysis of polypropylene at 438 °C [23].

	Yield (weight percent)	
Product	**Atactic PP**	**Isotactic PP**
Methane	0.1	0.09
Ethane	1.2	1.4
Propylene	12.2	10.7
Isobutylene	3.2	2.5
2-Pentene	15.8	15.5
3-Methyl-1-pentene	11.4	10.2
3-Methyl-3,5-hexadiene	2.0	1.7
4-Methyl-3-heptene; 2,4-dimethyl-heptadiene; 2,4-dimethyl-heptene	39.3	42.6
$C_{10}H_{20}$	2.4	2.1
4,6-Dimethyl-3-nonene	1.3	1.6
2,4,6-Trimethyl-8-nonene	9.5	10.2
$C_{13}H_{24}$	1.7	1.6

Note that the production of some of these compounds requires considerable rearrangement of the molecular structure – an indication of the complexity of the reactions that occur during this process.
Source: Adapted with permission from Chien, J. C. W.; Kiang, J. K. Y. In *Stabilization and Degradation of Polymers*; AMERICAN CHEMICAL SOCIETY, 1978; Vol. 169; pp 175–197. Copyright 1978 American Chemical Society.

depolymerized by using an oxidized copper scrubber as well as table salt to promote this reaction. This depolymerization provided a styrene monomer content of >83% in the liquid fraction, and the recovered monomer could be repolymerized without the need for further purification prior to the repolymerization reaction.

However, the depolymerization process is quite different for condensation polymers. The depolymerization of PET is shown in Figure 2.13. Here, the ester linkages of PET break down by reactive solvents such as water, methanol, or glycol, to produce monomer terephthalic acid, dimethyl terephthalate, and bis(2-hydroxyethyl) terephthalate, respectively [25].

Depolymerization via the unzipping approach is common for polymers such as polyacetals and poly(methyl methacrylate) (PMMA).

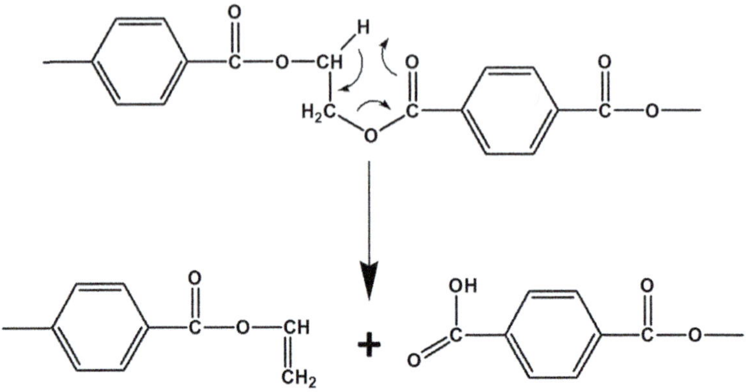

Figure 2.13 PET degradation through depolymerization.

Elimination of Low-molecular-weight Fragments Other Than Monomers In some cases, the cleavage of small fragments occurs, but the fragment is not a monomer. This type of reaction also results in very little change in the polymer's molecular weight during the early stages of this process.

For example, the elimination of low-molecular fragments that are not monomers, and which do not involve main chain scission, is the predominant process involved in the thermal degradation of PVC, as shown in Figure 2.14.

It is important to note that the presence of a double bond in the main chain facilitates the elimination of any adjacent HCl group, so once the reaction gets started, subsequent reactions tend to occur nearby, producing extended conjugation (as is similarly shown for polyvinyl acetate in Figure 2.15). This extension of the polymers' conjugation leads to color changes as the reaction progresses. As the reaction proceeds the polymer is initially yellow in color and then becomes orange and eventually its color changes to red brown.

One way to examine the relative dominance of random chain scission and depolymerization reactions during pyrolysis of a polymer is to look at the proportion of monomer generated in comparison to the generation of other small molecules, as shown in Table 2.5.

2.12.1.2 Some Polymers Degrade by Unzipping and Others by Random Scission

Unzipping in PMMA and PS As we have discussed earlier, thermal degradation (in the absence of oxygen) can lead to depolymerization and generation of the

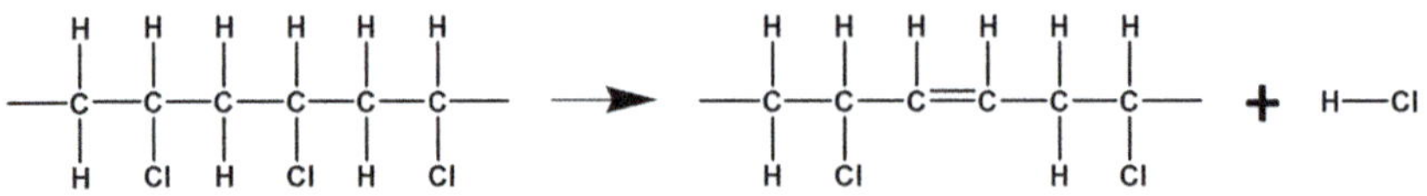

Figure 2.14 Thermal degradation of polyvinyl chloride.

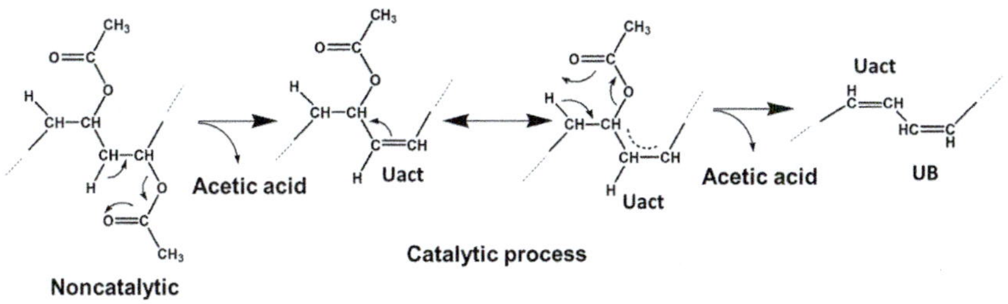

Figure 2.15 First step in the thermal degradation of polyvinyl acetate (rate-controlling step); elimination of acetic acid. The term Uact denotes the double bond catalyzing the deacetylation of the neighboring VAc entity, while UB denotes an inactive double bond [26] / Elsevier.

Table 2.5 Monomer yield in the pyrolysis of selected polymers.

Polymer name	Polymer structure	Monomer yield (%)	Predominant degradation pathway
Polyoxymethylene	$-[CH_2-O]_n-$	100[27]	Depolymerization
Polytetrafluoroethylene	$-[CF_2-CF_2]_n-$	96	Depolymerization
Poly(methyl methacrylate)	$-[CH_2-C(CH_3)]_n-$, with $CH_3O-C{=}O$ on the C	97[28]	Depolymerization
Poly(methyl acrylate)	$-[CH_2-CH]_n-$, with $CH_3O-C{=}O$ on the CH	1	Random chain degradation
Polystyrene	$-[CH_2-CH]_n-$, with C_6H_5 on the CH	41	Depolymerization
Poly(-deutero styrene)	$-[CH_2-CD]_n-$, with C_6H_5 on the CD	68	Depolymerization

(Continued)

Table 2.5 (Continued)

Polymer name	Polymer structure	Monomer yield (%)	Predominant degradation pathway
Poly(-methyl styrene)	$-[CH_2-C(CH_3)(C_6H_5)]_n-$	100	Depolymerization
Polyisoprene	$-[CH_2-C(CH_3)=CH-CH_2]_n-$	11	Random chain degradation
Polyethylene	$-[CH_2-CH_2]_n-$	1	Random chain degradation
Polyvinyl chloride	$-[CHCl-CHCl]_n-$	0	Random chain degradation

monomer or to the generation of a variety of decomposition products. Poly(methyl methacrylate) (PMMA) gives us about 95% monomer through the sequence of reactions shown in Figure 2.16, which begins with random chain scission. Note that the molecule on the bottom right of the figure is the methyl methacrylate monomer. The ceiling temperature is the temperature at which the polymerization and the depolymerization rates reach equilibrium. Of course, one should not heat a polymer to this temperature as it will decompose with significant monomers and degradation unless you want to intentionally degrade the polymer. For PMMA, the ceiling temp is 195 °C.

For PS, the ceiling temperature has been reported to be 310 °C.

Random Degradation in Polypropylene: PP does not give a large yield of its parent monomer when it degrades because PP does not unzip to its monomers to any significant extent. In fact, PP degrades via chain scission, hydrogen abstraction, and chain transfer processes. The various reactions involved in the thermal degradation of PP are shown in Figure 2.17a–c.

As mentioned above, one such reaction is chain scission (e.g., cleavage of C—C bonds in the polymer chain) as shown in Figure 2.17a.

Other reactions include hydrogen abstraction and chain transfer (the radical in one polymer chain abstracts H from another polymer chain as shown in

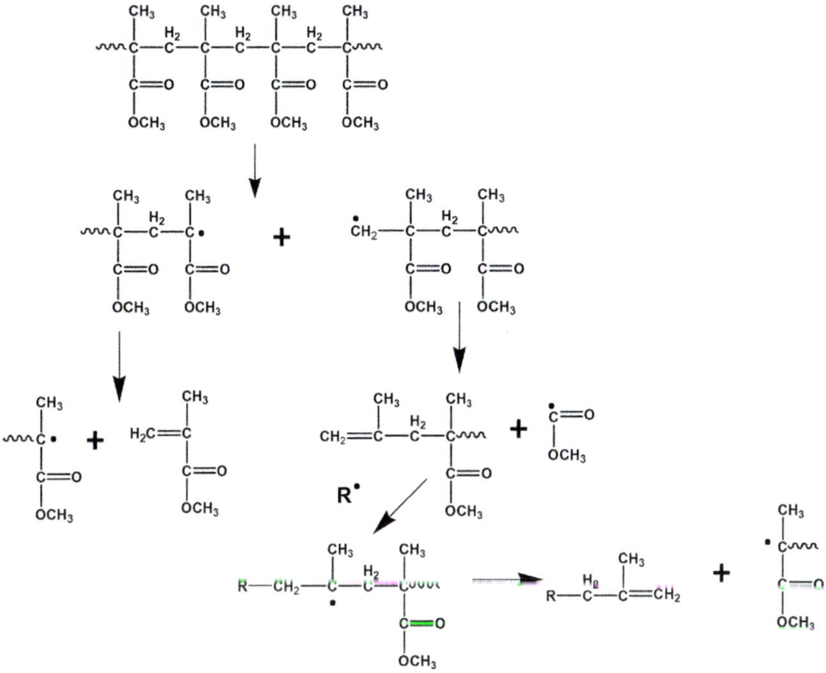

Figure 2.16 Thermal degradation of poly(methyl methacrylate).

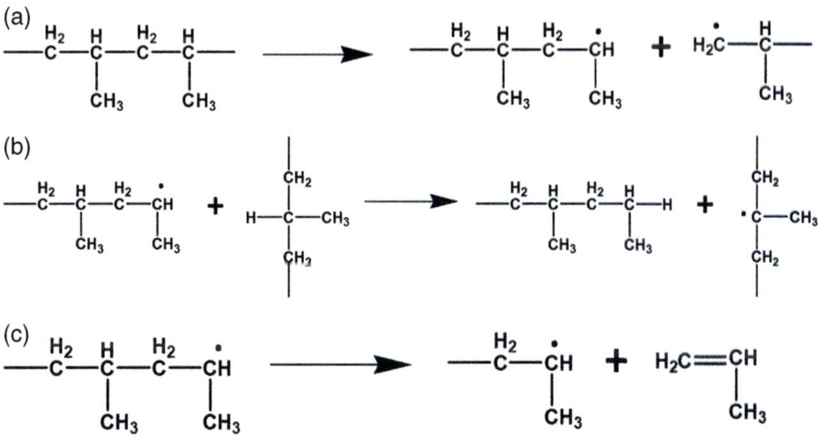

Figure 2.17 Thermal degradation of PP via: (a) chain scission, (b) hydrogen transfer and chain transfer, and (c) depolymerization into its parent monomers.

Figure 2.17b. Thus, the radical moves to another polymer chain where a new degradation reaction then begins)

PP can undergo degradation into its parent monomers, as shown in Figure 2.17c. Although this is not the main pathway of degradation, when it occurs, it leads to the formation of the parent monomer from the end of the polymer chain or from anywhere in the polymer chain.

The reason that PMMA and PS primarily produce their respective monomers during their thermal decomposition while PP produces a variety of compounds relates directly to the relative stability of the reactive intermediates (as well as the final molecules) that are formed. Likewise, the relative stability of these intermediates has an effect on the propensity of the polymers to undergo degradation reactions.

The guiding concept is perhaps somewhat counter-intuitive: the more *stable* the reactive intermediates (free radicals, in the cases we are dealing with), the *faster/greater* the degradation. This occurs because the slow step in the process is the formation of free radicals. Once these free radicals are formed, they react very quickly. Thus, while a less stable free radical will itself react somewhat more quickly than a more stable free radical, it takes more energy to produce the less stable radical in the first place and it will, therefore, be formed at a slower rate. The second part of this process – the reaction of the radical with other species – will be faster, but that has virtually no influence on the overall reaction rate. The first part, the formation of the radical, is the rate-limiting step (also called the rate-determining step).

In the last section, we mentioned the general rule that tertiary bonds are weaker than secondary bonds. This is because tertiary carbon free radicals are more stable than secondary ones, and secondary free radicals are, in turn, more stable than their primary counterparts [29]. Therefore, if a degradation route is available that will result in the formation of tertiary free radicals, it will occur more often than degradation steps that result in the formation of secondary or primary free radicals. Since the tertiary free radical is more stable, the bond that is broken to form a tertiary free radical is less stable than a comparable bond that would form a secondary or primary free radical – the reaction with that bond is favored energetically so that the reaction pathway is favored.

If we now look back at the PMMA example, the initial scission results in the production of one tertiary and one primary free radical. The tertiary free radical can easily "spit off" a monomer, thereby leaving another tertiary free radical. This is a relatively low-energy course of reaction, as it requires only that the carbon with an unpaired electron pair up with the adjacent carbon, disrupting that carbon's bond with the adjacent carbon (which is also tertiary).

If we compare this scenario with that for PP, the initial scission results in the production of one secondary and one primary free radical. While the secondary free radical can act in the same way as described above (spitting off a monomer), the resultant free radical is also secondary. Another route of reaction is for that free radical to grab off hydrogen from a CH moiety that is attached to a CH_3 group as well as to two main-chain carbons. That creates a tertiary free radical, so it is energetically more favorable and is exactly what is shown in the hydrogen abstraction reaction. It is also precisely what happens during an intermediate step (not shown) in the chain transfer reaction, before a main-chain bond is broken and a double bond formed. If we look at PMMA, we don't have any H atoms on the tertiary carbons that would be easy targets for these types of hydrogen abstraction reactions – so we don't get nearly as many of these competing reactions.

Note here also that a number of reactions are possible. Energy is not uniformly spread over the different polymer molecules, so in some places at some times, there will be sufficient energy for the somewhat-higher-energy reactions to occur. This is one of the main reasons that polymer degradation usually involves a variety of reactions, which in turn form a variety of products. This point has recently been explored by Rabnawaz et al. [24], with PS, and mixed polyolefins, where more uniform heat distribution resulted in high yields and fewer side products.

We can also look back at the half weight loss temperatures for PP and PE (from the figure shown above) and think about the relationship to free-radical stability. When we have random chain scission in PP, we produce a primary and a secondary free radical. When we have random chain scission in PE, we produce two primary free radicals. Since secondary free radicals are more stable, less energy is required to produce them, so PP loses half its weight at a lower temperature than is required for PE to do so.

To conclude this section, the thermal degradation of a polymer is dependent primarily on its chemical structure (bond types). Fluorinated polymers are the most stable because they have the largest dissociation energies. Peroxides are the least stable because of their weaker bonds. In fact, peroxides are used as initiators for polymerization because of their lower decomposition temperatures.

2.12.2 Oxidative Degradation

In contrast to pure thermal degradation, which is almost never encountered in packaging (exceptions include some recycling processes that use pyrolysis to produce fuels or small feedstock molecules from end-of-life plastics), **oxidative degradation** occurs in all commercially important polymers, including those used in packaging. The extent and significance of oxidation varies, but it occurs with virtually all polymers, at varying rates, during various stages of their life cycle.

All polymers (with very rare exceptions – polytetrafluoroethylene comes to mind) are susceptible to oxidation when they are exposed to a combination of oxygen and high temperature. This typically occurs during extrusion, molding, and thermoforming. It is one of the reasons why the "thermal history" of a resin sample is an important indicator of the properties that you can expect from it. As a general rule, each heat exposure event results in additional oxidation; therefore, steps such as compounding, pelletizing, and regrinding treatment tend to increase the amount of oxidation of the polymer, thus affecting its performance.

Highly sensitive polymers are very susceptible to oxidation during processing, as described above, but in addition, they are usually susceptible to oxidation during synthesis (polymerization) and even during storage or during use at ambient temperatures. **We should also note that** once oxidation is started, it continues (though generally slowly) throughout the lifetime of the polymer.

Oxidation is accelerated by exposure to heat, UV light, mechanical energy, and the presence of pro-oxidants (contaminants or deliberate additives, which will be discussed in more detail later). Oxidation tends to be an auto-accelerating process – the more oxidation has occurred, generally the faster the rate of subsequent oxidation.

The major changes that occur during oxidation are:

1) **Initiation:** the formation of free radicals ($\dot{R}$) denoted by a small dot while R denotes a polymer chain. These are highly reactive, and they can trigger new reactions. As we will discuss later, hydroperoxides are the most important source of free radicals.
2) **Chain Propagation:** This process proceeds via the reactions shown below:

$$R^{\cdot} + O_2^{\cdot} \rightarrow RO_2^{\cdot} \left(\text{this } O_2 \text{ comes from the atmosphere}\right)$$

$$RO_2 + R1 \rightarrow ROOH + R1 \left(\text{another radical generation}\right)$$

Note that we have formed another free radical $R\dot{1}$, which can then react with oxygen as shown in the first of these two steps. Also, note that ROOH is a hydroperoxide, which will react as shown next.

3) **Oxidative Chain Reaction:** This process proceeds via the reactions shown below:

$$R_1OOH \rightarrow R_1O^{\cdot} + {}^{\cdot}OH$$

$${}^{\cdot}OH + R2 \rightarrow R2^{\cdot} + H_2O$$

or

$$R_1O^{\cdot} + R_2^{\cdot} \rightarrow R_1 - O - R_2 \left(\text{Crosslinking}\right)$$

These chain-branching reactions occur because peroxides tend to be unstable; the —O—O— bonds are weak and are therefore easily disrupted. Note that we now have generated additional free radicals that can go on to react with other species.

The result, then, is a cascading series of reactions where the initial presence of a single free radical leads to a chain of events producing a large number of free radicals. That is what makes the reaction autoaccelerate.

The predominant reaction pathway differs for heat and UV energy sources:

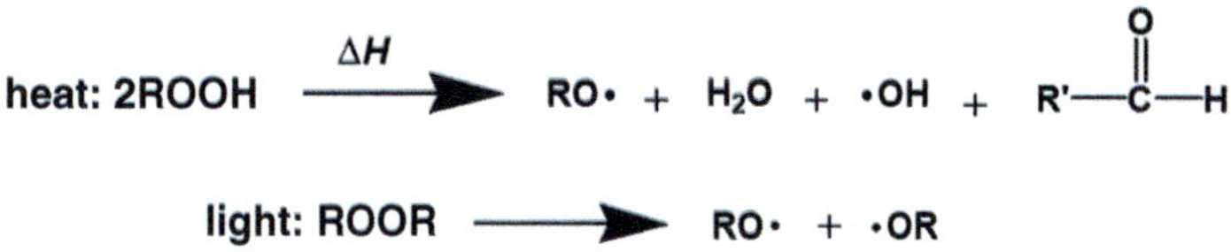

Cobalt, iron, manganese, copper, and other transition metals are important catalysts for hydroperoxide decomposition.

One of the most important consequences of oxidation occurs through chain scission reactions that are different from the typical reactions that we have described above. For example, in polypropylene, the reaction shown in Figure 2.18 can occur after hydroperoxide decomposition:

Note that this reaction results in chain scission – and consequently in a large decrease in molecular weight. Although this reaction occurs much less frequently than the more typical reaction described above:

$$RO^{\cdot} + RR \rightarrow ROOR + R^{\cdot}$$

Its practical significance is substantial. An average of a single chain scission per molecule can destroy the technological usefulness of plastic. The usual way to monitor such changes is their effect on **polymer viscosity** (since viscosity is an indirect measure of the average molecular weight).

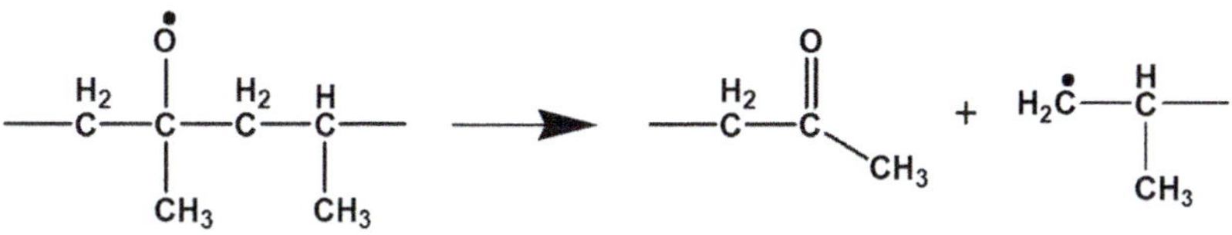

Figure 2.18 Chain scission in PP following hydroperoxide decomposition. (Note: the O in the first compound after the arrow is attached to the C by a double bond).

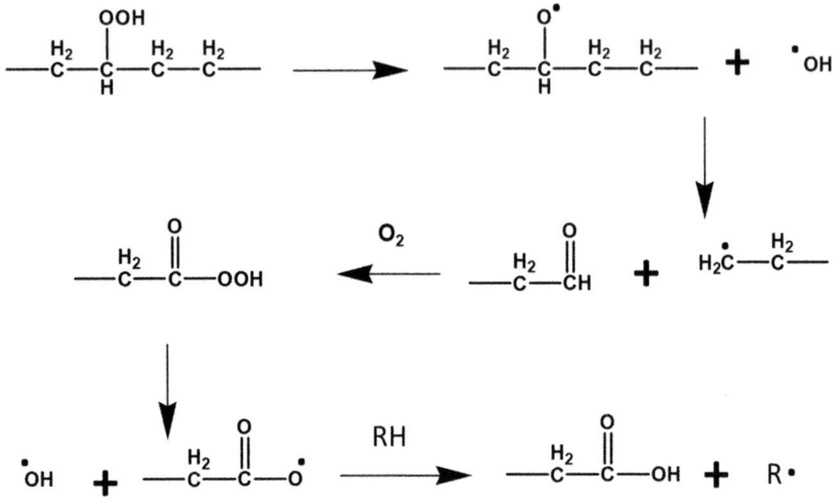

Figure 2.19 Oxidative chain scission in polyethylene.

Note that the precise structure shown above is not required for chain scission to occur. This is simply an illustration of the type of molecular rearrangement that results in chain scission. As another example, oxidative chain scission in PE proceeds, as shown in Figure 2.19, starting with the production of a free radical of the type shown above.

We will see later that the formation of carbonyl groups (which have the structure shown below):

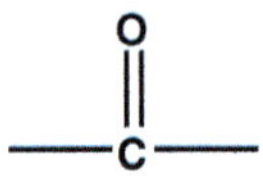

has a crucial role in accelerating subsequent **photodegradation reactions because carbonyl groups absorb UV light.**

Another kind of reaction that can and does occur, with or without oxidation, is the recombination of radicals, as shown in Figure 2.20.

This type of reaction generally occurs during crosslinking processes. Again, these reactions are typically less likely than the oxidation reactions described

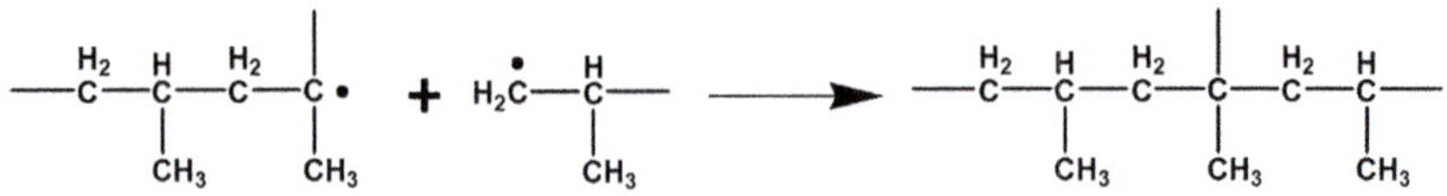

Figure 2.20 Example of a reaction involving the recombination of free radicals.

above, but they have appreciable practical significance. In some polymers, either chain scission or crosslinking predominates. In other polymers, both occur to roughly the same extent and to some degree counter balance each other. Oxidation reactions can compete with both chain scission and with crosslinking reactions.

As has been discussed earlier, oxidation can result from exposure to either heat or light energy, in the presence of oxygen. The mechanisms of the two reactions can be affected by the energy source (as described above). There are additional differences between thermal and photooxidation. Thermal oxidation processes tend to be bulk processes – they occur throughout the polymer sample (although, as we have discussed, crystallinity has an effect). In contrast, photooxidation tends to be very much a near-surface phenomenon, at least initially. Ultraviolet light energy tends to not penetrate very far into most polymers. Therefore, UV-induced oxidation reactions are primarily a surface phenomenon.

Both oxidative and nonoxidative reactions can occur in plastics. Thus, as one might expect, there is a large variety of degradation reactions that can and do occur. Some of the reactions that occur in PE oxidation are shown in Figure 2.21.

Note that reactions (2) and (3) in Figure 2.21 involve crosslinking, while reaction (6) results in chain scission. Reaction (4), which leads to (6), is favored when oxygen is abundant, while (2) and (3) are favored when there is not much oxygen present. Also, note that not all reaction products are shown. For example, in reaction (1), an ROH is also formed, and in (4), an $R^{\cdot}$ is formed.

Polypropylene undergoes a similarly complex set of reactions, as shown in Figure 2.22.

Reaction (2) in Figure 2.22 tends to occur if oxygen is deficient, and reaction (3) is observed if oxygen is plentiful. While recombination to form crosslinks is possible in PP, it is not favored because of the increased stability of the tertiary, as compared to primary, free radicals.

So, we can see that just as for nonoxidative reactions, the rate of oxidation is affected by the relative stability of the reactive intermediate species that are formed. However, in the case of oxidation, we must also keep in mind that the availability of oxygen is an additional important factor. We can get rid of the effects of physical structure and of oxygen availability, so that we can then pick out the effects of the stability of the intermediate species, by examining rates of oxidation of polymers in solution.

There are two types of influences that are important. One of these is known as the "electron delocalizing" effect [31]. The other effect is that of polarity. While both of these may be operating at the same time, frequently, one predominates over the other. If we have only C and H atoms, polarity is not typically significant – since we do not have any functional groups that have any substantial polarity. In this case, we can look back at our concepts of primary/secondary/tertiary carbons, and add to it a bit.

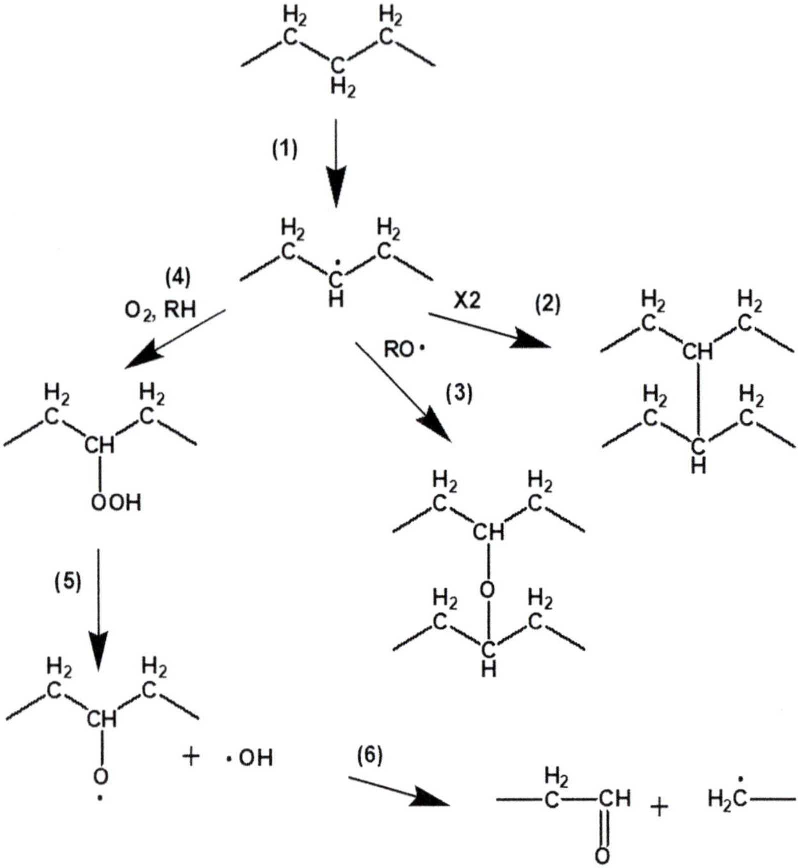

Figure 2.21 Polyethylene oxidation reactions [30].

The reason that a tertiary free radical is more stable than a secondary free radical, and secondary more stable than primary, is because of the ability this provides for the unpaired electron to kind of "roam around" a bit – the electron delocalizing effect. Think of the following species:

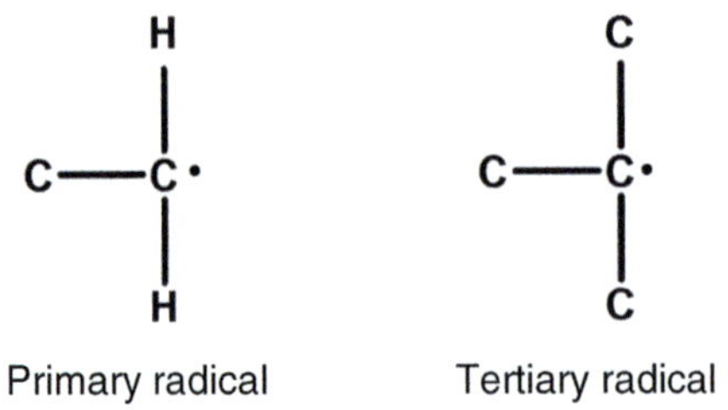

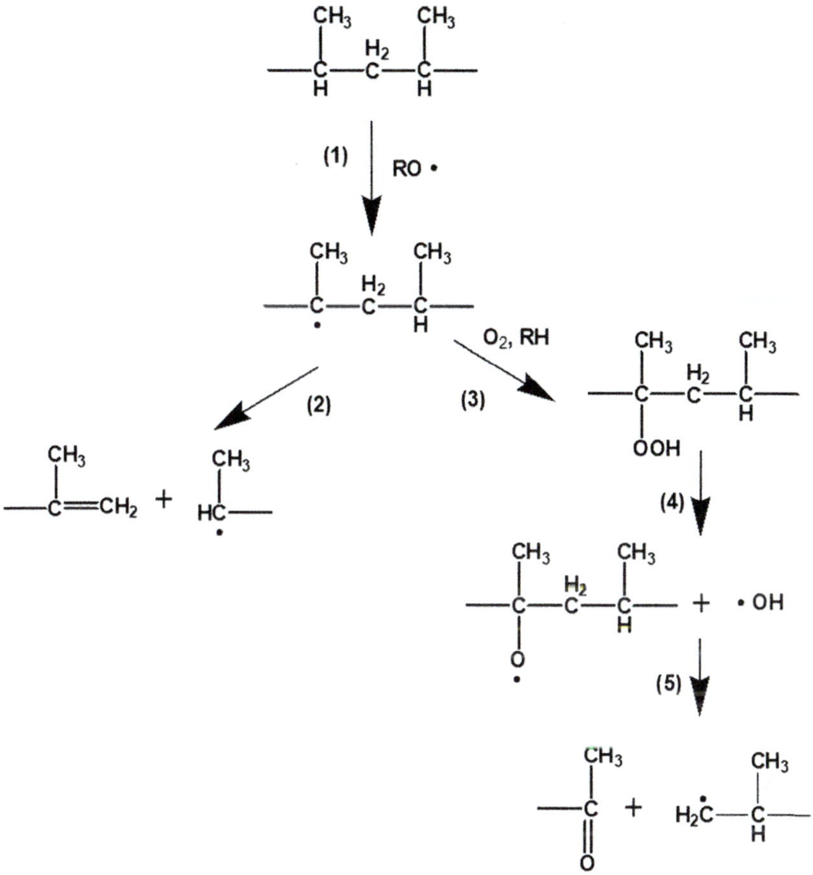

Figure 2.22 Polypropylene oxidation [30].

In the primary case, the unpaired electron on the carbon of interest cannot spend any time on the hydrogens because these atoms cannot accommodate any electron presence. On the other hand, the unpaired electron can spend a little time with the attached carbon. In the tertiary case, there are three attached carbons that the unpaired electron can spend a bit of time with – that is essentially what we mean by an electron delocalizing effect. Thanks to the presence of these neighboring carbons, the electron is not "trapped" in one location as much. That lowers the energy level of the radical species, thus making it somewhat more stable. As we have discussed earlier, a more stable intermediate species leads to a greater overall rate of reaction since the energy "hump" or barrier to get over (the activation energy) is lower, so a greater proportion of molecules will have sufficient energy to react.

We have discussed above the free-radical formation involving the cleavage of C—C bonds in the main chain in some random chain scission process. For oxidation, it is even more important in the context of the hydrogen abstraction reactions that lead to the formation of hydroperoxides. These hydroperoxides decompose, thereby generating even more free radicals to add on oxygen and then participate in a hydrogen abstraction reaction that in turn leads to the formation of the hydroperoxides. Therefore, this process continues on and on as repeated cycles.

When we apply this concept to oxidation rates, we find, as we have discussed earlier, that PP is more susceptible to oxidation than HDPE. In other words, the rate of oxidation of PP is greater than the rate of oxidation of HDPE. We can extend this same concept beyond the three categories of free radicals, and we will find that the **rate of oxidation increases** in the following series:

$$\text{PE} < \text{PP} < \text{PS} < \text{PB}\left(\text{polybutadiene}\right) < \text{PI}\left(\text{polyisoprene}\right)$$

The presence of the benzene ring in PS provides even more opportunity for electron delocalization than is offered by the methyl group in PP, so PS oxidizes more rapidly than PP. The presence of the double bond in polybutadiene provides even more ability for the electron to delocalize. To understand this, we need to remember the concept of resonance and see that the free radical can actually change its position, as shown below:

Therefore, although this looks like a secondary free radical, it is actually more stable (with a lower energy) than the tertiary free radical shown earlier. If we now add a methyl group to this carbon, in place of the hydrogen, to obtain polyisoprene:

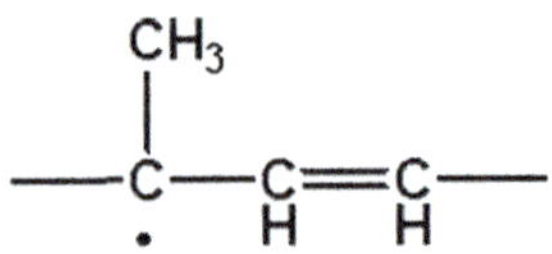

we will have an even more stable free radical and a still greater rate of oxidation. (Remember – if the free radical is MORE stable, it takes LESS energy for it to form, so removing a hydrogen to form a free radical that will then go on to react with oxygen is easier, so that this process occurs MORE rapidly.)

When noncarbon atoms are involved, polarity is generally more influential than electron delocalizing effects. The presence of a polar atom

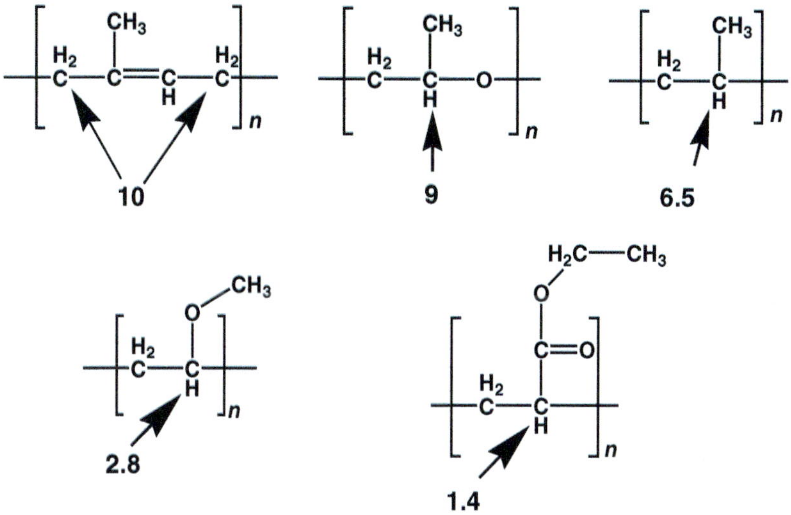

Figure 2.23 Relative rates of oxidation of various polymers. **A higher value denotes faster oxidation.** The arrows show the point where oxidation will most probably occur (the lowest energy location) in each structure.

influences how strongly the carbon holds onto an attached hydrogen. If the carbon is attached to an electronegative atom (Cl, for example), that atom draws two electrons more towards itself that is involved in that (polar) covalent bond. Consequently, the carbon will have a partial electron deficiency, so it can be thought of as holding on tighter to the other electrons that it has a share of, and specifically, to the electron belonging to any hydrogen that it is attached to, thus making it harder to sever that C—H bond. If it takes more energy to break a C—H bond when the C is attached to an electronegative species, that means the rate of oxidation will be lower, as shown in the following series:

> **Rate of oxidation:** PE > nylon > PET > PVC > PMA > PMMA > Polyacrylo nitrile (PAN) >> PTFE

Combining these effects, F. Rodriguez (***Principles of Polymer Systems*, 4th ed., Taylor & Francis, 1996**) presents a set of relative rates of oxidation as shown in Figure 2.23.

2.12.3 Photodegradation

Just as is the case for thermal degradation, we can, conceptually, differentiate between "pure" photodegradation and photochemical oxidation, but in packaging applications, we will have a mixture of reactions being initiated through photodegradation but propagating primarily through photochemical oxidation.

We already saw one example of photochemical oxidation, namely, hydroperoxide decomposition:

$$\mathrm{ROOH} \xrightarrow{h\nu} \mathrm{RO}\cdot + \cdot\mathrm{OH}$$

with subsequent hydrogen abstraction and oxidation leading to a cascade of events. This reaction requires energy of about 42 kcal/mol. UV light can also lead to the reaction shown below:

$$\mathrm{ROOH} \xrightarrow{h\nu} \mathrm{R}\cdot + \cdot\mathrm{OOH},$$

which requires energy of about **70 kcal/mol.**

Still, another possibility is:

$$\mathrm{ROOH} \xrightarrow{h\nu} \mathrm{ROO}\cdot + \cdot\mathrm{H}$$

but this reaction requires about 90 kcal/mol. Thus, the first of these reactions is the most energetically favorable and therefore predominates.

Photodegradation of PVC involves scission of the C—Cl bond, forming free radicals, followed by hydrogen abstraction by the Cl, thus leaving a double bond in the main chain. While thermal degradation has the same end result, it can occur by both free-radical and non-free radical mechanisms. In contrast, photodegradation occurs almost entirely through a free-radical process.

For photodegradation of a polymer to occur, we must have radiation reaching the molecular species involved, the radiation must be absorbed by the molecule, and it must be sufficiently energetic to result in changes in chemical bonds. Light energy decreases with increasing wavelength. Therefore, short wavelength, high energy radiation such as UV is very damaging, while longer wavelength IR does not break polymer chains. Another factor that must be considered is the quantum nature of light. It is not possible for only part of a "packet" of light energy to be absorbed. The closer the amount of the energy in a quantum of absorbed radiation is to the energy of a chemical bond, the more likely it is that that energy will be dissipated by fracturing that bond. (It should also be noted that this quantum nature also influences the absorption of radiation at energy levels that will not result in bond fracture but rather in changes in the vibrational motion of atoms or groups of atoms. The resulting differences in the ability of polymers with different chemical structures to absorb varying wavelengths of light are the basis for IR spectroscopy as a plastics identification technique).

If we examine a portion of the electromagnetic spectrum, we find the wavelengths and approximate energies that are shown in Figure 2.24.

We can compare these energy ranges with some typical values for the dissociation of single bonds of 150–400 kJ/mol, as shown in Table 2.6.

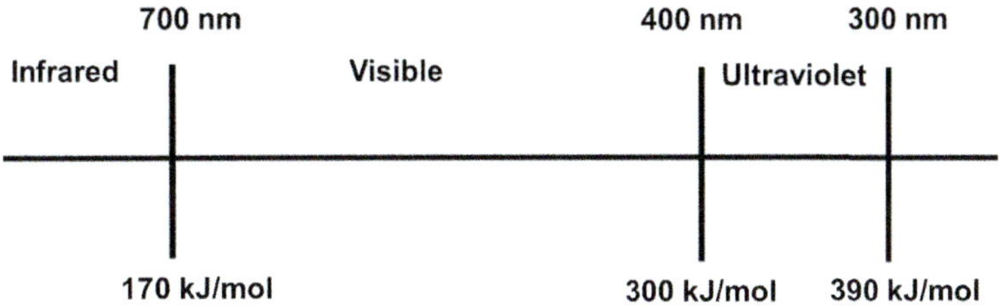

Figure 2.24 Scale of wavelengths and energies in the electromagnetic spectrum.

Table 2.6 Typical single-bond dissociation energies.

Bond type	Bond dissociation energy (kJ/mol)
C—C bonds	260–400
C—H bonds	320–420
O—O bonds	147
C—O bonds	330

We can see that energies that fall in the infrared or visible part of the spectrum are not sufficient to break polymer bonds, while energies in the ultraviolet portion can be sufficient. Since the main source of ultraviolet light is sunlight, exposure to sunlight is the most damaging in terms of the initiation of photodegradation. Since glass blocks most ultraviolet light, indoor light exposure is seldom a significant concern, but outdoor exposure can lead to rapid color change and embrittlement. **NOTE: Light bulbs used indoors do not emit any UV waves.**

As indicated above, for incident light energy to result in degradation, the light must be absorbed by the molecule. Saturated bonds (such as C—C, C—H, O—H, C—Cl, etc.) are transparent to light with wavelengths above about 200 nm, and thus these structures do not absorb light in the UV (or visible) regions. Therefore, it is tempting to conclude that plastics such as PE, polypropylene, and PVC are not susceptible to photodegradation. Unfortunately for our ability to use these materials outdoors without stabilization, that is not the case. Remember that anything we manufacture out of plastic has been exposed to heat and oxygen during processing – so it has already suffered some degree of thermal oxidation. This history leaves in the polymer molecule some chromophoric groups (groups that are able to absorb incident light energy). These groups are usually present in some form of unsaturated structure – in particular ethylenic groups (C—C double bonds), aromatic groups (benzene rings), or carbonyls (C=O groups). Of course, polymers with more chromophoric groups in their structures tend to absorb UV light more strongly than those with fewer of these groups.

Absorption of UV light alone is not sufficient to determine the amount of photodegradation (or photochemical oxidation) that will occur. The other major factor is what happens to that energy once it has been absorbed. Here, carbonyl groups play a particularly important role. Light energy absorbed by carbonyls can lead to three major different types of reactions, all of which result in main chain scission. These are known as Norrish I, Norrish II, and Norrish III reactions, as shown in Figures 2.25, 2.26, and 2.27, respectively.

Note that the free radicals produced in a Norrish I reaction (Figure 2.25) will go on to react with oxygen or engage in other degradation reactions, as previously discussed.

Note that in the case of a Norrish II reaction (Figure 2.26), main chain scission occurs through reactions that do not result in the formation of free radicals.

Note once again, we have main chain scission without leading to the formation of free radicals. This type of reaction can only occur if there is a hydrogen atom available on a side group adjacent to the carbonyl group, so it is relatively rare.

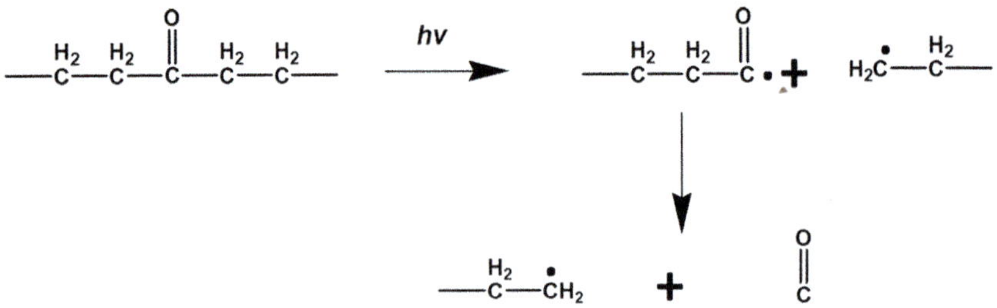

Figure 2.25 A Norrish I reaction [30].

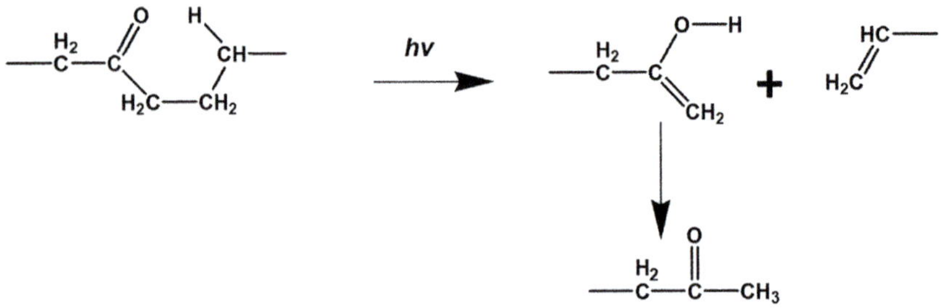

Figure 2.26 A Norrish II reaction.

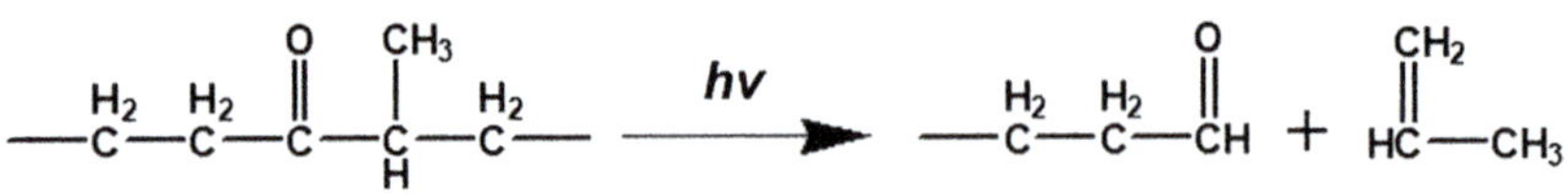

Figure 2.27 A Norrish III reaction.

The activation energies of Norrish I and Norrish II reactions are different, with a Norrish II reaction having an activation energy (E_a) of about 0.85 kcal/mol and a Norrish I reaction having an E_a of about 4.8 kcal/mol. That means that an increase in temperature speeds up a Norrish I reaction much more than it speeds up a Norrish II reaction. Consequently, while Norrish II reactions tend to predominate at room temperature, Norrish I reactions tend to predominate at elevated temperatures.

It should be noted that the "activation energy" associated with a reaction rate refers to the temperature dependence of the rate. Most reaction rates have an Arrhenius-type dependence on temperature, which is expressed by the classic Arrhenius equation:

$$k_T = k_o e\text{-}^{\,Ea/RT}$$

where k_T is the reaction rate constant at temperature T; k_0 denotes the pre-exponential factor, E_a is the activation energy, R is the gas constant, and T denotes the (absolute) temperature.

We also must keep in mind that photodegradation tends to occur mainly at and near the polymer surface. If there are strongly absorbing groups, little UV light will penetrate into the interior of the polymer – and consequently, there will be little light-induced degradation in the interior. In contrast, if the UV light is not strongly absorbed, little energy will be absorbed anywhere in the polymer, and again, there will be little degradation.

With some compounds, particularly those containing aromatic structures, we get strong UV absorption but much less photodegradation than we might expect. PS, for example, absorbs strongly due to the many benzene rings that are present in its structure.

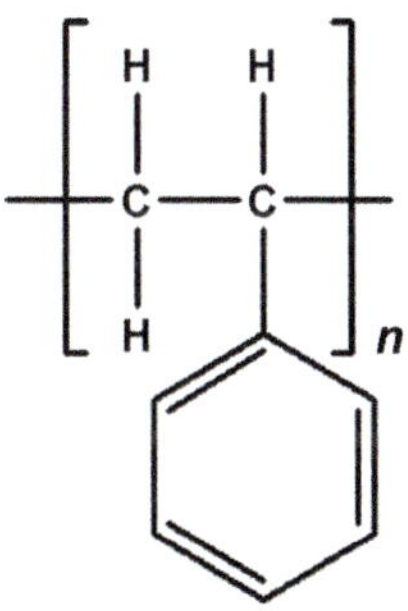

However, PS is comparatively stable to photooxidation since the ring acts as an "energy sink," thus allowing the radiation energy to be dissipated without resulting in bond rearrangement or dissociation.

With polyolefins, polyacrylonitrile, PVC, polyvinyl acetate (PVA), and polyvinyl alcohol (PVOH), we tend to get both chain scission and crosslinking reactions upon exposure to light. In the absence of oxygen, crosslinking will predominate.

Polycarbonate, like PS, is relatively stable to photooxidation, again due to the "energy sink" provided by the aromatic rings, even though it has additional absorbance from the carbonyl group present in its repeat unit:

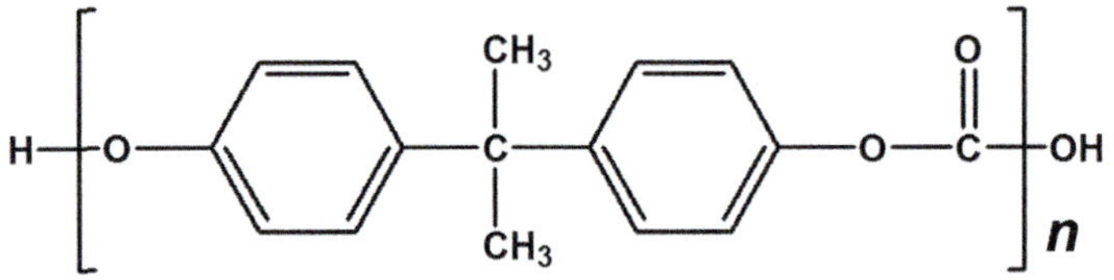

In the case of PET, we have different relative proportions of sensitive and stable absorbing groups than exist in PC and PS. Consequently, even though PET has aromatic rings, it still tends to undergo chain scission on exposure to UV light, with scission occurring at any of the points marked by the numbers 1, 2, and 3:

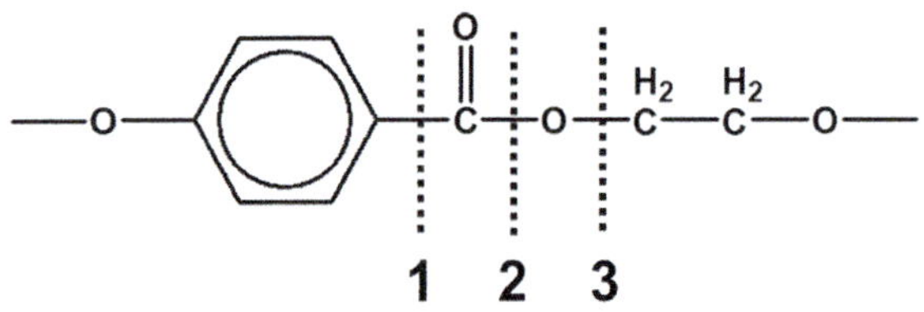

Polyamides undergo both crosslinking and chain scission reactions, but crosslinking usually predominates.

Since the surroundings of the bond, as well as the bond itself, influences absorption, different polymers tend to have different characteristic regions in the electromagnetic spectrum at which they absorb most strongly – and thus characteristic parts of the UV region that are more likely to cause degradation. Table 2.7 shows characteristic values for selected polymer families.

2.12.4 Chemical Degradation

When we discussed oxidation, we really had already talked about one type of chemical degradation – just one that is so common that we usually put it in a category of its own. Other chemical reactions also occur in susceptible polymers. Temperature is a critical variable in all chemical degradation processes.

Table 2.7 Characteristic maximum absorption wavelengths and associated energy for selected polymer families [32].

Polymer	Activation spectra maxima (nm)	Energy (kcal/mole)
Polyester (PET)	325	88
Polystyrene	318	90
Polyethylene	300	96
Polypropylene	310	92
Polyvinyl chloride	310	92
Polycarbonate	295	97
Poly(methyl methacrylate)	290–315	99–91

2.12.4.1 Hydrolysis

One important type of chemical degradation, especially for the family of condensation polymers, is hydrolysis. The end result of hydrolysis is often, although not always, the reverse of polymerization, giving us back the starting monomers. Often, the mechanism involves initial production of free radicals followed by reaction of water with the free radicals. Alternatively, it may involve ionic intermediates. One example of a free-radical mechanism leading to chain scission is the hydrolysis of polyamides:

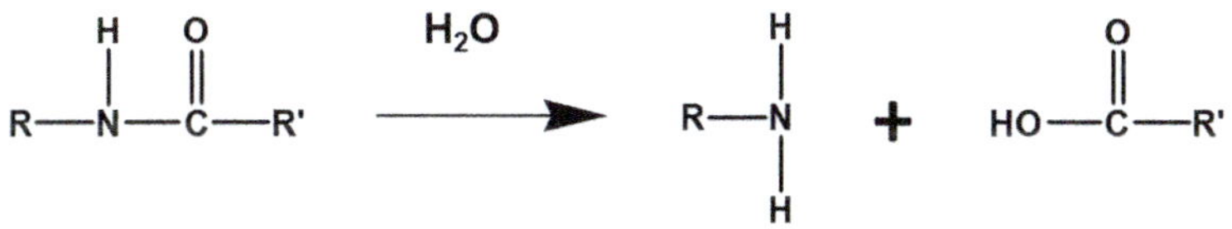

Notice that this reaction is simply the reverse of the condensation polymerization process between an amine and a carboxylic acid, one of the common routes for producing nylons.

Hydrolysis can also take place by reactions that do not involve chain scission. For example, the hydrolysis of polyvinyl acetate is actually used to produce polyvinyl alcohol:

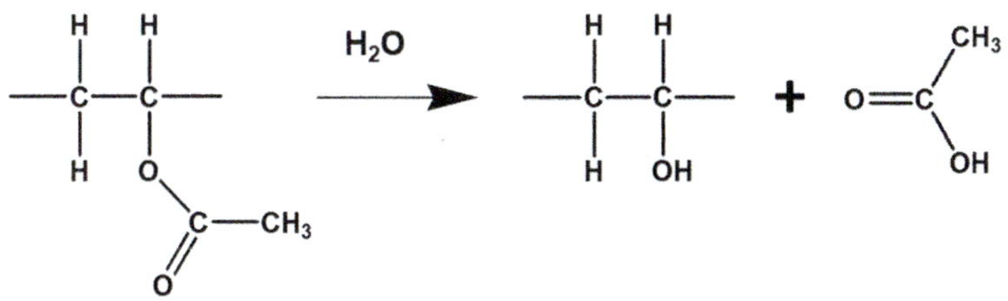

Meanwhile, polyacrylonitrile hydrolysis produces carbonyl groups that can then initiate thermal or photodegradation:

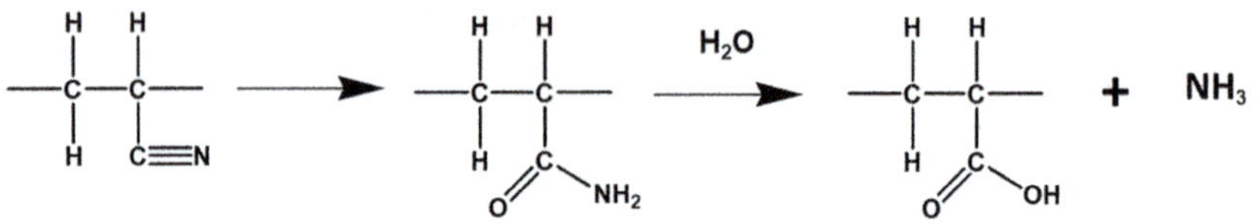

Besides polyamides, polyvinyl acetate, and polyacrylonitrile, other polymers such as polyesters, polycarbonates, polysaccharides, polyurethanes, and others are also susceptible to hydrolysis. In most cases, these reactions are catalyzed by acids or bases. The rate of hydrolysis is generally controlled (limited) by the amount or rate of diffusion of water into the bulk polymer and primarily occurs in the amorphous regions of the polymer. In general, condensation polymers are more susceptible to hydrolysis than are addition polymers. For many of these condensation polymers, it is critical that the polymers be dry when they are processed (extruded, molded, etc.) so that hydrolysis can be avoided during the forming process.

In some cases, hydrolysis is also used as a recycling mechanism. If the hydrolysis reaction produces monomers, these monomers can be recovered, purified, and subsequently repolymerized. There are some commercial facilities in the United States that recycle nylon 6 (primarily from discarded carpets) through hydrolysis [33].

2.12.4.2 Environmental Stress Cracking

Environmental stress cracking (ESC) is a very different type of chemical degradation [34, 35]. It is defined classically as a situation where a reagent (chemical) that does not appreciably attack or dissolve a polymer when the polymer is unstressed will cause catastrophic failure when the polymer is stressed, usually by the initiation and propagation of cracks or crazes.

In other words, if we test the packaging material that is in contact with the product, we do not find an appreciable deterioration of its properties. If we subject the package, either unfilled or filled with an innocuous product, and expose it to stress, we do not experience failure. However, when we combine exposure to the stress-crack agent with moderate stress levels, we can encounter catastrophic failure. Certain materials are known to be stress-crack agents for certain plastics. The mechanisms involved in stress cracking are complex, and predictions are difficult. Some stress-crack agents are known to not be a solvent for the polymer in question (detergents with HDPE); others are known to be solvents but very poor ones (zinc chloride solutions with nylon, alcohols with PS). (Obviously, a material that would be a good solvent for plastic would not be packaged in that plastic in the first place and would also cause a decrease in properties on exposure without any stress being applied.)

The rate of stress-cracking depends on the reagent, its concentration, the particular plastic resin, temperature, stress, and other factors – including the polymer morphology and surface roughness.

The most common mechanism for stress-cracking involves some source of localized weakening on the surface of the plastic due to the chemical action of the stress-cracking agent. Stress then results in localized failure, which opens up a crack and exposes additional surface to penetration by the reagent. The crack acts as a stress concentrator, amplifying the effect of the reagent and causing further failure. The result is propagation of the crack through the material and catastrophic failure of the package.

Crack Initiation Can Occur Through a Variety of Mechanisms

- Selective absorption of the agent in localized regions (for example, amorphous regions of a semicrystalline polymer), resulting in plasticization, lowering of the T_g, softening, weakening, and consequently localized failure.
- Selective solvation of the polymer by the reagent in localized areas, such as low molecular weight fractions and oligomers, leading to a weak point in the surface, perhaps amplified by stress-concentrating surface defects (for example, scratches), thus leading to the initial failure.
- Complexing of the reagent with active sites along the chain (for example, polar groups in semicrystalline polymers) leads to localized destruction of crystallinity and consequent reduction of strength, again leading to localized failure.

The general pattern for a given polymer and reagent is for no stress-cracking to occur below a certain level of stress or strain.

2.12.4.3 Other Chemical Reactions

Chemical degradation can also occur through other chemical reactions, as well as through effects related to solubility. Certain chemical structures in polymers are more susceptible than others (susceptibility is related to bond strength). For example, hydroxyl groups are generally reactive, as they are readily dehydrated.

2.12.4.4 Summary of Chemical Degradation Susceptibilities of Various Polymers

Polyolefins (including LDPE, HDPE, etc.):

Resistant to most corrosives (acids, bases, etc.), but exceptions are:

- Cl_2 at elevated temperatures or in the presence of UV radiation.
- F_2 at moderate temperatures.

- Cl_2 plus either sulfur dioxide (SO_2) or sulfuryl chloride (SO_2Cl_2) leads to the sulfonation of nitric acid at high temperatures, thus resulting in the formation of nitroalkanes.

High-density PE:
- Resistant to acids except nitric and some concentrated acids.
- Resistant to commercial petroleum products such as fuel oil, hydraulic fluid, etc., at room temperature, but not at elevated temperatures.
- Resistant to alcohol.
- Significantly affected by carbon tetrachloride, cyclohexane, *n*-heptane, toluene, *p*-xylene, benzene, and chlorine.
- Discolored by concentrated acetic and chromic acid at 140 °F (60 °C).
- Susceptible to stress cracking by detergents, especially in the case of highly crystalline homopolymers of HDPE.

Low-density PE:
- Susceptible to stress cracking by acetone, citric acid, ethylene glycol, glycerin, soap, carbon tetrachloride, chloroform, ethanol, methanol, and turpentine.
- Possible stress cracking by carbon disulfide, corn oil, ethyl acetate, formaldehyde, olive oil, aqueous phenol, butyl acetate, ethylene dichloride, gasoline, motor oil, toluene, and xylene.
- Markedly swollen by carbon disulfide, carbon tetrachloride, chloroform, gasoline, toluene, turpentine, xylene, ethylene dichloride, heptane, acetone, ethyl acetate, butyl acetate, and so on.

Polypropylene:
- Affected by acetone, butyl acetate, carbon disulfide, carbon tetrachloride (CCl_4), chloroform, ethyl acetate, ethylene dichloride, gasoline, heptane, toluene, turpentine, xylene, concentrated acids or acids at high temperatures, benzene, household detergents, dibutyl phthalate, ethanol, ethyl ether, kerosene, oils, and ketones

Polystyrene:
- If the PS is pure, it will be relatively resistant to oxidation.
- Susceptible to exposure to Cl_2 and sunlight or high temperature.
- Affected by concentrated H_2SO_4, resulting in sulfonation.
- Affected by hot fuming nitric acid, leading to nitrated PS, as well as to mixtures of sulfuric and nitric acid.

Polyvinyl chloride:
- Affected by Cl_2, leading to additional chlorination.
- Resistant to most chemicals.

Condensation Polymers:

- Affected by a variety of chemicals (acid and bases, etc.).

2.12.5 Mechanical Degradation

Mechanical degradation involves mechanically induced decline in a polymer's performance. It includes friction, abrasion, irreversible deformation, crazing (fine cracks/voids), cracking, fracture, and fatigue under static or dynamic loading. Much mechanical degradation, although not all, is due to mechanical scission of bonds due to imposed stress, followed by oxidation. Temperature is an important variable, with increased temperature often leading to higher rates of degradation. Once bond cleavage (dissociation) occurs, we can have the same types of reactions as those we have already discussed, with free-radical mechanisms typically predominating.

In crystalline polymers, mechanical degradation is usually initiated within the amorphous regions of the polymer, as shown in Figure 2.28, where the solid blocks represent crystallites, and the colored lines represent segments of molecules that extend through the amorphous regions between the crystallites (these are called tie molecules).

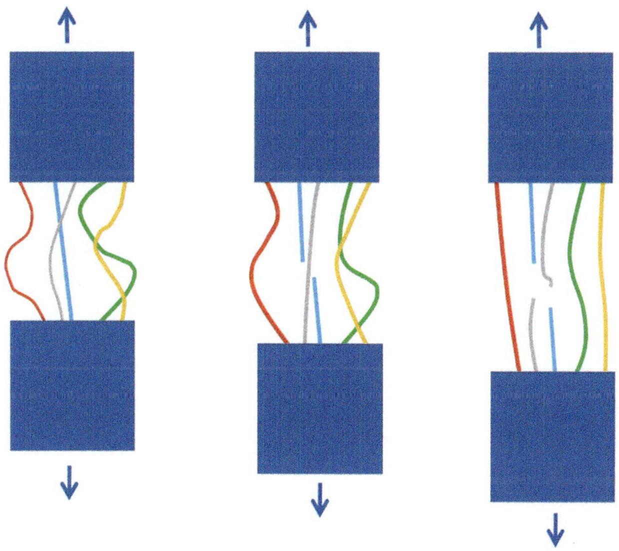

Figure 2.28 Mechanical degradation. Notice that in the first picture at the left, the blue molecule is carrying the load. In the second picture in the middle, it has stretched beyond its limit so that it has ruptured, and the purple molecule is now carrying the load. In the third picture at the right, the purple molecule has also ruptured, and now the red and orange molecules are carrying the load.

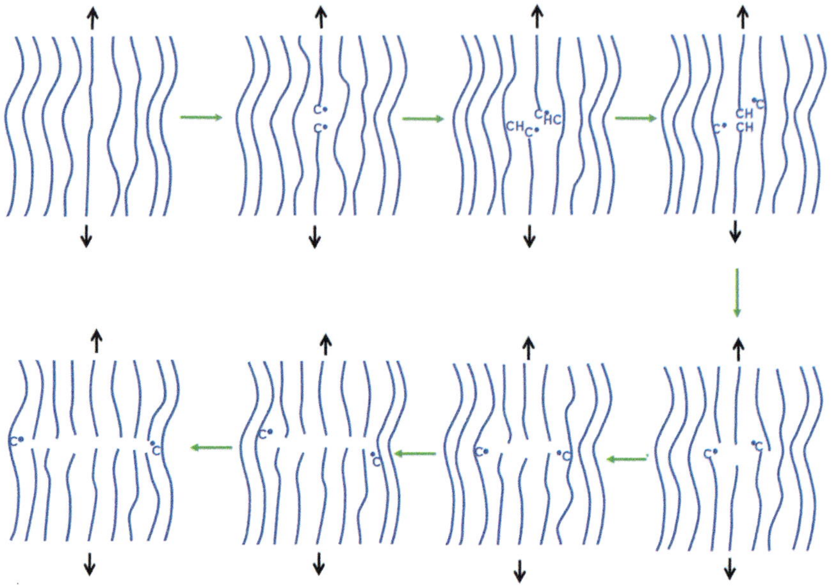

Figure 2.29 Development of a microcrack due to propagation of free-radical reactions with hydrogen abstraction.

Once a free radical is formed by mechanical stress, microcracks can be formed via hydrogen abstraction, thus leading to the formation of new free radicals and additional chain scission, as shown in Figure 2.29. Of course, if oxygen is present, oxidation reactions can also occur, and recombination leading to crosslinking can also occur.

In most cases, mechanical degradation is accompanied by chemical reactions – often oxidation. In the case of polymers such as PVC, both oxidative and nonoxidative chemical changes can occur following initial random chain scission, as shown in Figure 2.30.

Note that in reaction (5) in Figure 2.30, the free radical is transferred to another PVC molecule, which will go on to react further. The Cl· radicals can also react to remove an H from a polymer molecule, thus forming HCl.

2.13 Methods for Studying Polymer Degradation

A variety of methods are used to evaluate the degradation of polymers. As is the case for the determination of shelf life, such tests often employ accelerated conditions to speed up the degradation process and thereby enable faster evaluation. We will not delve deeply into these methods, but here is a brief overview.

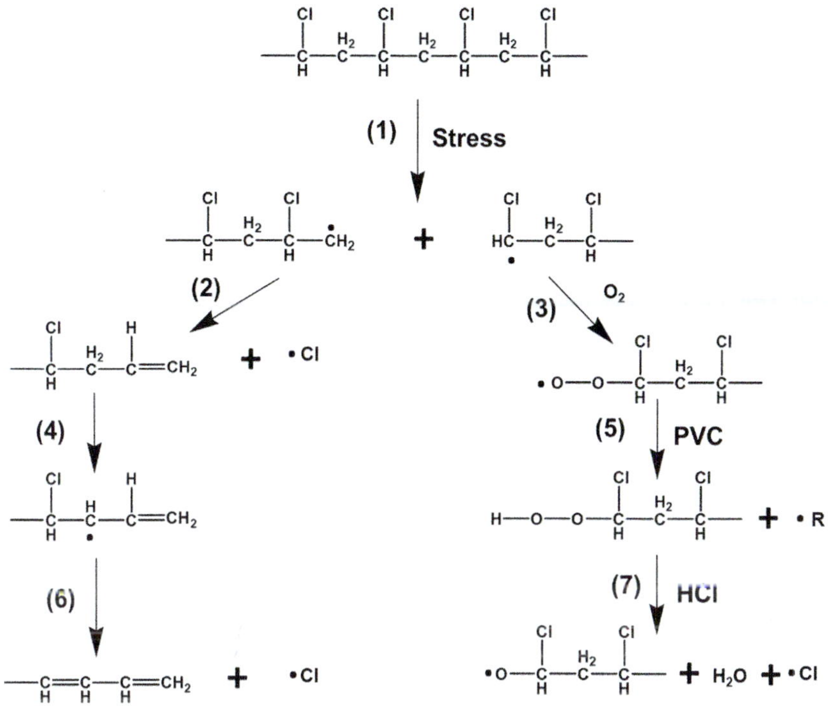

Figure 2.30 Mechanical degradation of PVC.

As is also the case for shelf-life studies, extrapolation from measures of degradation during accelerated conditions to degradation expected under actual exposures conditions is often quite difficult.

2.13.1 Weathering

Evaluation of photodegradation is often characterized as "weathering" and typically includes exposure to both UV light and moisture. There are also tests that involve UV exposure alone, without moisture.

Such evaluation may be done outdoors or under controlled laboratory conditions. If it is conducted outdoors, it is susceptible, of course, to normal variations in weather. There are standard test locations in Florida and in Arizona where these types of investigations are conducted. Obviously, the moisture exposure differs between these two locations. The ASTM D1435 protocol prescribes sample mounting and placement procedures.

In the lab, if moisture is included, the equipment (often known as a "weatherometer") usually provides cycles of light and dark, with fluctuating temperatures that result in daily condensation/evaporation cycles.

It has proven to be difficult to correlate lab-based and outdoor exposure. Some of this is because the outdoor exposure is likely to lack consistency; some is due to the inherent difficulty in correlating results from accelerated and "normal" exposure.

2.13.2 Thermal Degradation

There are two basic types of methods that are used to measure thermal degradation. One evaluates isothermal weight loss, monitoring the loss of weight at a specified temperature. More often, anisothermal methods are used, where the temperature changes during the test. There are a variety of these methods; some of the most common are:

- Thermogravimetric analysis (TGA) – measures the change in weight with temperature (also called thermogravimetry and abbreviated TG).
- Differential thermal analysis (DTA) – compares the temperature between the sample and a reference.
- Differential scanning calorimetry (DSC) – measures the relationship between heat flow and sample temperature.
- Thermal volatilization analysis (TVA) – measures increased pressure from the production of volatiles with an increase in temperature.
- Thermomechanical analysis (TMA) – measures the amount of deformation under a static load with increasing temperature.

Another approach to thermal degradation is to look not at the polymer but rather at the volatile products of the thermal degradation reactions. That is how yields of volatiles during pyrolysis are evaluated, for example. (Of course, this general concept can be extended to the evaluation of volatile products of oxidation or of other chemical reactions, as well.)

2.13.3 Photochemical Degradation

There are various irradiation devices that can be used in the evaluation of light-initiated chemical changes in polymers.

2.13.3.1 Change in Molecular Weight

One important measure of polymer degradation is the change in molecular weight. This can be evaluated by the same techniques that are used to evaluate the molecular weight of any polymer. The most important measures are viscometry (measurement of polymer viscosity, either in solution or in the melt) and GPC.

2.14 Stabilization of Polymers

Stabilizers are added to plastics to enhance their longevity and prevent degradation from oxygen, heat, and light, thereby ensuring improved product safety, durability, and performance. These additives allow plastics to retain their properties during manufacturing and use. Although there are exceptions, it should be noted that in most cases, additives are not chemically bound with the polymer but rather are physically mixed with it or its precursors.

There are two main ways through which polymers are stabilized:

1) stabilization using additives, and
2) stabilization via structural modification of the polymer.

Among these, the use of additives is the most common approach because of its lower cost, and the most important category of additives is antioxidants. Structural modification of polymers is a less popular way to stabilize polymers because it is not only costly but also can drastically affect the properties of the polymers. Thus, additives blending with plastics is the most common approach to stabilize plastics [36].

2.14.1 Antioxidants

Antioxidants act as inhibitors and impede polymer oxidation. Typical classifications of antioxidants include:

- Heat Stabilizers
- Melt Stabilizers
- Light Stabilizers
- Antifatigue Agents
- Antiozonants

The classification system shown above is based on where in the system the antioxidant exerts its primary effect. **All antioxidants act primarily by interfering, in some way, with the free-radical reactions that lead to the incorporation of oxygen into the polymer structure, thus interrupting the oxidation cycle.** Note that oxidation tends to be an autocatalyzing process, so interrupting this continually accelerating set of reactions is the key to successfully inhibiting oxidation. There are several points at which the oxidation cycle can be interrupted, as shown in **Figure 2.31**.

The oxidation cycle can be considered to start from the bold **R•** free radical near the middle of **Figure 2.31**. The red arrows then depict the passage of this free radical around the cycles, with the reactions that it undergoes, culminating in the formation of the ROH compound at the end. Note that along the way, it can react

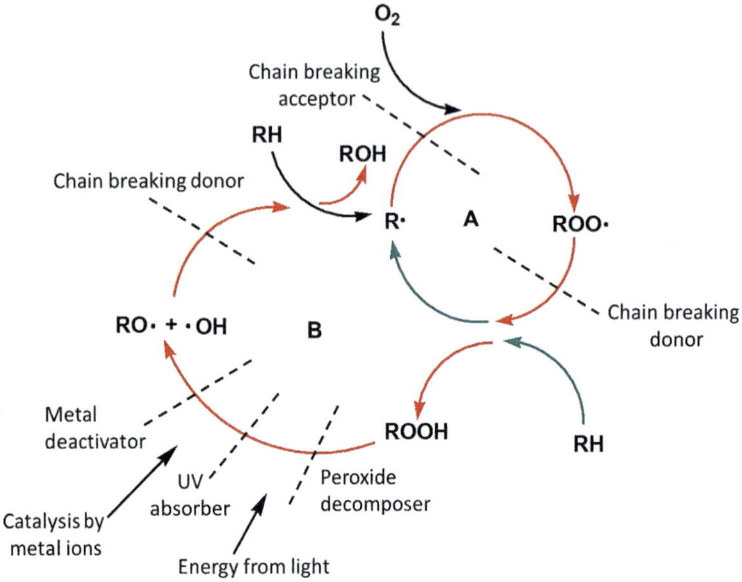

Figure 2.31 Schematic depiction of the oxidation cycle. The figure is modified from Ref. (37).

with two more polymer molecules, the RH at the beginning of the blue arrows, and the one at the beginning of the purple arrow. Hydrogen abstraction from these molecules results in the formation of a new R• free radical, thereby beginning the cycle all over again. The heavy dotted lines indicate the points at which the cycle can be interrupted, the places where antioxidants or other stabilizers can have an effect. It should be noted that heat can also provide energy for peroxide decomposition, although this is not shown in the diagram.

Antioxidants acting on the "A" cycle shown in Figure 2.31 are sometimes categorized as primary antioxidants because they directly interfere with the oxidation process itself, while antioxidants acting on the "B" cycle are categorized as preventive antioxidants because they are aimed at preventing the generation of the free radicals that lead to oxidation.

One major category of antioxidants is kinetic chain-breaking agents. These have two basic subtypes, including chain-breaking donors (CB-D) and chain-breaking acceptors (CB-A).

2.14.1.1 Chain-breaking Electron Acceptors (CB-A Antioxidants)

CB-A reacts with polymeric carbon radicals. Chain-breaking acceptor antioxidants are substances that act as electron acceptors (they can also be thought of as hydrogen acceptors and/or as oxidizing agents). The typical mechanism for these

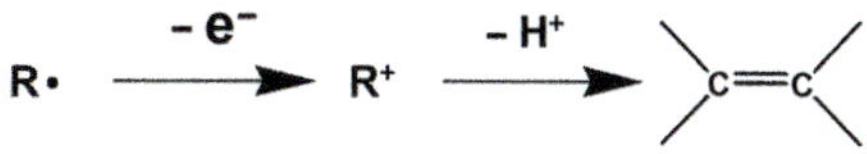

Figure 2.32 Illustration of how chain-breaking electron acceptors (CB-A antioxidants) help to prevent plastic degradation by converting alkyl radicals and subsequent formation of a stable unsaturated polymer.

Figure 2.33 Addition of a stabilizer to a free radical. S stands for stabilizers, and RS˙ stands for stable radical.

antioxidants is to remove the alkyl radical to give a carbonium ion (R+), subsequently yielding stable reaction products such as unsaturated polymers (Figure 2.32). In other words, this process allows polymer radicals to form a relatively stable polymer (C=C bond is more stable than free radical), and this polymer retains its integrity and prevents further chain events.

If we look at this reaction, we can see where the two names come from. The antioxidant can be thought of as an electron acceptor if we concentrate on the first reaction step, where the free radical gives up an electron, thereby leaving all the electrons paired up but resulting in an ion with a net positive charge. Alternatively, if we concentrate on the overall reaction, we can think of the antioxidant as a **hydrogen acceptor** because the net result is the loss of a hydrogen atom (along with its electron) to the antioxidant, generally leaving a double bond in what used to be the free radical.

Often, the reaction is more realistically described as adding the stabilizer to the free radical, thereby giving a more stable free radical, as shown in Figure 2.33:

where RS˙ is not very reactive and cannot abstract a hydrogen atom from another polymer. In other words, rather than shifting the hydrogen from the polymer molecule to the stabilizer molecule, we have added these molecules together and repositioned a hydrogen atom so that we have greater stability. **Some specific examples of this stabilization behavior will be described below.**

CB-A antioxidants are generally effective only in oxygen-deficient conditions. Otherwise, the activities of these antioxidants are rivaled by the addition of oxygen to R˙, which generates ROO˙. Examples of CB-A antioxidants are **benzoquinone** and related compounds, **aromatic nitro and nitroso compounds**, and substances generating "stable" free radicals such as **galvinoxyl and nitroxyl radicals.**

The route through which benzoquinone functions as an antioxidant is shown in Figure 2.34.

Note that one molecule of benzoquinone can stabilize two free radicals, but then it is "used up" and has no more stabilizing activity. It should be noted that the key to its ability to act as a stabilizer is that the free radical produced in the first reaction above is not energetic enough to abstract a hydrogen atom from a polymer

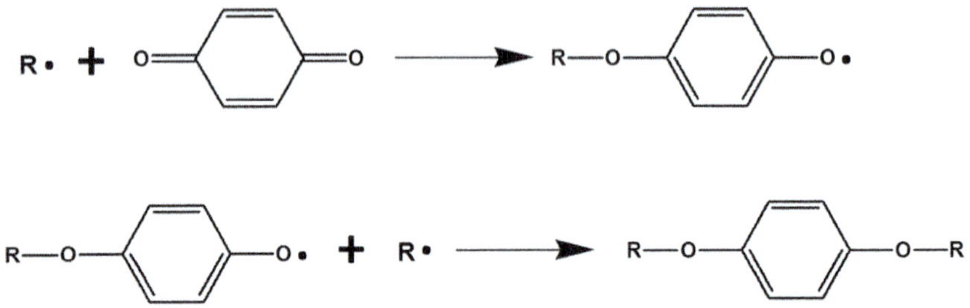

Figure 2.34 Reaction route through which benzoquinone acts as an antioxidant.

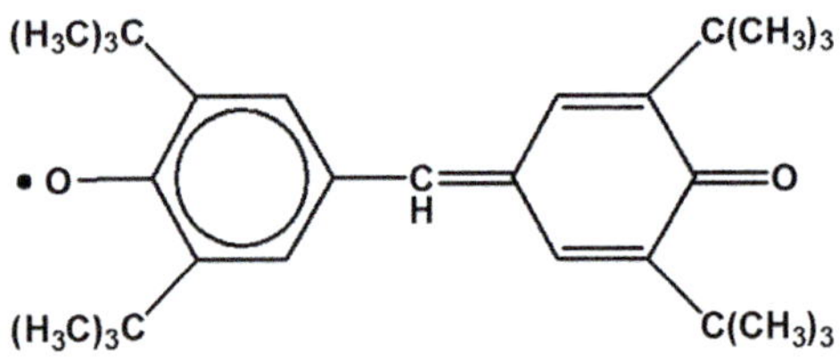

Figure 2.35 Structure of a galvinoxyl free radical.

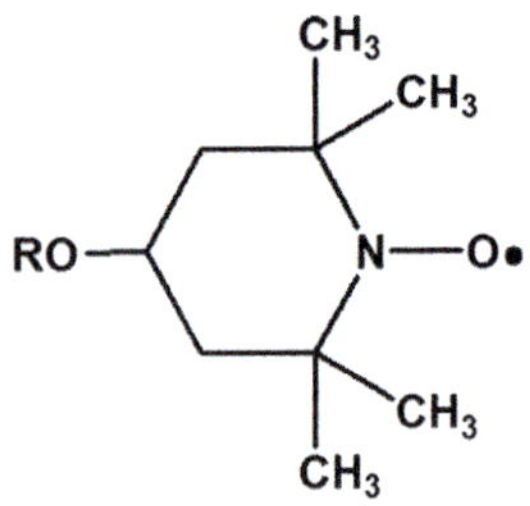

Figure 2.36 Structure of a nitroxyl free radical.

molecule. If it did so, it would not be effective as an antioxidant. Because it is relatively stable, it can combine with (react with) existing free radicals, but it does not participate in the creation of additional free radicals.

A galvinoxyl free radical looks like this (Figure 2.35):

And a **nitroxyl free radical** looks like this (Figure 2.36):

Just as was the case for benzoquinone, these free radicals have the ability to react with polymer free radicals, thereby stabilizing them, but do not participate in hydrogen abstraction, so they do not create new polymer free radicals. So, substances that produce these free radicals exhibit a CB-A antioxidant effect.

2.14.1.2 Chain-breaking Electron Donors (CB-D Antioxidants)

CB-D reacts with polymer peroxide radicals. CB-D antioxidants are to a certain extent the reverse of CB-A antioxidants in their mode of action; they act as **electron donors (or as hydrogen donors or reducing agents).** The typical mechanism for these antioxidants is to remove an alkylperoxide radical to give a hydroperoxide, as shown in Figure 2.37:

Again, if we concentrate on the first step, we see these as electron donors; if we look at the overall reaction, we see them as hydrogen donors.

$$ROO\bullet \xrightarrow{+\,e^-} ROO^- \xrightarrow{+\,H^+} ROOH$$

Figure 2.37 Removal of an alkylperoxide by a CB-D antioxidant to yield a hydroperoxide.

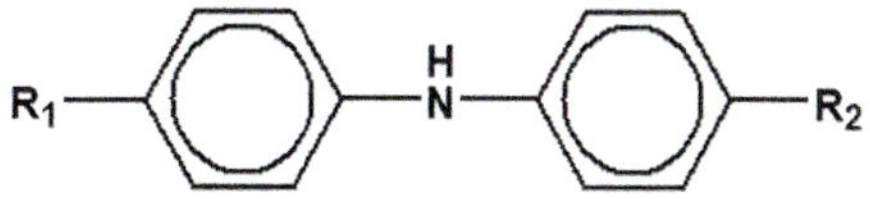

Figure 2.38 Structure of an arylamine.

CB-D antioxidants are a much more widely used category than CB-A antioxidants.

Two primary groups of CB-D antioxidants are arylamines (aromatic amine) and substituted phenols.

The structure of an arylamine is shown in Figure 2.38. Note that in this structure, if $R_1 = R_2 = H$, we have diphenylamine; if $R_1 = R_2 =$ an octyl group (C_8H_{17}), we have dioctylamine, etc.

Arylamines act as stabilizers by participating in reactions such as those shown in Figure 2.39. As you can see, the amine can participate in several CB-D steps

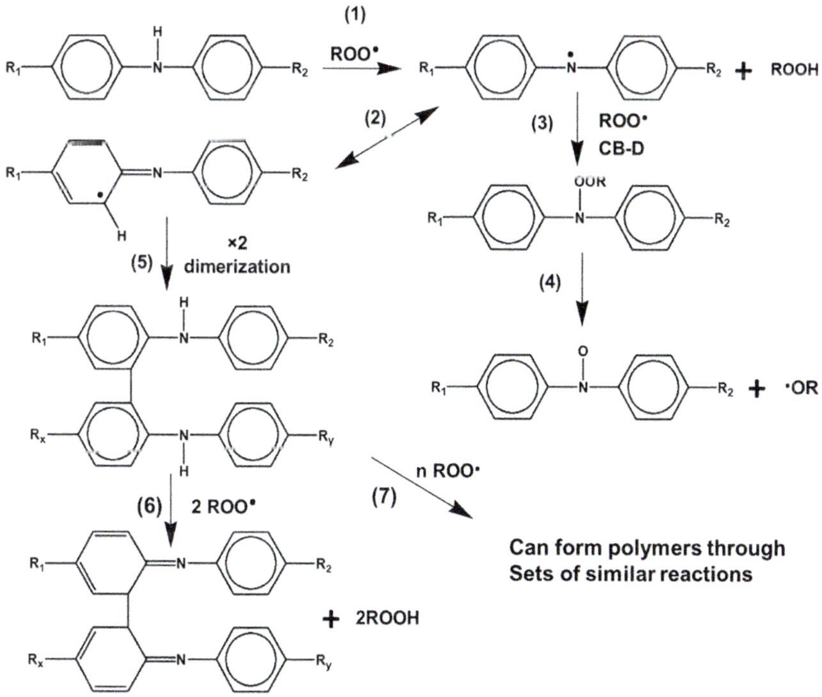

Figure 2.39 Reactions involving arylamine antioxidants.

(reactions 1, 3, 6, and 7). It should also be noted that one disadvantage of these compounds is **that the product of reaction 6 is colored due to its extensive conjugation**. This can pose unpleasant color issues.

The key to the activity of these substances is that they have hydrogen that can be removed relatively easily because the resulting free radical (e.g., the product of reaction 1) is stabilized by resonance involving the ring structures.

Substituted phenols are compounds such as BHT (butylated hydroxytoluene, 2,6-di-*tert*-butyl phenol), which can react as shown in Figure 2.40. The reactions shown above are only a few of the many reactions that are possible. As only one example, note that the product of reaction (2) is a peroxide, which can decompose to produce two free radicals through cleavage of the O-O bond – and will do so under appropriate conditions. The presence of heat or light energy increases the chance of this decomposition. The resultant free radicals can then participate in further oxidation reactions. We will see in other places, as well, that under certain conditions, antioxidants can become pro-oxidants. Notice also that once again we see compounds that are colored being formed as a consequence of the stabilization reactions. In general, hindered phenols such as BHT do not present as significant color problems as do hindered amines. Once again, hindered phenols

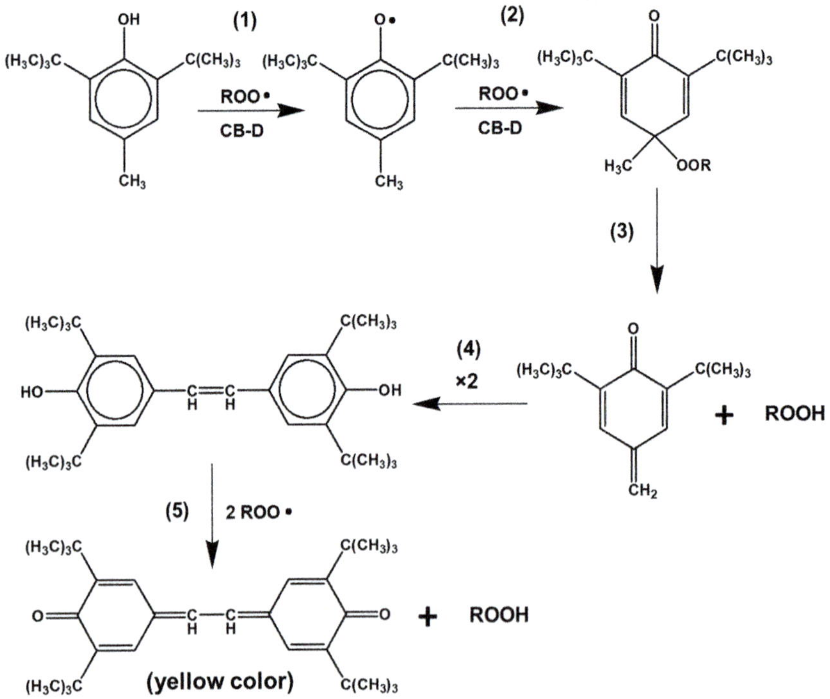

Figure 2.40 Antioxidant reactions of BHT.

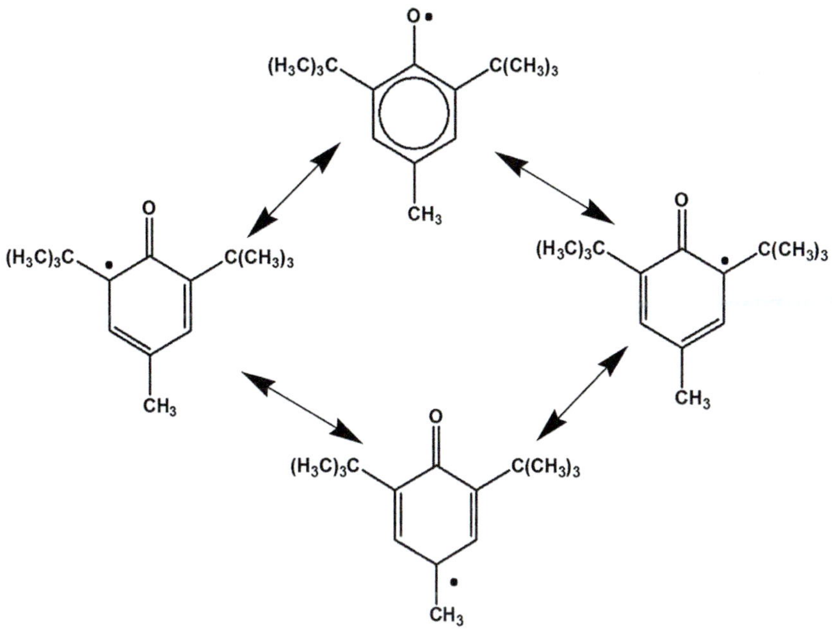

Figure 2.41 Resonance structures of a hindered phenol free radical.

are effective as antioxidants because the resultant free radicals (such as the product of reaction 1 above) are stabilized by resonance. Figure 2.41 shows the four resonance structures that this species can alternate between.

Another thing to notice about BHT is that it has a relatively low molecular weight, so it can be lost through volatilization. When a long-acting antioxidant is needed, or one that is less volatile at elevated temperatures, compounds related to BHT but with higher molecular weights can be used as alternatives. Some examples of commercial antioxidants in this category are shown in Figure 2.42.

Another well-known antioxidant that falls in the hindered phenol group category is vitamin E (α-tocopherol, $C_{29}H_{50}O_2$), which is shown in Figure 2.43:

This antioxidant also undergoes a complex set of reactions that we will not try to illustrate. (Moreover, to further complicate matters, vitamin E is actually a mixture of eight slightly different compounds, all of which are similar to the α-tocopherol shown above.)

2.14.2 Peroxide Decomposers

As mentioned earlier, the other major category of antioxidants is preventative oxidants, which act in the B cycle in Figure 2.31 to prevent the generation of **additional free radicals.** The most significant category of these antioxidants is **peroxide decomposers** (PD). These compounds act to remove **hydroperoxides**

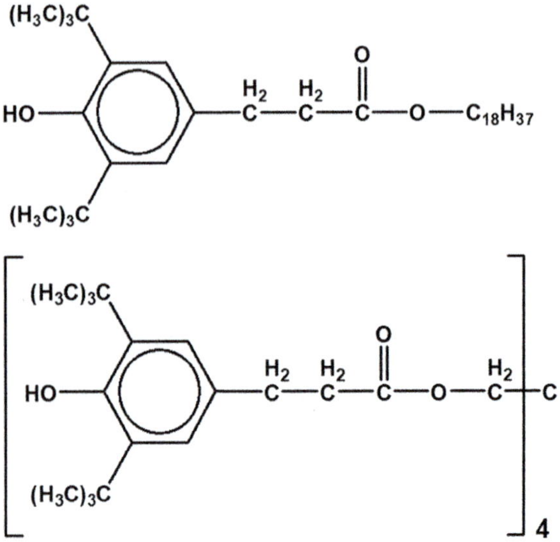

Figure 2.42 Derivatives of BHT are used as commercial antioxidants.

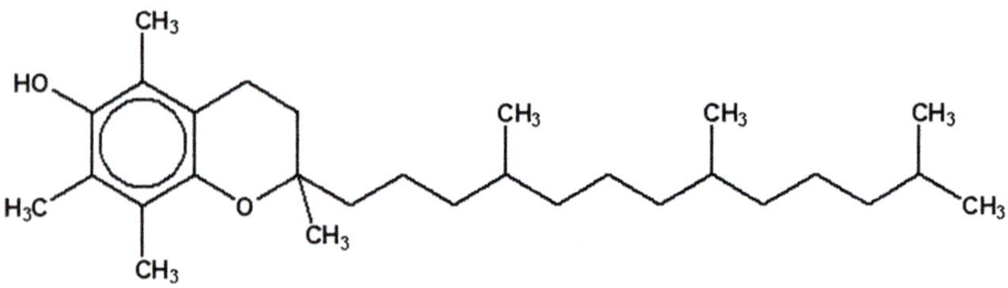

Figure 2.43 Structure of vitamin E.

(**ROOH**), producing stable molecules in a way that does not generate free radicals or involve thermal hydrolysis. Typically, they act by reducing the hydroperoxides (ROOH) so that they become converted to alcohols (ROH). Here, too, we have two major groups:

- catalytic peroxide decomposers, and
- stoichiometric peroxide decomposers

2.14.2.1 Stoichiometric Peroxide Decomposers (PD-S)

Stoichiometric peroxide decomposers, as the name implies, have functional groups that can react with the peroxides and are consumed during the reaction. Once all of the available functional groups have reacted, these compounds lose their stabilizing capabilities.

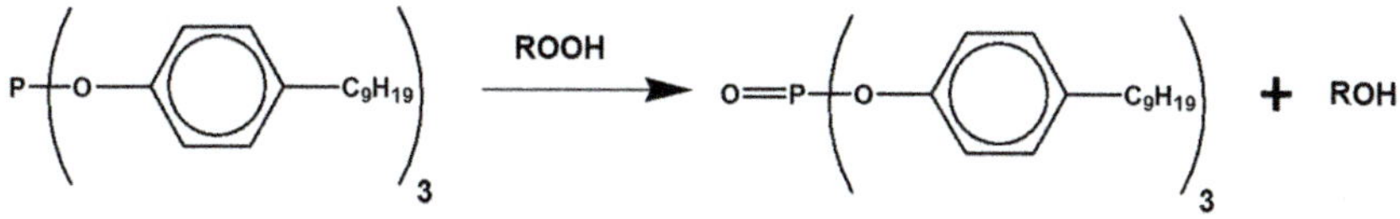

Figure 2.44 Reaction of tris(nonylphenyl) phosphite, with the phosphorous undergoing oxidation from a phosphite to a phosphate with decomposing ROOH into ROH.

A typical category of PD-S additives is phosphite esters, such as tris(nonylphenyl) phosphite, which reacts in the manner shown in **Figure 2.44**. A key to this reaction is that the phosphorus atom oxidizes from a +3 to a +5 oxidation state (from a phosphite to a phosphate), and ROOH reduces to ROH.

2.14.2.2 Catalytic Peroxide Decomposers (PD-C)

Catalytic peroxide decomposers, as the name implies, act as catalysts for peroxide decomposition, emerging from the sequence of reaction steps essentially unchanged, and therefore, they do not become "used up" as the reaction proceeds. One example is sulfur compounds such as dialkyl thiodipropionates:

$$ROCOCH_2CH_2SCH_2CH_2COOR$$

The reaction mechanisms are generally quite complex (and thus we are not going to try to illustrate them here). A disadvantage of these compounds is that while they are designed to inhibit oxidation, in some conditions, they may instead catalyze peroxide decomposition in a way that leads to the generation of free radicals. If this happens, they become prooxidants rather than antioxidants.

2.14.3 Metal Deactivators (MDs)

Stabilizers such as MDs and UV absorbers (UVA) do not directly target reactions leading to oxidation but instead are designed to influence reactions that tend to accelerate oxidation. Transition metal ions (especially cobalt, iron, manganese, and copper) catalyze the oxidation of plastics, primarily by catalyzing peroxide decomposition. Compounds that coordinate with these metal ions and thereby prevent them from catalyzing oxidation reactions serve as MDs. In general, these fall in the category of chelating agents. One example is the oxamide family:

$$RNHCOCONHR$$

such as oxanilide:

$$C_6H_5NHCOCONHC_6H_5$$

It should be noted that recent information suggests that some of these compounds can function as **chain-breaking antioxidants** in addition to serving as complexing agents.

Question: what is the source of these metals in plastics? Primarily, these metal residues arise from catalysts used in the synthesis of polymers; for example, Tin is used as a catalyst in PET synthesis, while TiO_2 is used for PE.

2.14.3.1 UV Light Absorbers and Other Light Stabilizers

UV light absorption can degrade polymers. To address this issue, UV light absorbers can be used to interrupt the absorption of UV light by the polymer and its consequent degradation/oxidation. In general, they can be divided into two categories which include UV light absorbers and quenchers of photo-excited states.

2.14.3.2 UV Light Absorbers (UVA)

One mode of action of UV light absorbers is to absorb incoming UV light energy, thus preventing it from being absorbed by sensitive groups in the polymer. **The absorbed energy must, of course, be dissipated in a way that does not lead to the generation of free radicals.** Common examples include carbon black (which also acts as a free-radical scavenger), titania, and compounds with phenolic or quinonoid groups. Carbon black imparts a black color, and titania imparts a white color. If colorless additives are desired, common choices are 2-hydroxy-4-octoxybenzophenone (Figure 2.45):

and members of the benzotriazole family (Figure 2.46):

It should be noted that the presence of the —OH group next to the C=O and C—N groups is critical. If it is replaced by an alkyl group, the compound no longer has activity as a stabilizer. In fact, replacing the —OH group in 2-hydroxy-4-oxtoxybenzophenone with an H gives us a compound that is a photosensitizer, which promotes rather than inhibits photooxidation.

The ways in which benzophenones and benzotriazoles act as light stabilizers are shown in Figure 2.47. In both cases, the term $+h\nu$ denotes light absorption, while the symbol **-Δ** indicates the evolution of heat. Notice that hydrogen bonding (indicated by the dotted lines) plays a crucial role in these transformations. If the —OH group is not present, these compounds act as excellent UV absorbers because of their unsaturated bonds, but they tend to form free radicals rather than safely dissipating the energy.

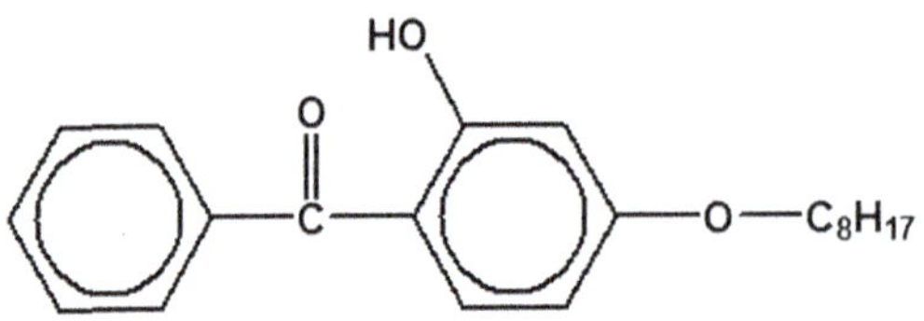

Figure 2.45 Structure of octoxybenzophenone.

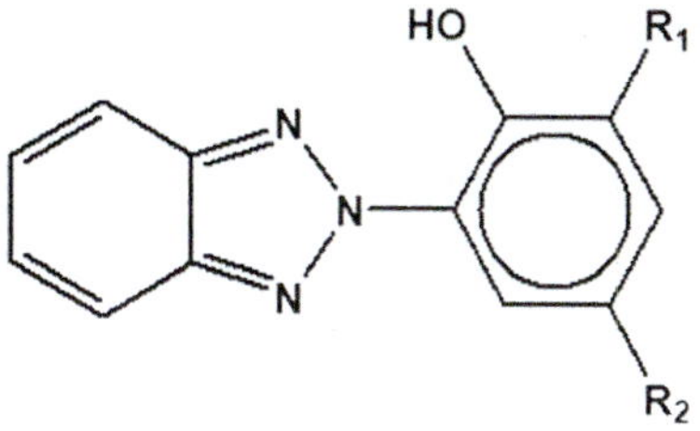

Figure 2.46 General structure of the benzotriazole family.

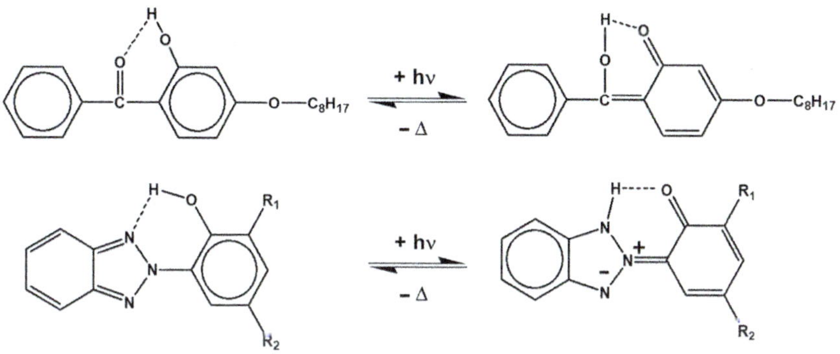

Figure 2.47 Schematic illustration of the ways through which benzophenones and benzotriazoles behave as light stabilizers.

2.14.4 Quenchers of Photo-excited States

Another mode of action of light stabilizers is to quench **photoexcited states of the polymer** by accepting the excess energy from UV absorption, again in a way that does not lead to the generation of free radicals. Commercial stabilizers in this category include nickel-organic compounds such as nickel dibutyl dithiocarbamate (NDC, which also acts as a hydroperoxide decomposer) and the very important family of hindered amine light stabilizers (HALS), which also act as free radical scavengers. As these examples show, many commercial stabilizers function by using a combination of mechanisms. In fact, there is some controversy about whether the actual energy-transfer quenching action is of much significance. Nonetheless, we will use this categorization to separate light stabilizers targeting photooxidation from those targeting thermal oxidation (which are often themselves unstable to UV degradation, so are not effective as photostabilizers). The structure of NDC is shown in Figure 2.48.

The comparative effectiveness of some HALS stabilizers (Figure 2.49) is shown in Table 2.8.

The stabilization mechanism for these compounds involves the removal of the H from the NH group and its combination with an ROO˙ free radical.

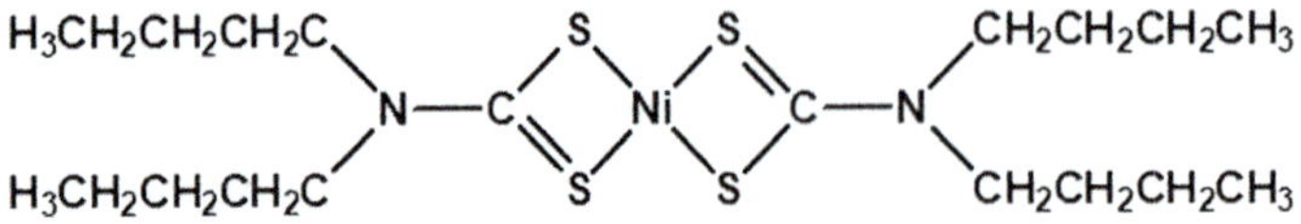

Figure 2.48 Structure of NDC.

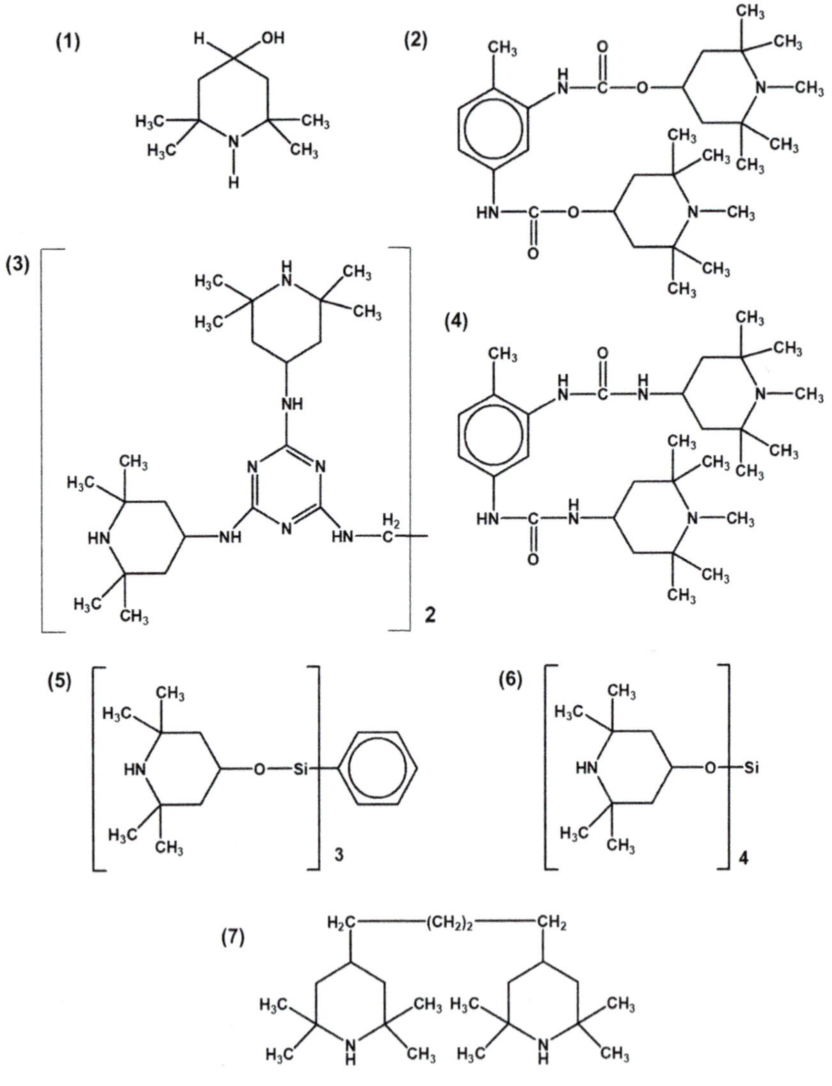

Figure 2.49 Structures of the HALS stabilizers listed in Table 2.8.

Table 2.8 The effect of various photostabilizers at 1 wt% concentration on the length of the induction period for photooxidation of polyisoprene.

Stabilizer (See Figure 2.49)	Induction period (h)	Stabilizer (See Figure 2.49)	Induction period (h)
none	3	#4	32
#1	4	#5	51
#2	21	#6	54
#3	23	#7	63

The corresponding structures for these stabilizers according to their identification number are shown in Figure 2.49. Data is from: V. Y. Shlyapintokh, *Photochemical Conversion and Stabilization of Polymers*, Hanser Pub., Munich, 1984.

In addition, a complex set of other reactions result in the regeneration of active species, and consequently, continuing stabilizing activity.

2.14.5 Special Stabilizers

Other types of stabilizers are also used, especially for PVC which undergoes a self-catalyzed dechlorination reaction, as we have discussed earlier. The major activity of such stabilizers is either to remove HCl that is generated in the degradation reaction (so it cannot catalyze further degradation) or to remove allylic chlorides (those adjacent to double bonds, and hence more likely to be reactive).

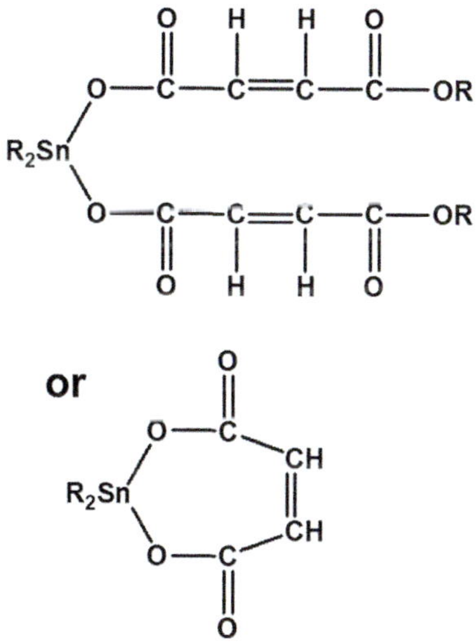

Figure 2.50 Chemical structure of dialkyl tin dimaleate. (top) and dialkyl tin maleate (bottom).

Some examples of PVC stabilizers include dialkyl tin maleates (Figure 2.50), dialkyl tin thioglycolates (Figure 2.51), and metal carboxylates (metal salts of carboxylic acids). The metal carboxylates react in the ways shown in Figure 2.52. The elimination of conjugation by a metal salt and capture of HCl is also shown in Figure 2.53. Examples of some stabilizers that are used for various polymers are provided in Table 2.9.

Until recently, lead-based stabilizers such as lead carbonate used to be commonly used for PVC (although not often in packaging applications), but they have been phased out because of increasing awareness and concern regarding lead toxicity. Fatty acid salts of Ca and Zn are commonly used to stabilize PVC packaging film and

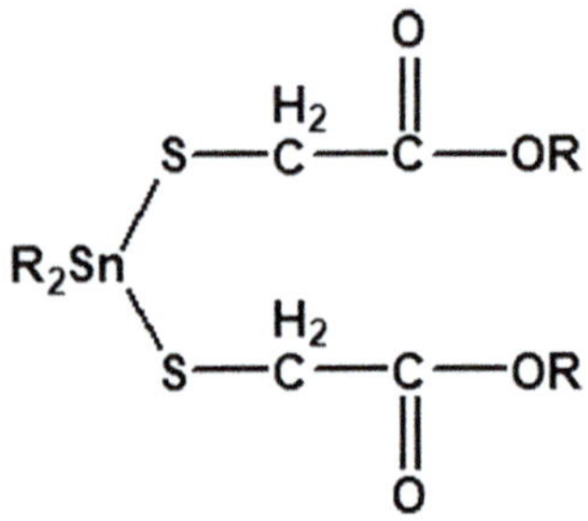

Figure 2.51 Chemical structure of a dialkyl tin thioglycolate compound.

bottles. These stabilizers are nontoxic and FDA approved. Various ratios are used for particular applications. A general guideline is to prevent discoloration, a Ca:Zn ratio of 2:3 is optimal, while for the greatest extension of lifetime, a 4:1 ratio is best. One important thing to note is that amine stabilizers cannot be used with PVC, as they promote dehydrochlorination.

2.14.6 Drawbacks of Stabilizers

While stabilizers are very useful tools that can help extend the lifetime of plastics, thus providing economic and even environmental benefits (by helping to minimize the amount of plastic required due to its longer lifetime and lifespan between recycling processes), they also can pose a number of issues. For example, they can interfere with the recycling of plastics. In the case of mechanically recycled plastics, antioxidants cause issues in waste sorting, migration (where the additive may end up contaminating a packaged product, for example), and aesthetics [38].

Carbon black is employed as a pigment to change the color of plastics, and it also serves as a filler to provide structural reinforcement and protection against UV light [39]. Although carbon black provides the aforementioned benefits, it also leads to issues during the sorting of polymer waste due to its high absorption rates of infrared (IR) radiation (particularly when IR or near-infrared (NIR) systems are used in the sorting process) [40]. In addition, carbon can cause

$$HCl + MY_2 \text{ (metal carboxylate)} \rightarrow MYCl + RY$$

Figure 2.52 Ways through which metal carboxylates can react (M denotes a metal in the reaction shown above). One goal is to convert the byproduct HCl into noncorrosive salt.

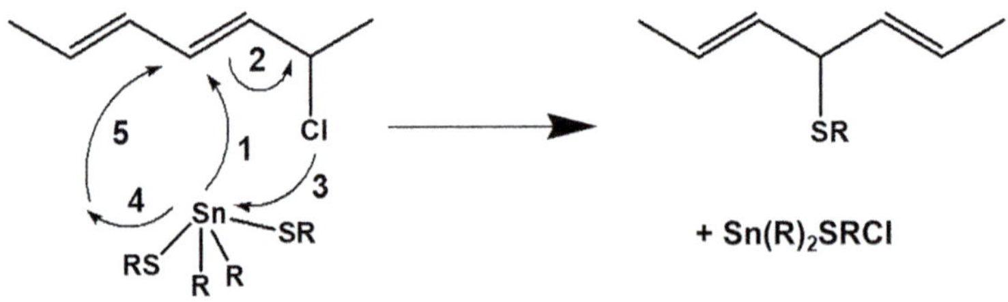

Figure 2.53 Metal salt can eliminate conjugation (alternating double bonds) and capture HCl.

Table 2.9 Examples of some stabilizers that are used for various polymers [38, 39, 51].

Polymer	Stabilizers	Function
Linear low-density polyethylene (LLDPE), high-density polyethylene (HDPE), high impact polystyrene (HIPS), acrylonitrile-butadiene-styrene (ABS)	Bisphenol A	Antioxidants and UV stabilizers
PVC	Cadmium and lead compounds	Heat stabilizer (note that the lead compounds are being phased out
PVC	Nonylphenol (barium and calcium salts)	Heat stabilizer
PVC	Arsenic compounds, organic tin compounds, triclosan	Biocides
LLDPE, HDPE, HIPS, ABS	Nonylphenol compounds	Stabilizers, antioxidants, and UV stabilizers
PVC	Phthalic esters (such as DEHP)	Plasticizers
PET	Dipentyl phthalate (DPP), di(2-ethylhexyl) adipate (DEHA), dioctyl adipate (DOA), diethyl phthalates (DEP), diisobutylphthalate	Plasticizers
Polyvinylidene chloride (PVDC)	Acetyl tributyl citrate (ATBC)	Plasticizers
Plastic food packaging (PET, etc.)	Arylamines, phenolic compounds, organophosphites	Antioxidants (the phenolic and organophosphate compounds act by reducing hydroperoxides that are formed when alcohols are oxidized)
PVDC, PVC blends, vinyl chloride copolymers	**Primary heat stabilizers**: mixed metal salt blends, lead compounds, organotin compounds: **Secondary heat stabilizers:** epoxy compounds, organophosphites, beta diketone compounds	Heat stabilizers

(Continued)

Table 2.9 (Continued)

Polymer	Stabilizers	Function
HIPS	Montmorillite clay	Filler: Helps prevent chain scission reactions during degradation, and increases the viscosity [38]
Cloisite 15A clay	PS	Filler: Protects against chain degradation and also improves mechanical properties after recycling [38]
Polyethylene	Curcumin, caffeic acid, vitamin E, quercetin, *b*-carotenes	Naturally occurring antioxidants [39]

discoloration of the recycled material and adversely affect its aesthetic appearance, thereby lowering its value.

Antioxidants are designed mainly to offer protection during the first processing cycle for polymers, but there has been a lack of research regarding antioxidant stability during repeated extrusion cycles. **The migration of contaminants that are degraded/oxidized additives from the previous use is a concern if the plastics are to be used again for food contact applications.**

While additives can pose issues during the recycling process, in some cases, additives such as compatibilizers are used to facilitate recycling processes. Although we have focused on stabilization, it should be noted that additives are often added to polymers for other purposes besides stabilization, such as to provide reinforcement or enhance their mechanical strength, to act as fillers, plasticizers, curing agents, foaming agents, to modify their viscosity, and so forth.

Case Study – Braskem

One trend that has been developing in recent years has been a shift toward biobased plastics as alternatives to their petroleum-based counterparts. These types of polymers can be produced from biomass, such as agricultural products or byproducts. While avoiding the consumption of non-renewable petroleum resources, the production of bio-based plastics can also generate fewer emissions than is the case with the production of non-renewable counterparts.

Braskem is the leading manufacturer of biobased plastics, using sugarcane as a feedstock to produce ethanol, which is then used for the production ethylene monomer for polyethylene (I'm green™) bio-based plastics product line [41]. The sugarcane that is used as a source for Braskem's absorbs carbon dioxide from the atmosphere as it grows, thus leading to emission reductions of approximately 4 kg of carbon dioxide for every kg of bio-based plastics that are produced, in comparison with those produced from petroleum feedstocks. Once the sugarcane has grown, it is collected and then crushed at a mill, thus producing ethanol and sugar. The bagasse cane used as a fuel for power plants for electricity generation.

Recently, Braskem has invested $87 million USD to expand the production capacity of its plant in Rio Grande do Sul, Brazil, thereby expanding its bio ethylene production capacity by 30%, up to 260,000 tons per year from its previous capacity of 200,000 tons per year. The bio-based plastics are used in a wide range of products, including packaging, plastic toys, bags, reusable water bottles, industrial cables and wires, and numerous other applications. In addition, Braskem's bio-based plastics are widely exported to over 30 countries and are used in over 250 product brands [42].

Bio-based plastics that offer similar performance to their petroleum-based counterparts and which can be processed in the same way as conventional polymers while offering better sustainability are known as drop-in bioplastics [43, 44]. Bio-based PE, which would offer similar performance yet better sustainability in comparison with petroleum-based PE would be one such example of a drop-in bioplastic.

Since Braskem began large-scale production of bio-based PE in 2010, they have also expanded to the production of other bio-based polymers such as bio ethylene-vinyl acetate (bio-EVA) in 2018, which was also derived from sugarcane [45, 46]. More recently, Shell and Braskem have partnered to facilitate the production of bio-based polypropylene for the U.S. marketplace [47]. In this arrangement, Shell is providing Braskem with bio-attributed and biocircular propylene feedstocks, to be used for bio-propylene production.

Coca-Cola has been developing plant-based PET bottles, having recently developed 100% plant-based PET bottles [48, 49]. This represents the culmination of efforts by Coca-Cola's adoption of completely bio-based bottles since they launched their PlantBottle beverage bottles in 2009, which were 30% bio-based but 70% petroleum-based. Coca-Cola had announced a 100% bio-based PET bottle in 2015, which was produced at a laboratory scale, but that plant-based bottle demonstrated that a 100% bio-based PET bottle could indeed be produced, but it was not yet considered to be ready for commercial-scale production.

While the commercial sector has been very active in advancing biobased polymer technologies, cooperation between government, industry, and academia also has a role. One such example is the NSF Center for Sustainable Polymers at the University of Minnesota. This research center focused on three key areas, including the conversion of biomass to polymer ingredients, the development of high-performance plastics and elastomers, as well as the development of strategies for sustainable polymer degradation, chemical recycling, and compatibilization.

2.15 Summary

In this chapter, we have provided an overview of plastics as well as their properties and characterization. In addition, we have also discussed the various ways through which these materials degrade and the strategies that are employed to stabilize these materials. In terms of stabilizers or other additives, it is highly desirable to use naturally occurring materials when possible [40]. Although plastics pose a number of environmental challenges, they also have numerous unique properties that will likely ensure that they remain in use for decades to come. The use of biobased plastics holds great promise, especially for applications such as packaging [50].

Problem Set

1 Why do recycled plastics have weaker mechanical properties compared to virgin plastics?

2 How can we determine the thermal degradation of plastics?
A X-ray studies.
B Thermogravimetric analysis.
C Microscopic methods (e.g., SEM).
D None of the above.

3 Which polymer will thermally oxidize faster among the following?
A Linear PE.
B Lightly branched PE.
C Densely branched PE.
D All degrade at the same rate.

4 What are the two potential causes of environmental stress cracking (ESC) in plastics?

5 Based on the figure below, what is the impact of CB-D on polymer stability?

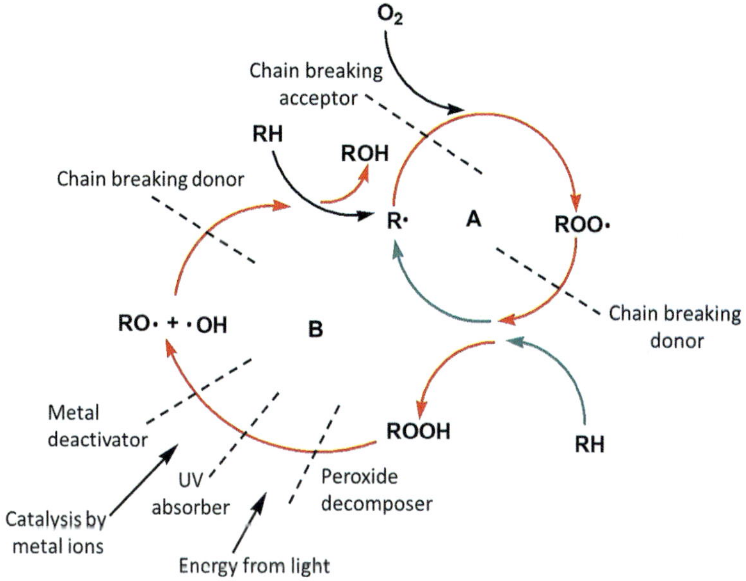

References

1 Carothers, W.H. (1929). Studies on polymerization and ring formation. I. An introduction to the general theory of condensation polymers. *J. Am. Chem. Soc.* 51 (8): 2548–2559. https://doi.org/10.1021/ja01383a041.

2 Flory, P.J. (1953). *Principles of Polymer Chemistry*. New York: Cornell University Press. Available from: https://archive.org/details/dli.ernet.286013.

3 Odian, G. (2004). *Principles of Polymerization*, Chapter 1, p. 23, where it appeared as Fig 1–4. Wiley https://doi.org/10.1002/047147875X.ch1.

4 Stepto, R.F.T. (2010). Dispersity in polymer science (IUPAC Recommendation 2009). *Polym. Int.* 59 (1): 23–24. https://doi.org/10.1002/pi.2748.

5 Stepto, R.F.T. (2009). Dispersity in polymer science (IUPAC Recommendations 2009). *Pure Appl. Chem.* 81 (2): 351–353. https://doi.org/10.1351/PAC-REC-08-05-02.

6 Günther, H. (2013). *NMR Spectroscopy : Basic Principles, Concepts and Applications in Chemistry*. John Wiley & Sons, Incorporated.

7 Atta-ur-Rahman (1986). Chemical Shift in ^{1}H-NMR Spectroscopy. In: *Nuclear Magnetic Resonance: Basic Principles* (ed. Atta-ur-Rahman), 1–33. New York, NY: Springer US https://doi.org/10.1007/978-1-4612-4894-1_1.

8 Yang, R. (2018). *Analytical Methods for Polymer Characterization*. Boca Raton: Taylor & Francis Group, LLC.

9 King, S.B. and Wong, Y.L. (2023). *1H NMR Spectroscopy for CHEM 222L*. Wake Forest University. Available from: https://users.wfu.edu/ylwong/chem/nmr/h1/splitting.html.

10 Questions and Answers in MRI (2023). T2 Relaxation: Definition. Available from: https://mriquestions.com/what-is-t2.html

11 Ahamed, B.A. and Lakshmaiah, M.V. (2016). Measurement of viscosity for synthetic polymers. *Int. J. Adv. Res. Electr. Electron. Instrum. Eng.* 5 (5): 3866–3870.

12 UMass Chan Medical School Electron Microscopy Facility. What is Electron Microscopy? [Internet]. Available from: https://www.umassmed.edu/cemf/whatisem/

13 Inkson, B.J. (2016). 2 - Scanning electron microscopy (SEM) and transmission electron microscopy (TEM) for materials characterization. In: *Materials Characterization Using Nondestructive Evaluation (NDE) Methods [Internet]* (ed. G. Hübschen, I. Altpeter, R. Tschuncky, and H.G. Herrmann), 17–43. Woodhead Publishing. https://www.sciencedirect.com/science/article/pii/B978008100040300002X.

14 Khan, M.K., Wang, Q.Y., and Fitzpatrick, M.E. (2016). 1 - Atomic force microscopy (AFM) for materials characterization. In: *Materials Characterization Using Nondestructive Evaluation (NDE) Methods* (ed. G. Hübschen, I. Altpeter, R. Tschuncky, and H.G. Herrmann), 1–16. Woodhead Publishing https://www.sciencedirect.com/science/article/pii/B9780081000403000018.

15 Jalili, N. and Laxminarayana, K. (2004). A review of atomic force microscopy imaging systems: application to molecular metrology and biological sciences. *Mechatronics* 14 (8): 907–945. https://www.sciencedirect.com/science/article/pii/S0957415804000455.

16 Odian, G. (2004). *Principles of Polymerization*, Chapter 1, p. 25, where they appear as Fig. 1--5 (fringed micelle) and 1–6 (folded chain lamella), respectively. Wiley.

17 Zhang, M., Guo, B.H., and Xu, J. (2016). A review on polymer crystallization theories. *Crystals (Basel)* 7 (1): 4.

18 University of Washington. Young's Modulus. https://depts.washington.edu/matseed/mse_resources/Webpage/Biomaterials/young%27s_modulus.htm (accessed 16 June 2023).

19 SpecialChem SA. Strength at Break (Tensile). https://omnexus.specialchem.com/polymer-properties/properties/strength-at-break-tensile

20 MechaniCalc, Inc. Mechanical Properties of Materials. https://mechanicalc.com/reference/mechanical-properties-of-materials

21 JohnR, R. (ed.) (2023). *CRC Handbook of Chemistry and Physics, 104th Edition (Internet Version 2023)*. Boca Raton, FL: CRC Press/Taylor & Francis.

22 Miandad, R., Rehan, M., Barakat, M.A. et al. (2019). Catalytic Pyrolysis of Plastic Waste: Moving Toward Pyrolysis Based Biorefineries. *Front. Energy Res.* 7: 27. https://doi.org/10.3389/fenrg.2019.00027.

23 Chien, J.C.W. and JKY, K. (1978). Pyrolysis and oxidative pyrolysis of polypropylene. *Adv. Chem.* 169: 175–197.

24 Kumar, V., Khan, A., and Rabnawaz, M. (2022). Efficient depolymerization of polystyrene with table salt and oxidized copper. *ACS Sustain. Chem. Eng.* 10 (19): 6493–6502. https://doi.org/10.1021/acssuschemeng.1c08400.

25 Khan, A., Naveed, M., Aayanifard, Z., and Rabnawaz, M. (2022). Efficient chemical recycling of waste polyethylene terephthalate. *Resour. Conserv. Recycl.* 187: 106639.

26 Rimez, B., Rahier, H., Van Assche, G. et al. (2008). The thermal degradation of poly(vinyl acetate) and poly(ethylene-co-vinyl acetate), Part I: Experimental study of the degradation mechanism. *Polym. Degrad. Stab.* 93 (4): 800–810.

27 Berkowicz, G., Majka, T.M., and Żukowski, W. (2020). The pyrolysis and combustion of polyoxymethylene in a fluidised bed with the possibility of incorporating CO_2. *Energy Convers. Manag.* 214: 112888.

28 Kaminsky, W. and Franck, J. (1991). Monomer recovery by pyrolysis of poly(methyl methacrylate) (PMMA). *J. Anal. Appl. Pyrolysis* 19: 311–318.

29 Liu, X. Organic Chemistry I. Pressbooks. https://kpu.pressbooks.pub/organicchemistry/.

30 Gardette, M., Perthue, A., Gardette, J.L. et al. (2013). Photo- and thermal-oxidation of polyethylene: comparison of mechanisms and influence of unsaturation content. *Polym. Degrad. Stab.* 98 (11): 2383–2390.

31 Reuter, L. and Lüchow, A. (2021). Real space electron delocalization, resonance, and aromaticity in chemistry. *Nat. Commun.* 12 (1): 4820.

32 Deanin, R.D., Orroth, S.A., Eliasen, R.W., and Greer, T.N. (1970). Mechanism of ultraviolet degradation and stabilization in plastics. *Polym. Eng. Sci.* 10 (4): 228–234. https://doi.org/10.1002/pen.760100408.

33 Mihut, C., Captain, D.K., Gadala-Maria, F., and Amiridis, M.D. (2001). Review: recycling of nylon from carpet waste. *Polym. Eng. Sci.* 41 (9): 1457–1470.

34 Koch, S., Meunier, M., Hopmann, C., and Alperstein, D. (2020). A combined experimental and computational study of environmental stress cracking of amorphous polymers. *Polym. Adv. Technol.* 31 (2): 297–308.

35 Jansen, J.A. (2004). Environmental stress cracking - the plastic killer. *Adv. Mater. Process.* 162: 50–53.

36 Hahladakis, J.N., Velis, C.A., Weber, R. et al. (2018 Feb). An overview of chemical additives present in plastics: migration, release, fate and environmental impact during their use, disposal and recycling. *J. Hazard. Mater.* 344: 179–199.

37 Rowe, R.K. and Sangam, H. (2002). Durability of HDPE geomembranes. *Geotext. Geomembr.* 20: 77–95.

38 Schyns, Z.O.G. and Shaver, M.P. (2021). Mechanical recycling of packaging plastics: a review. *Macromol. Rapid Commun.* 42: 2000415.

39 Al-Malaika, S., Ashley, H., and Issenhuth, S. (1994). The antioxidant role of α-tocopherol in polymers. I. The nature of transformation products of α-tocopherol formed during melt processing of LDPE. *J. Polym. Sci. A Polym. Chem.* 32 (16): 3099–3113.

40 Neo, E.R.K., Yeo, Z., Low, J.S.C. et al. (2022). A review on chemometric techniques with infrared, Raman and laser-induced breakdown spectroscopy for sorting plastic waste in the recycling industry. *Resour. Conserv. Recycl.* 180: 106217.

41 Braskem. Sustainably sourced bio-based plastic. https://www.braskem.com/imgreen/all-about-being-bio-based (accessed 25 March 2024).

42 Braskem (2023). Braskem expands its biopolymer production by 30% following an investment of US$ 87 million. https://www.braskem.com.br/usa/news-detail/braskem-expands-its-biopolymer-production-by-30-following-an-investment-of-us-87-million (accessed 25 March 2024).

43 Balart, R., Garcia-Garcia, D., Fombuena, V. et al. (2021). Biopolymers from natural resources. *Polymers* 13: 2532. https://doi.org/10.3390/polym13152532.

44 Barret, A. (2018). What are drop-in bioplastics? *Bioplastics News* (28 August). https://bioplasticsnews.com/2018/08/28/drop-ins-bioplastics/ (accessed 25 March 2024).

45 Braskem. About our Portfolio. https://www.braskem.com/imgreen/portfolio (accessed 25 March 2024).

46 Bioplastics (2021). Braskem to invest US$61 million to increase biopolymer production. *Bioplastics Magazine* (25 February). https://www.bioplasticsmagazine.com/en/news/meldungen/20210225-Braskem-invests-US-61-million-to-increase-biopolymer-production.php (accessed 25 March 2024).

47 Shell Chemicals. Shell Chemicals and Braskem bring certified bio-attributed and bio-circular propylene to the U.S. market. https://www.shell.com/business-customers/chemicals/media-releases/2024-media-releases/shell-chemicals-and-braskem-bring-certified-bio-attributed-and-bio-circular-propylene-to-the-us-market.html (accessed 25 March 2024).

48 The Coca-Cola Company (2021). Coca Cola Collaborates with Tech Partners to Create Bottle Prototype Made from 100% Plant-Based Sources. https://www.coca-colacompany.com/media-center/100-percent-plant-based-plastic-bottle (accessed 25 March 2024).

49 Packaging World (2022). Coca-Cola, Suntory Cross the 100% Plant-Based Bottle Finish Line,. https://www.packworld.com/sustainable-packaging/article/22171823/cocacola-suntory-debut-100-plantbased-bottles (accessed 25 March 2024).

50 Rabnawaz, M., Wyman, I., Auras, R., and Cheng, S. (2017). A roadmap towards green packaging: the current status and future outlook for polyesters in the packaging industry. *Green Chem.* 19 (20): 4737–4753.

51 Hahladakis, J.N., Velis, C.A., Weber, R. et al. (2018). An overview of chemical additives employed in plastics: migration, release, fate and environmental impact during their use, disposal and recycling. *J. Hazard. Mater.* 344: 179–199.

3

Wood

3.1 Introduction

Wood packaging material, or WPM, is also called **non-manufactured** wood packing. It is defined as hardwood and softwood packaging other than that comprised wholly of wood-based products such as plywood, particle board, oriented strand board, veneer thin slices of wood), wood wool, etc., which have been created using glue, heat, and pressure or a combination thereof, used to support, protect, or carry a commodity. In packaging, WPM applications include pallets, skids, pallet collars, containers, crates, boxes, cases, bins, reels, drums, load boards, or dunnage (Figure 3.1).

3.2 Wood Degradation

We can divide the degradation of wood into three major categories:

- weathering,
- chemical degradation, and
- biological degradation (decay).

Weathering refers to degradation that involves the breakdown of wood materials **due to exposure to natural weather conditions** such as sunlight, temperature fluctuations, and moisture [1, 2].

Chemical degradation involves the alteration or destruction of the chemical structure of wood caused by exposure to certain chemicals.

Biological degradation (decay) corresponds to the breakdown of wood by living organisms such as fungi, bacteria, insects, or marine borers.

Degradation, Stabilization, and Recycling of Packaging Materials, First Edition.
Muhammad Rabnawaz, Susan E.M. Selke, and Ian Wyman.

Figure 3.1 Photographs of a stack of wood pallets (left) and a wood crate for the packaging of citrus fruit (right).

3.2.1 Weathering

Weathering can be defined as the reaction of wood to exposure to the outdoors above ground. In particular, **weathering results from exposure to a combination of moisture, light, and heat**. The chemical changes are most often initiated by ultraviolet (UV) light, although atmospheric pollutants can also play a significant role in weathering. (Note: Atmospheric pollutants, especially ozone, can also have significant roles in accelerating the degradation of plastics. Further discussion regarding the degradation of plastics is provided in Chapter 2.)

The general pattern of behavior is that wood surfaces become rough as the grain rises, and the wood forms small **cracks or defects that are called "checks"** (Figure 3.2). These checks grow into large cracks, so that the grain of the wood may loosen, boards of wood may become warped and pull away from fasteners. **As this weathering progresses,** the roughened surface **changes color** and **gathers dirt and mildew (mold or fungus)**, and eventually the surface coherence is lost. At this point, the wood becomes friable – it splinters, and fragments come off.

Figure 3.2 Illustration of checks that have formed on wood due to weathering. *Source:* Image used with permission from RONA Inc.

3.2.1.1 Effects of Various Influences on Weathering

If we examine the primary effects of the individual factors, we find the following:

Heating and Cooling Cycles Moisture affects the wood as rain or dew is absorbed deeply into the wood by capillary action. Consequently, the wood swells or shrinks, and stresses arise due to the moisture gradients that exist within the wood. The stresses are largest where the gradients are the steepest, which is usually at the surface. If these stresses are unbalanced or unevenly distributed, they can cause the wood to undergo warping (bending) and checks to form on its surface. **If temperature variations are sufficient to produce freeze/thaw cycles,** the freezing and thawing (melting) of absorbed water (especially from expansion as the water freezes and to a lesser extent from contraction as it melts) produces additional stress. Note: Ice has 10% more volume than the equivalent weight of liquid, which results in this expansion during freezing.

Light Light acts to induce *photochemical degradation*. As is the case for plastics, the **UV portion of the spectrum is largely responsible** for this type of degradation. **Photochemical degradation results in yellowing or browning of the wood,** which proceeds to eventual graying. The effect of light is mostly due to **photochemical decomposition of the lignin component of the wood** and typically affects only the **top 0.05–2.5 mm of the wood surface**. Since wood is an excellent light absorber, **UV light penetrates only about 75 μm (~0.003 inches) into the wood**. Visible light penetrates farther, about 200 μm, as this light is not absorbed well by lignin. ***Propagation of free radicals*** is responsible for the damage in the lower surface layers. **Note:** Because lignin is an excellent UV absorber, researchers are investigating if lignin can be used as a UV stabilizer for polymers and coatings [3]. The chromophores in lignin give composites their UV-protective quality, but they also result in an unwanted dark hue.

3.2.1.2 Effects of Wood Composition

As discussed in Chapter 2, polymer chemistry (chemical composition and types of bonds) has a critical role in the degradation of the polymer. Wood is a complex mixture of polymers, and thus, its chemistry determines its stability. So, to understand the effects of weathering, we need to look at both the chemical and physical structures of wood.

Wood is composed primarily of **cellulose, hemicellulose, and lignin**, but it also contains minor amounts of various extractives and minerals. A typical wood sample is composed of about **50% cellulose**. Figure 3.3 shows the composition of wood from Bigtooth Aspen trees. Different types of wood will have somewhat differing percentages of the major components.

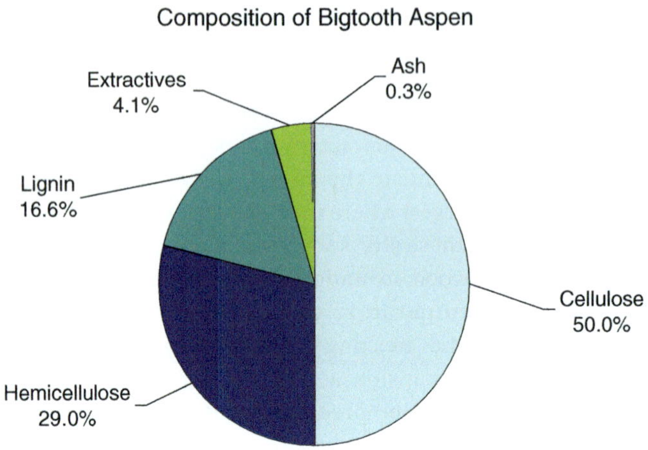

Figure 3.3 Composition of Aspen. *Source:* Drawn by Dr. Selke.

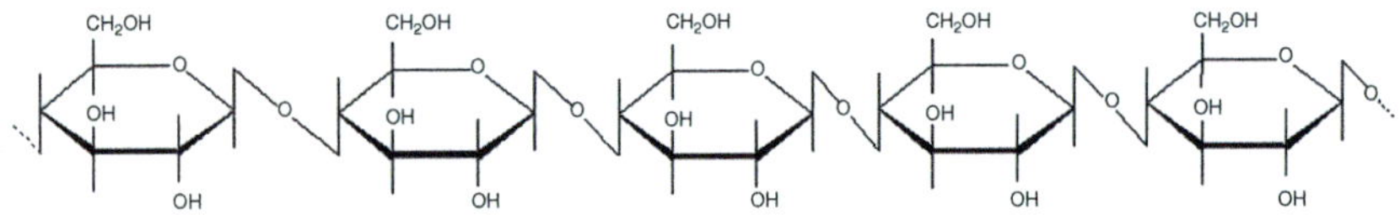

Figure 3.4 Structure of cellulose.

Cellulose (Figure 3.4) is made by the polymerization of glucose in a very regular repeating arrangement that allows the cellulose to pack into a crystalline array. The large number of hydroxyl (–OH) groups in cellulose results in **very strong hydrogen bonding, rendering the cellulose crystallites quite resistant to numerous types of chemical reactions**. The strong hydrogen bonding is also responsible for the great propensity of wood to sorb and otherwise interact with water, primarily in the amorphous regions and on the boundaries of the crystallites.

It should be noted that the chemical structures of some types of starch are identical to that of cellulose, except that the glucose units (also called glucose residues) are joined by α-1-4 linkages instead of the β-1-4 linkages found in cellulose. As a consequence, the starch molecule takes a helical rather than a linear shape, and this reduces starch crystallization. This phenomenon causes starch to be much more susceptible to chemical and biological attack than is the case with cellulose.

Hemicellulose: Hemicellulose is a copolymer of various sugars (five- and six-member rings), and it also has a branched rather than a linear structure. Consequently, hemicellulose is amorphous and is susceptible to attack. Hemicellulose is a shorter-chain polysaccharide that fills the space between cellulose fibers and lignin.

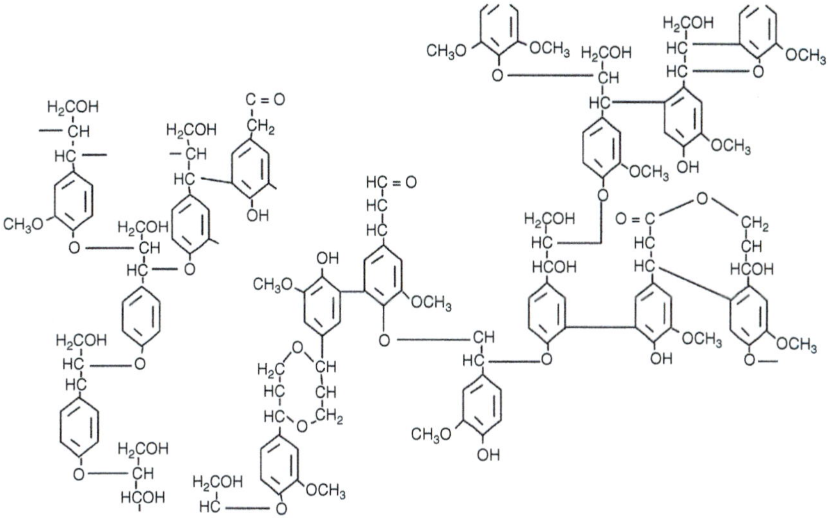

Figure 3.5 Examples of structures found in lignin molecules. This shows only a portion of a molecule.

Lignin: As mentioned above, lignin (Figure 3.5) is another major component of wood. Lignin has a complex molecular structure, which is based predominantly on phenylpropane units. In contrast to cellulose and hemicellulose, which are not strong UV absorbers, lignin absorbs UV light readily due to its abundance of aromatic rings and ether bonds. The absorbed UV energy tends to cause the production of free radicals and subsequent reaction of the phenolic hydroxyls with oxygen – this process is called photochemical oxidation. (Removal of acidic groups also occurs, resulting in an increase in acidity, which can accelerate oxidation.) Lignin also has abundant hydroxyl groups, so it, too, interacts strongly with water. In fact, photochemical oxidation of lignin tends to produce degradation products that are water-soluble, and hence, they can be washed out of the wood if it is exposed to running water, such as rain. **Softwood tends to have a higher lignin content than does hardwood.**

Further, the distribution of the various components of wood is not uniform throughout the structure. As you all know, wood has a cellular structure. The cells have a hollow center (the lumen) and a relatively thick cell wall, as illustrated in Figure 3.6. Individual cells are joined together by a substance called the middle lamella. These cellular sub-structures are described in further detail below.

Lumen refers to the interior of a wood cell, which is enclosed by the cell wall. This cellular interior usually becomes an empty or void space following the death of the cell after its contents have decayed or dissipated. The size and shape of the lumen can influence the physical properties of the wood, such as its density and strength.

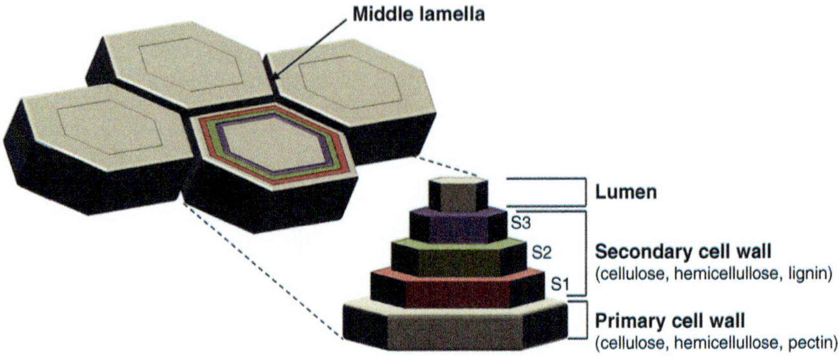

Figure 3.6 Diagram showing the structures of wood cells. *Source:* [4] / Springer Nature. / CC BY 4.0.

The **primary cell wall** refers to the outermost layer of the cell. This wall is relatively thin and flexible so that it can expand as the cell grows. The main components of the primary cell wall are cellulose, hemicellulose, and pectin.

The **secondary cell wall** is formed on the interior of the primary cell wall. This layer is much thicker than the primary cell wall, and it imparts structural rigidity and strength to the cell. It is composed of multiple layers (S1, S2, and S3), with the S2 layer being the thickest and providing most of the mechanical strength to the cell. The main components of the secondary cell wall are cellulose, hemicellulose, and lignin.

The **middle lamella** is the very first layer that is formed during cell division (also called mitosis), and this layer is **located between the cell walls of adjacent cells.** The middle lamella is **rich in lignin**, which helps to **bind the cells together.** It acts somewhat like a glue, maintaining the structural integrity between adjacent cells. After cell division is complete, the primary and secondary cell walls form inside the boundary established by the middle lamella.

In a typical softwood, the middle lamella is comprised of about 70% lignin, 20% hemicellulose, and only about 10% cellulose, as shown in Figure 3.7. The cell wall itself consists of an extremely thin primary cell wall and a thick secondary cell wall with three distinct layers, S1, S2, and S3 in Figure 3.7, which are differentiated from each other by the orientation of the cellulose microfibrils within them. The middle layer of the secondary wall is the thickest, and consists of approximately **20% lignin, 30% hemicellulose, and 50% cellulose**.

Therefore, photochemical oxidation primarily attacks the middle lamella because of its rich lignin content. As water washes out solubilized lignin degradation products, the remaining material is predominantly comprised of **whitish-to-gray**

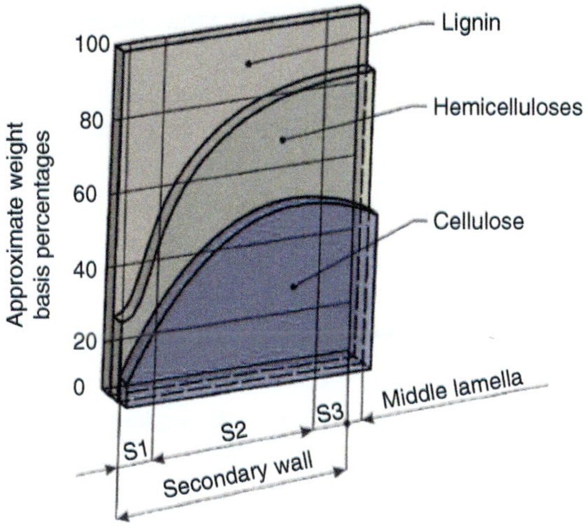

Figure 3.7 Relative proportions of lignin, hemicellulose, and cellulose across a wood cell wall. The *Y*-axis denotes percent by wt. [5]. *Source:* Image reprinted from ref.(6) This article is an open access article distributed under the terms [6] / MDPI. / CC BY 4.0.

fibers, which are high in cellulose, rather than having the light brown appearance that is characteristic of lignin. The graying of wood that occurs with weathering results from a combination of this exposure of the whitish fibers in the wood with the growth of mildew and other fungi.

The extent to which weathering erodes the surface is determined by exposure conditions and by the characteristics of the wood. Within a piece of wood, there are variations in the strength of the bonding between cells, leading to differences in erosion rates and, therefore, to an increase in surface roughness. **Wood produced early in the growing season** (called earlywood) erodes more rapidly than the thicker-celled wood that is produced later in the growing season (called latewood). Moreover, some species of wood are more resistant to weathering than others. In general, **hardwoods** erode at a slower rate than softwoods.

Protection Against Weathering (Natural Degradation) There are various ways to protect wood against weathering, and some key examples are briefly described below.

Surface treatment is one way to prevent or minimize weathering by protecting the wood against exposure to light and water. Paints, varnishes, and stains can all provide such protection. In some cases, water repellants are used, which consist of plastic resin, solvent, wax, and a **preservative (fungicide).**

Fungicides are also used alone as preservatives, but they target decay rather than photochemical processes.

Inorganic surface treatments with chromium, copper, or zinc salts can also be used to prolong wood life. For example, copper has biocidal properties and is toxic to fungi, bacteria, and many insects. Copper also acts as a radical scavenger and being a transition metal, can exist in different oxidation states. This allows it to donate or accept electrons, which can be used in the neutralization of free radicals.

3.3 Chemical Degradation

Wood is also susceptible to chemical degradation. Under acidic or alkaline conditions, all the components of wood can be hydrolyzed. Hydrolysis occurs faster under acid conditions since acidic groups are released by hydrolysis, which act to neutralize alkaline chemicals but enhance the action of acids.

Lignin hydrolyzes very easily, even in hot water. Lignin reacts readily with oxygen, as well as with chlorine, bromine, iodine, nitrogen dioxide, and a variety of other agents. **While we may think of the chemical degradation of wood as something to be avoided, it is actually the basis of chemical pulping processes**. The goal of pulping is to degrade the wood sufficiently to allow separation of the fibers. Alkaline conditions are generally used because they are milder and hence permit the cellulose fraction to be recovered largely intact while the lignin is extensively degraded to liberate the fibers. Acid hydrolysis is too harsh for this process, resulting in considerable hydrolysis of the cellulose along with the lignin. Even many largely mechanical pulping processes benefit from a limited amount of lignin hydrolysis to help liberate the fibers – recall that lignin hydrolyzes in hot water. Therefore, a steam explosion process, for example, while depending mostly on mechanical force from the expansion of the steam, also results in some lignin hydrolysis [7].

3.4 Biological Decomposition (Decay)

In many cases, the lifetime of wood is limited not by weathering but rather by **decay.** Since weathering is a surface phenomenon, **the wood may maintain its overall strength even when its surface integrity has been greatly reduced**. Depending on the application, the loss of surface integrity may or may not render the article unfit for continued use. **Decay, on the other hand, can cause massive destruction throughout the wood**. Decay results primarily from the action of fungi, insects, and marine borers, depending on the exposure conditions. **A key to preventing fungal decay is to keep the wood below the fiber-saturation moisture content (about 27%**

of the wood's dry weight), as the fungi need free water to be readily available in order to grow and reproduce.

The relative susceptibility of the major wood components to decay is quite different than their susceptibility to UV-induced oxidation. **For decay, cellulose and hemicellulose are much more susceptible than lignin.** A wide variety of bacteria and fungi can metabolize cellulose and hemicellulose. In contrast, lignin is degraded by relatively few microorganisms, most of which are fungi. **The three types of fungi** that are most capable of **degrading lignin are known as white-rot, soft-rot, and brown-rot fungi** [8]. Brown-rot fungi produce the most rapid decay. Larger organisms such as insects and marine borers also digest cellulose. They can breach the lignin barrier mechanically (by chewing through it) but do not actually metabolize the lignin.

The way wood has been treated (processed) can significantly affect its susceptibility to decay. Two major types of cells are produced in the cambium (the growing layer) of a living tree (see Figure 3.8): **thick-walled fiber cells**, which give the tree its strength, and thin-walled **parenchyma cells**, which provide food storage.

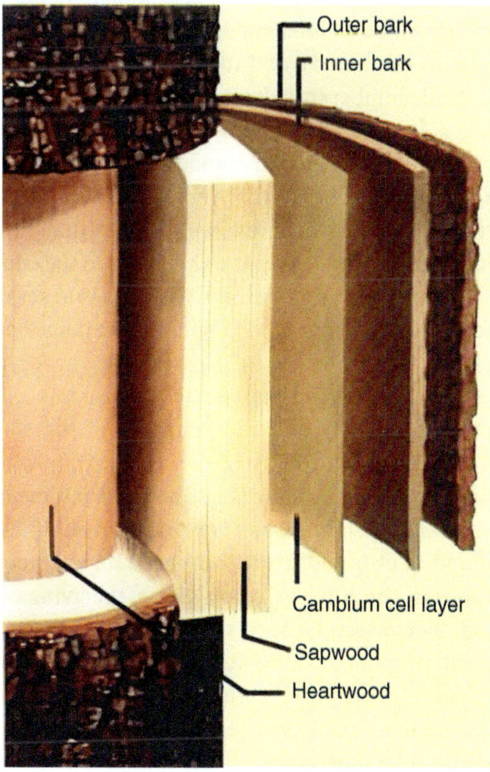

Figure 3.8 Structure of a tree. *Source:* Picture from Durability by Nature, Canadian Wood Council, 2022. Image used with permission from the Canadian Wood Council (cwc.ca) [9].

(In hardwoods, an additional type of cell, vessels, provide sap flow.) Mature fiber cells die within a few days or weeks and lose their cytoplasmic contents (the contents occupying the hollow space inside the cells). At this point and thereafter, they consist almost entirely of cell wall polymers: cellulose, hemicellulose, and lignin. In contrast, **the parenchyma cells remain alive for years**. They only lose their cytoplasmic contents when **sapwood** is transformed into **heartwood**. These cytoplasmic contents include sugar, starch, amino acids, and proteins. Therefore, sapwood is highly susceptible to invasion by the large number of fungi and bacteria that can use these reserve food materials but that are not able to attack the more complicated cell wall polymers. In contrast, heartwood (which comprises the core of a tree trunk) does not contain these materials and is more resistant to decay than sapwood.

Another factor that increases the decay resistance of heartwood is that in many trees, extractives accumulate in the heartwood, which is toxic to decay organisms. These extractives, which serve a protective function, **are not present in sapwood**. Such substances include phenols, terpenes, and alkaloids, among others. Trees differ by species in the amounts and types of such substances that they produce and therefore differ in decay resistance. Highly resistant species include cypress, many cedars, and Osage orange. White oak, Douglas fir, and certain pines are classified as moderately resistant. **This accumulation of extractives often causes the heartwood to exhibit a darker color, as shown** in Figure 3.9.

While lumber from heartwood is more decay resistant than sapwood, the opposite is true in a living tree. When a living tree is attacked, it has active defense mechanisms that can be brought to bear – delivering protective chemicals to the site of the attack. This delivery depends on active sap flow. Therefore, in a living tree, sapwood is more resistant to decay than heartwood since there is no sap flow in heartwood to deliver the chemicals to the site of the attack.

When lumber is cut, the first step is to season it (dry it out). If lumber is air-seasoned, the stored food materials are depleted by the **respiration** of the parenchyma cells themselves. However, if the drying treatment is delayed, fungi, bacteria and molds can invade and use the cell contents as food. This does not usually seriously affect the strength of the wood, but it does cause stains and affects the permeability of the wood. If the wood is kiln-dried immediately, the living cells are killed by heat, and these reserve foods are retained. This provides a substrate for the growth of microorganisms if the wood becomes wet again. **If the wood is chipped, that increases its surface area and exposes the reserve food; in the piles of wood chips**, the living cells along with fungi and bacteria rapidly convert the food reserves to CO_2, H_2O, and heat. This phenomenon can result in spontaneous combustion of the wood pile.

The keys to the prevention of wood decay are to keep the **wood dry, use resistant species, and/or use preservatives**. The wood preservatives that are used most widely are creosote, pentachlorophenol, chromated copper arsenate, and ammoniacal copper arsenate. However, there is growing pressure to avoid the use

Figure 3.9 Darkening of heartwood. The heartwood toward the interior of this piece of wood is significantly darker than the sapwood that is adjacent to it toward the exterior. Photograph taken by the authors.

of these chemicals because of environmental concerns, and the arsenic-based preservatives are being phased out [10, 11].

Problem Set

1 In the case of wood, what are the factors that affect weathering?
  A Moisture.
  B Temperature.
  C Light.
  D All of the above.

References

1 Acıcan, T., Alibaş, K., and Özelkök, I.S. (2007). Mechanical damage to apples during transport in wooden crates. *Biosyst. Eng.* 96 (2): 239–248.
2 Niklewski, J., van Niekerk, P.B., and Marais, B.N. (2023). The effect of weathering on the surface moisture conditions of Norway spruce under outdoor exposure. *Wood Mater. Sci. Eng.* 18 (4): 1394–1404.

3 Zhang, Y. and Naebe, M. (2021). Lignin: a review on structure, properties, and applications as a light-colored UV absorber. *ACS Sustain. Chem. Eng.* 9 (4): 1427–1442.
4 Kim, S.M., Dien, B.S., and Singh, V. (2016). Promise of combined hydrothermal/chemical and mechanical refining for pretreatment of woody and herbaceous biomass. *Biotechnol. Biofuels* 9 (1): 97.
5 Ali, I. (2012). Study of the mechanical behavior of recycled fibers. Applications to papers and paperboards. PhD thesis. King Abdulaziz University. Advisors: Bloch, J.-F. and Passas, R.
6 Mitaľová, Z., Klichová, D., Botko, F. et al. (2023). Green machining of NFRP material. *Machines* 11 (7): 692.
7 He, Q., Ziegler-Devin, I., Chrusciel, L. et al. (2020). Lignin-first integrated steam explosion process for green wood adhesive application. *ACS Sustain. Chem. Eng.* 8 (13): 5380–5392.
8 Kirk, T.K. and Cowling, E.B. (1984). Biological decomposition of solid wood. In: *The Chemistry of Solid Wood* (ed. R. Rowell), 455–487. Washington, DC: American Chemical Society.
9 Canadian Wood Council. (2024). Durability by Nature. https://cwc.ca/en/why-build-with-wood/durability/durability-by-nature/ (accessed 19 November 2024).
10 United States Environmental Protection Agency. (2023). Overview of wood preservative chemicals. https://www.epa.gov/ingredients-used-pesticide-products/overview-wood-preservative-chemicals (accessed 13 June 2023).
11 CBC News. (2002). Canadian companies to phase out CCA wood. https://www.cbc.ca/news/canada/canadian-companies-to-phase-out-cca-wood-1.352525 (accessed 4 April 2002).

4

Paper Degradation and Stabilization

4.1 Introduction

Paper is primarily composed of:

Cellulose: Cellulose fiber is one of the main components of wood, which is used to prepare paper. Further details regarding cellulose are provided in Chapter 3.

Binders: These have an important role in paper fabrication as they help enhance the printability of paper as well as its physical properties, such as its flexibility, strength and its resistance against water. They also facilitate the binding of fillers and pigments to the paper. The main purpose of a binder is to ensure that the cellulose fibers that comprise the paper adhere together so that the paper exhibits the desired characteristics. Starch is the most common material that is used as a binder for papermaking processes, but other materials such as latex, synthetic polymers, casein, or soy are sometimes used as binders when coated paper is being prepared [1]. The structure of starch is quite similar to that of cellulose as it is a complex carbohydrate that is composed of long chains of glucose molecules, and starch can act as a glue to hold cellulose fibers together in the finished paper. When coated paper is being produced, binders are selected so as to ensure that there is firm adhesion between the paper substrate and the coating.

Lignin: As is the case with cellulose, lignin is a natural component of wood, and further details regarding lignin are provided in Chapter 3.

Fillers: These are used to enhance various characteristics of paper, such as its printability and appearance. Fillers help smoothen the texture of paper, and enhance its brightness and opaqueness, while also helping to prevent ink from bleeding through to the other side. Common types of fillers used in paper production include calcium carbonate, clay (also called kaolin), talc,

Degradation, Stabilization, and Recycling of Packaging Materials, First Edition.
Muhammad Rabnawaz, Susan E.M. Selke, and Ian Wyman.

titanium dioxide, and starch [2]. Among these fillers, calcium carbonate is the most widely used, and helps provide good brightness along with a smooth surface to enable better printability. Moreover, calcium carbonate helps to protect paper against acid degradation due to its alkalinity. Clay is a silicate-based or aluminosilicate material that is highly abundant and occurs naturally. Clays are often used to enhance the glossiness of coated paper. In addition, they help improve the printability and smoothness of paper surfaces. Talc (magnesium silicate) is a soft and hydrophobic material that is used to reduce "tackiness" and facilitate the coloring of papers as it is compatible with dyeing processes and can facilitate ink transfer [3]. Titanium dioxide is used when high brightness and high opacity is required, although its high cost limits its use to specialty paper applications. Starch enhances the printing characteristics as well as the strength of paper. As we have seen above, starch is also used as a binder.

Surface coatings: Paper is also coated with materials such as per- and polyfluoroalkyl substances (PFAS), latexes, wax, polyethylene (PE), and other materials.

Among the materials listed above, **cellulose and binders** are mandatorily found in paper. Lignin is also present, albeit in less or more quantity depending on the type of paper involved, how the pulping was performed, and whether or not any bleaching was performed. Lastly, fillers and surface coatings are optional, and their amount varies depending on the application.

4.2 Durability and Permanence

Two important criteria for paper are "durability" and "permanence." Durability of paper refers to the ability to resist the effects of wear and tear when in use. In contrast, permanence refers to the ability of paper to last hundreds of years without significant deterioration under normal use and storage conditions. In other words, the ability to remain chemically and physically stable over long periods of time is called permanence. Lack of durability is most often evidenced by bursting or tearing of the paper. In contrast, lack of permanence is often manifested by discoloration, loss in strength (embrittlement), or changes in chemical properties. Although the passage of time or storage is the key aging aspect (as opposed to usage) that permanence refers to, it should be noted that the conditions (temperature, humidity, pH, etc.) under which the storage is undertaken can affect the permanence of the stored paper.

To understand these changes in paper performance, it is important to consider both changes in the chemical makeup of the paper (predominantly in the cellulose, at least in bleached paper) and changes in the physical structure and properties of the paper. It is possible to have considerable change in the physical properties of the paper with no discernible change in chemical structure.

The reasons for loss of strength on aging of paper are not fully understood. But at least two types of changes are known to be involved:

1) Weakening or embrittlement of fiber–fiber bonds either those derived from natural hemicelluloses in the fibers or from additives used during the papermaking process.
2) Structural changes in the fibers themselves, possibly involving cross-linking or increases in crystallinity.

The extent of change in paper performance upon aging depends on, among other variables:

- the type of fiber;
- the pulping process used;
- the amount and type of secondary ingredients present; and
- the atmospheric conditions to which the paper is subjected.

Paper is subject to many types of degradation. For cellulose, these can involve:

- hydrolysis;
- oxidation;
- cross-linking;
- changes in crystallinity;
- photolysis (ultraviolet [UV]-induced chain scission);
- photo-oxidation; and
- thermal decomposition.

Like cellulose, hemicellulose and lignin are subject to similar degradative processes. Mechanisms for these degradation processes are essentially the same as those that we discussed for polymers in the plastics section, since, after all, paper is composed of polymers.

Acidity factor: One very significant factor affecting paper performance is hydrolysis of cellulose caused by excessive acidity in the paper. This became evident as an issue after the introduction of wood pulp. **While it is not normally a serious issue for packaging papers, because of their relatively short useful life,** it is a serious problem for **libraries seeking to preserve printed materials**. There are widespread examples of books that were published in the 1600s now being in much better shape than books published in the 1900s, as a result of acid-induced hydrolysis.

There are numerous sources of acid in paper. One of the most common is alum, which has a general chemical formula of $XAl(SO_4)_2{\cdot}12H_2O$, where X is a monovalent cation such as potassium or ammonium. Alum is often used in rosin, a sticky material, used as a sizing agent for paper. Hydrolysis of alum results in the production of acid. The permanence of paper can be directly correlated to the pH of a water extract of the paper. If the pH is under 5, the paper will have poor stability. **A pH of 6–8 is desirable for maximum stability.**

Yellowing of paper is not necessarily correlated with deterioration of strength or other performance properties. For example, groundwood paper has very poor color stability due to its high lignin content, but generally has good strength retention. In general:

- high yield pulps have poor color stability due to their larger lignin content;
- sulfate (kraft) pulps have poorer color stability than sulfite pulps. For pulping, either sulfite or sulfate processes are used. Sulfite is an acidic pulping process used for rayon grade pulp for cloth manufacture. Sulfate or Kraft pulping is alkaline in nature and used mainly for paper making. Sulfate process paper has more lignin and thus has poor color stability.
- lightly bleached sulfite pulps have better color stability than highly bleached sulfite pulps. This is counter intuitive, but this is likely because in the case of highly bleached pulps, some unwanted chromophoric/photochromic structures are introduced into paper, leading to poorer color stability when exposed to light and heat and
- chlorine dioxide bleached pulps have better color stability than hypochlorite bleached pulps by enabling selective and effective removal of lignin by chlorine dioxide.

Of note, for packaging applications the main type of pulp used in paper and paperboard is kraft, both bleached and unbleached.

4.2.1 Quality of Paper Fiber

The quality of the paper fibers is very important. Cotton, cotton rags, and linen produce paper with better permanence than most wood pulp, in part because harsher chemical treatment is required to obtain purified cellulose fibers from wood. Therefore, the cellulose starts out with more extensive chemical degradation leading to chromophoric and photochromic structures. In addition, wood fibers are shorter. On the other hand, retention of hemicellulose tends to create problems with acidity. **Hemicellulose tends to contain many acidic groups which are released on hydrolysis**. Similar problems can result from the retention of lignin. Mechanical pulp, such as groundwood, and some unbleached pulps are very undesirable when permanence of paper is a goal. However, the situation is somewhat complicated, since wood itself is very stable, and papyrus sheets are extremely stable even though they contain lignin and hemicellulose along with cellulose.

General guidelines have been developed to ensure that paper will have high stability. For example, it should:

- have a low functional group content in the cellulose (typical functional groups include aldehydes, ketones, and carbonyls) and this will be controlled by the fiber source and the pulping process used for fiber recovery;
- use an alkaline filler or a filler with a neutral pH (such as calcium carbonate);
- use a stable sizing material;
- contain a high percentage of long fibers;
- contain a low percentage of fine fibers;
- contain no lignin; and
- contain no aluminum salts (alum).

Accelerated testing is often performed to predict the permanence and durability of paper. While there are a number of performance variables that can be tested, it has been found that folding endurance is a measure that tends to be fairly sensitive, much more sensitive than tensile strength, so it is often used as the measure. This test measures the number of times a paper can withstand folding and unfolding before it breaks under a standard condition.

Pulps containing lignin can undergo color changes due to UV-induced lignin oxidation, which causes darkening. Light also degrades cellulose, hemicellulose, other carbohydrates, fats, resins, and waxes that may be present in the pulp. **In pulps that do not contain lignin, color change is closely correlated to the content of carbonyl groups.** Therefore, chemical treatments that react with chromophoric carbonyl groups to convert them to nonchromophoric hydroxyl groups improve their color stability. These **carbonyl groups** are mostly found in the hemicellulose fraction, where they result from degradation reactions, such as those produced by bleaching. The wavelength of the light has an effect on the types of color change that result. **Wavelengths longer than 480 nm** tend to have a slight bleaching effect (photo-bleaching effect), while shorter wavelengths (especially those under 300 nm), tend to produce a dulling and yellowing of the pulp.

Other factors affecting paper degradation include exposure to atmospheric contaminants such as sulfur dioxide, nitrogen oxides, and ozone [1]. Ozone is a potent oxidizing agent. Sulfur dioxide in the presence of metals, especially iron, plus oxygen forms sulfur trioxide (SO_3), which in the presence of water forms sulfuric acid (H_2SO_4). **Sulfuric acid** contributes to hydrolysis of cellulose, and consequent reduction in molecular weight and in turn weakening the paper strength. **Degradation reactions generally increase with increasing temperature**, and they also tend to be **faster with higher moisture content.** However, the reaction rates are not totally linear with moisture, and they may reach a minimum at a humidity that is somewhat greater than zero. (A small amount of water can have a stabilizing effect through its interaction with sensitive groups.)

4.3 Biological Degradation of Paper

Paper is also subject to biological deterioration (biodegradation) through attacks by various organisms such as bacteria and fungi. **In general, microbiological attack is very dependent on moisture content.**

Paper protection during the paper making process: Since the highest moisture paper encounters is during the paper-making process, prevention against biological deterioration frequently begins with protection of the recirculating pulp mix. **Slimicides** are commonly used to prevent growth of slime organisms during the manufacturing process [5]. Various compounds can be used in this role, including **chlorine and pentachlorophenol.** Slimicides use different pathways to attack microorganisms. Slimicides can produce radical species that attack fats in the microorganism's cell membrane or act by altering the chemistry of the proteins inside the microorganism's body.

Paper protection after paper is made: Other agents are directed toward limiting the growth of fungi and bacteria in paper once it is produced, especially for those papers that can be expected to be exposed to high moisture contents for considerable periods of time, such as filters, building papers and boards, and multiwall bags. Compounds used in this role include **pentachlorophenol and other chlorinated phenols**, **copper sulfate, 8-hydroxyquinoline or its copper salt** as well as others. Many of these compounds can also be used as slimicides.

Active Packaging: A third use of antimicrobial compounds is to protect packaged products from the growth of microorganisms. Sodium alkylbenzene sulfonate is sometimes used in soap wraps. Calcium propionate or sodium propionate can be used in wrapping papers for butter, margarine, and bread to inhibit mold growth. Sorbic acid can be incorporated into paper used for wrapping cheese. Biphenyls can be used in paper for paperboard boxes or tissues for the packaging of fruits, especially citrus fruits, to prevent the growth of blue mold and stem-end rot organisms. These antimicrobial compounds can function both to protect the paper as well as the product inside.

Of note, while infestation with insects and rodents can be a significant problem, there has been little success with the use of insecticides and poisons to prevent such an attack. These substances tend to be highly toxic, and they are often effective only after major damage has already been done. There has been some limited use of pyrethrins with piperonyl butoxide as synergistic agents in the outer layers of multiwall bags for food products.

4.4 Wet-Strength Papers

A very different type of "degradation" of performance is that associated with the interaction of paper with moisture. In this context, we are not referring to chemical changes that might result in the paper due to hydrolysis, but simply to the **physical changes resulting from the interaction between the hydrophilic groups in**

the paper with water, and consequent changes in the mechanical strength and cohesiveness of the paper. At its extreme, this is the basis of paper-forming or paper-reforming processes, in which the wood fibers are suspended in water in order to facilitate their rearrangement into the desired paper structure. Hydrophilic groups in the fibers cause them to be readily wetted and swollen by water. When the paper sheet is formed, the swollen fibers are compacted into a web. As the water is removed via drying, inter-fiber bonding forces develop between adjacent surfaces, providing strength and coherence to the finished sheet. When the sheet is rewetted, the fibers again swell, the inter-fiber bonding is weakened, and the paper loses most of its strength. On complete saturation with water, the typical tensile strength retention of paper is between 1% and 10% of that of the dry strength of the paper. If we need the paper to retain a larger portion of its strength, we must in some way block some of the effect of water on the paper. Wet-strength papers are treated papers that can retain a higher proportion of their strength on complete saturation with water. Even these papers do suffer significant loss of strength, but rather than retaining only 1–10% of their original strength, they typically retain between 20% and 50% of their original strength when they are fully saturated with water.

It is important to note that wet-strength paper is defined as paper that shows extraordinary resistance to rupture or disintegration when it is saturated with water. This is *not* the same as water repellency or resistance to wetting, which interferes with water saturating the paper in the first place. Water repellence or water resistance are commonly achieved by application of wax or a polymer coating (usually polyethylene) to the paper, to serve as a water barrier. We will discuss coated paper for improved water resistance in Section 4.5 of this chapter.

Paper with enhanced wet strength is desirable for numerous applications, including the following:

- Paper products used as drying or wiping media, such as paper towels and facial tissues.
- Papers that will be exposed to weather, such as packing cases and paper bags.
- Paper used in wrapping wet materials, such as butcher wrap, and packaging for frozen and prepared food.
- Paper that will be immersed in water during a processing operation, such as photographic paper, filter paper, and tea bags.
- Paper used as disposable materials in place of textiles, such as hospital bed sheets and hospital gowns.

So, how one can improve the wet strength of paper? The following four approaches have been identified.

1) Employing a cross-linked resin network at the fiber interface to reduce fiber swelling. This is the most widely used in the industry.
2) Water-insoluble adhesive between the fibers.

3) Higher degree of interatomic attraction (hydrogen bonding) than normally occurs in paper.
4) Development of covalent bonding at the fiber interfaces with or without an additional phase. This is not really used for practical applications.

Commercial systems using the fourth alternative above have not yet been developed, but systems using the other three have been used. The reason why the fourth mechanism listed above has not been used is that attempts to produce covalent chemical bonds between fibers have usually resulted in a decrease in dry strength properties, which is obviously undesirable. The decrease in dry strength is likely due to embrittlement as well as a reduction in the number of hydrogen bonds between fibers.

The second mechanism (water-insoluble adhesive between the fibers) using animal glues, formaldehyde and cellulose xanthate solutions has been used in the past to improve the wet-strength of paper, but these materials are not currently used for this purpose. Latex is currently employed to some extent as an internal bonding agent in paper, but it is mainly used for other strength enhancing purposes rather than the development of wet strength, as it is less effective than the other methods that will be discussed below.

If we look at the **third mechanism,** the development of a **higher degree of secondary (hydrogen) bonding** can be achieved through physical or chemical modification of the fiber, or by use of additives. **Prolonged beating will gelatinize the fiber, yielding some increase in the wet strength.** This approach is used in greaseproof and glassine paper. The parchmentization of paper involves treating it in a bath of strong sulfuric acid. This method is used to some extent for the **packaging of frozen or moist fatty foods**. While vulcanization is typically used to treat rubber, it is also used to treat cellulose fibers. Vulcanized fiber produced through treatment with strong zinc salt solutions gelatinizes the cellulose [6, 7] and is used to some extent for electrical insulation, gaskets, tags, and luggage but not for packaging applications. Addition of modified starches (dialdehyde starches) increases the wet strength in neutral or acidic conditions, although not in alkaline conditions. There has been some use of polyethylene imine (PEI), a polymer of $[-CH_2-CH_2-NH-]_n$, for this purpose in unsized (containing no sizing agents) absorbent papers, but it tends to cause development of a yellowish color.

The **first alternative** listed earlier employs a cross-linked resin network at the fiber interface to reduce fiber swelling. This is the one that is the most commonly used route to produce paper with enhanced wet strength. The first wet-strength papers fabricated via this approach emerged in the 1930s and 1940s, when thermosetting resins were applied to finished paper sheets as an impregnating solution, and then cured. **When the resins were cured, the resulting cross-links**

provided a support structure for the paper fibers. However, this is a costly approach to increasing wet strength. Processes that permitted resins to be added to the paper stock before the paper was formed, absorbed by the fibers, and subsequently cured had obvious advantages in terms of process simplicity and economics.

Currently, the materials most extensively used in this role are those which are based on **urea-formaldehyde (UF), melamine formaldehyde, and polyamide-polyamine-epichlorohydrin resins**. It is estimated that between **4% and 8%** of all paper produced in the United States is treated for improvement of wet strength. In terms of tonnage, the largest markets are Kraft linerboard and bag and sack paper. **Other applications** include towels and sanitary tissues, butcher wrap, frozen food stock, hanging papers (wallpaper), tag stock, poster boards as well as construction paper and boards. **Melamine resins** [8, 9] are used in map papers, currency papers, laundry tags, food wraps, and some photo-production papers.

4.4.1 Major Categories of Wet-Strength Papers

4.4.1.1 Urea-Formaldehyde B-stage Derivatives

Urea-formaldehyde resins (sometimes called **UF resins**) are the simplest example of the aminoplast thermosetting type of wet strength agents. **Under neutral or alkaline conditions**, urea reacts with formaldehyde to produce dimethylolurea (DMU) as shown in Figure 4.1:

DMU is regarded as the **A-stage** (monomeric) resin. Under acidic conditions and at high temperatures, it condenses to three-dimensional insoluble polymers. If paper is impregnated with DMU solution and it is polymerized using acid catalysis, its wet strength will be enhanced. However, the problem with this approach is that **DMU is nonionic and monomeric**, so it tends not to be adsorbed by the pulp fibers from an aqueous solution. Therefore, it must be applied by some type of impregnation method such as surface sizing or spraying treatment. Furthermore, the wet strength tends to diminish during aging under damp moist conditions, due to hydrolytic degradation. **Therefore, this type of wet strength agent is not currently used**.

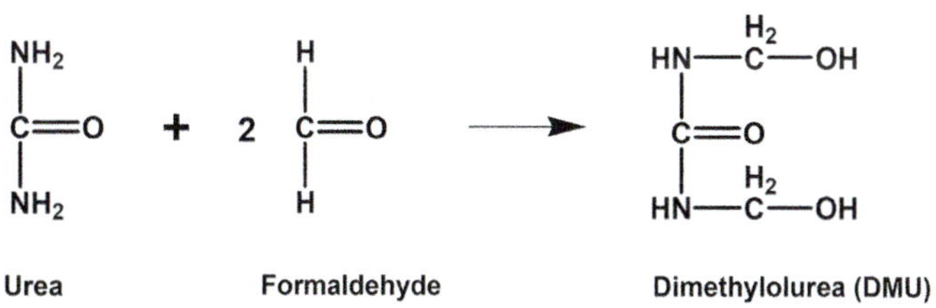

Figure 4.1 Reaction between urea and formaldehyde leading to DMU.

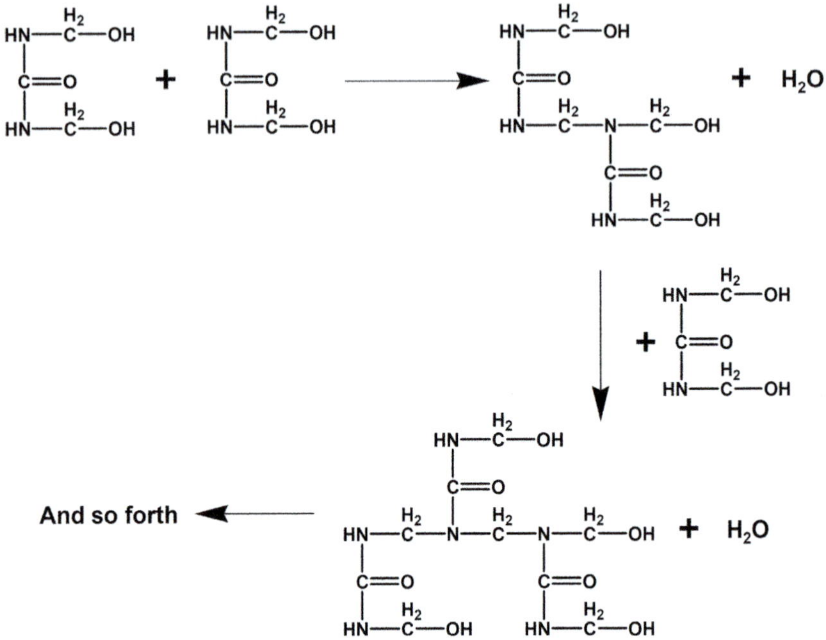

Figure 4.2 Condensation polymerization of DMU leading to a three-dimensional network.

To get better retention of the wet-strength agent in the paper mat/sheet, the polymerization can be partially accomplished before it is applied on the paper mat – getting to the B-stage (low molecular weight polymer), by the reactions shown in Figure 4.2.

Note that a branched and eventually cross-linked polymer is formed, with elimination of a water molecule occurring at each step. Now, however, a **new problem has emerged**. The elimination of many hydroxyl groups to form the stage B resin produces a syrupy material that is water immiscible. To retain water miscibility, highly polar groups are incorporated into the resin molecules. This is often done by the addition of bisulfite, thus forming sulfonate groups in the polymer. This bisulfite-modified resin is anionic (has a negative charge). To get good adsorption onto the pulp, alum is commonly added to the paper furnish (imparting a positive charge to the pulp). Different pulps react differently to this treatment. Unbleached kraft, with its abundance of **carboxyl groups**, is highly receptive to this type of wet-strength treatment, while more purified bleached pulps are much less receptive.

Once the **bisulfite-modified resin** has been incorporated into the pulp, subsequent heat treatment, such as during drying of the paper, leads to additional condensation polymerization, producing a cross-linked network, the **C-stage**. Because the cured resin is hydrophobic, these papers end up **highly nonabsorbent**. This is desirable for some applications, but not for products that are intended to be absorbent (such as sanitary absorbent products, paper towels, etc.).

A **major advance** in UF wet-strength agents was the discovery that polyfunctional amines could be used in place of bisulfite as modifying agents. The resins then remain water miscible but are cationic (positively charged) rather than anionic, so they are absorbed and retained without requiring the use of alum. This significantly improves the useful life of the paper. Widespread use of amine modified UF resins began in the late 1940s.

4.4.1.2 Melamine-Formaldehyde Cationic Colloids and Derivatives

The next advance was the development of melamine-formaldehyde resins (sometimes called MF resins). Melamine is a trifunctional amine that was first produced commercially from calcium cyanamide, as shown in Figure 4.3 (note that, for simplicity, not all reactants and products are shown in this reaction scheme):

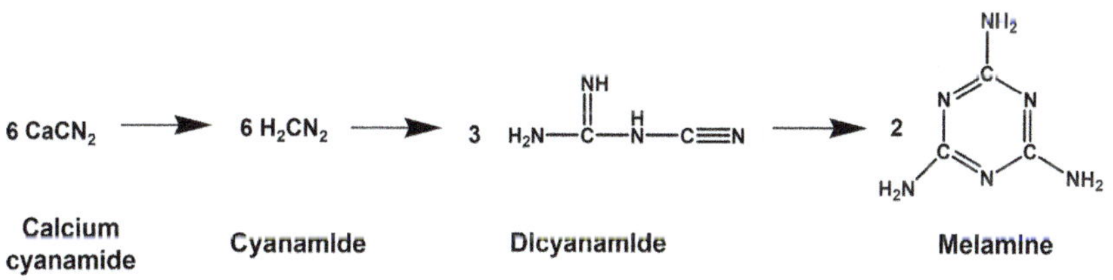

Figure 4.3 Synthetic route leading to melamine.

Melamine reacts with formaldehyde to form a series of methylol derivatives, ranging from mono- to hexamethylol melamine, depending on the number of amino hydrogens that have been replaced by methylol groups, as shown in Figure 4.4:

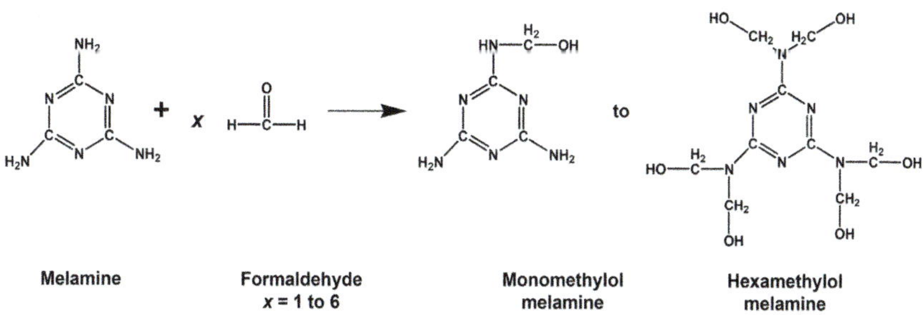

Figure 4.4 Synthetic route leading from melamine to monomethylol melamine and hexamethylol melamine.

In the presence of hydrochloric acid, methylol melamines undergo a self-limiting polymerization, forming colloidal particles about 20 monomer units in size. This is the B-stage, or gel stage. The hydrochloric acid dissociates into H^+ and Cl^- ions, and the H^+ ions associate with the nitrogen atoms of these colloidal particles, rendering them electropositive, and therefore they are readily absorbed by the pulp fibers. After a mat has been formed on the fibers via this adsorption process, as is the case for UF resins, subsequent additional condensation reactions produce a cross-linked structure: the water-insoluble C-stage polymer.

In comparison with urea-based resins, the greater abundance of functional groups in these melamine-based systems results in greater wet strength per unit of resin used, while also offering improved permanence during aging. Unfortunately, these resins are also higher cost materials.

Both the UF and MF resins require acidic conditions during curing – typically a pH of about 4–5. This acidic pH adversely affects the permanence of the paper and can lead to corrosion.

4.4.1.3 Polyamide-Polyamine-Epichlorohydrin (PPE) Resins

PPE resins [10, 11] were developed in 1957. Sometime PPE is also referred to as polyaminopolyamide-epichlorohydrin (PAE). Their principal advantage over other resins is their ability to be absorbed by the fiber and undergo curing to the C-stage in neutral or alkaline conditions. This leads to better retention of wet strength under acidic pH conditions, although they exhibit reduced retention at high pH. In addition, these resins allow the paper to retain its absorbency and softness. These advantages have led to widespread use of PPE resins. The reaction sequence is shown in Figure 4.5.

4.4.2 Overview of Wet-Strength Resins

Wet-strength resins that are commonly used today share the following characteristics:

1) The wet-strength agents are **readily absorbed** from water onto the pulp fibers. Adsorption is rapid, but not necessarily complete.
2) The resin is absorbed during the B-stage and does not produce wet-strength at this point. It must be cured to bring it to the C-stage. The heat in the drying oven is insufficient to completely cure the resin. The curing continues, albeit at a slower rate, after the paper leaves the machine. **It is typically complete in**

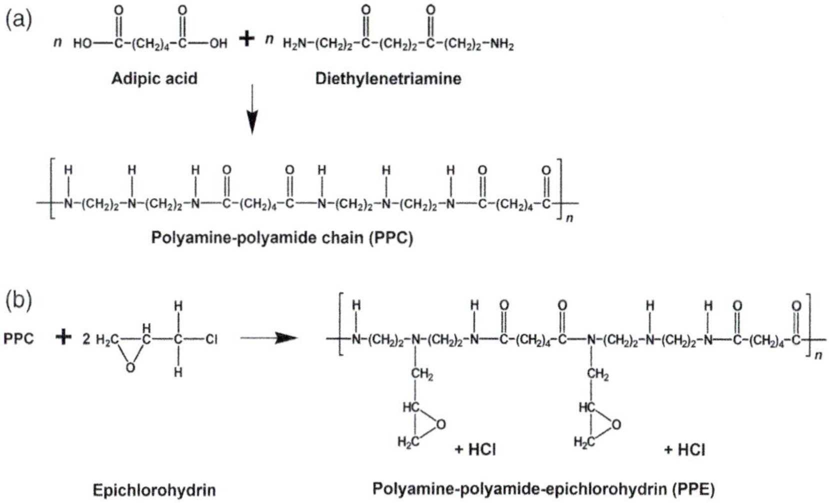

Figure 4.5 Synthetic route leading to a PPE resin. (a) Synthesis of PPC precursor; and (b) synthesis of PPE after PPC and epichlorohydrin reaction.

about two weeks, although the curing rate is affected by the temperature of the storage area.

3) Since all papermaking operations produce **broke or mill broke** (a term referring to waste, such as that obtained from trimming and joining operations), recycling is essential. For UF and MF resins, mildly acidic conditions and elevated temperatures are used to defiber the broke (resuspend the fibers so they can be recycled). For PPE resins, an alkaline pH of about 10 is used for this process. **Unfortunately, this requirement complicates the recycling of postconsumer wet-strength paper**. The fibers will not resuspend easily without acid or alkaline treatment. The recycled paper mill is not usually set up to provide either of these conditions, and certainly not to segregate the paper according to which condition it needs and provide appropriate treatments.

In general, the amount of wet strength increases as additional resin is applied, until a point is reached where the improvement slows down and then levels off. Figure 4.6 illustrates this phenomenon. The shaded area shows typical values of wet strength retention.

As you can see, the amount of wet strength levels out at about 2% wet strength resin by weight. Of course, this graph is qualitative. There are differences between resins of different types as well as differences associated with the type of paper involved and how it was produced.

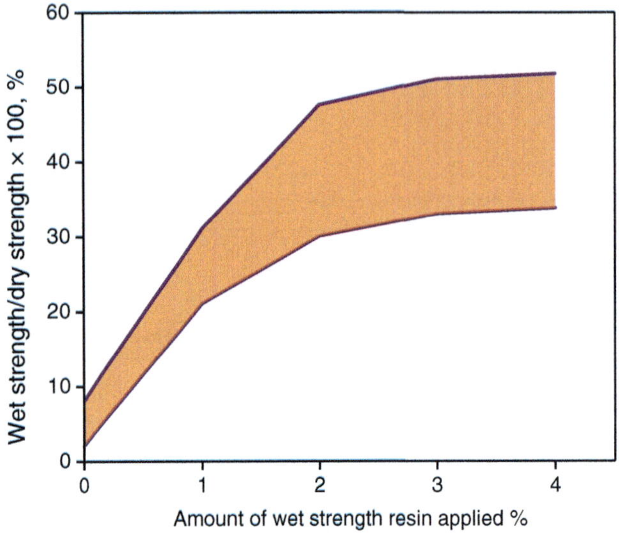

Figure 4.6 Relationship between the amount of wet strength resin applied and the resulting relative change in paper strength.

4.5 Sustainable Materials for Paper Coating for Packaging Applications

Paper is a highly desirable material for numerous applications such as packaging as it is a food safe, renewable, and biodegradable material with good mechanical properties. However, its main shortcoming is that it has poor resistance against liquids such as water and oils. A number of strategies have been developed to address this issue, such as lamination, the application of coatings onto paper substrates, and the use of fluorinated chemicals. We will briefly discuss these three strategies below.

In general, **lamination** refers to the application of a polymer-based film onto an uncoated paper substrate. Synthetic polymers such as low-density polyethylene (**LDPE**) have been commonly used for this purpose. Although they can enhance the antiwetting properties and grease-resistance of paper, these types of coatings have certain drawbacks. In particular, it is difficult to separate these coatings from the paper substrate, which impedes recycling efforts. In addition, LDPE as well as many other synthetic polymers are neither rapidly biodegradable nor compostable. In addition, these types of polymers are derived from nonrenewable petroleum resources. Due to environmental concerns, there has been a shift toward the use of bio-based polymers as laminates, with a common example being poly(lactic acid) (PLA). While these types of materials are derived from renewable resources such as agricultural by-products or other forms of biomass, they also have certain shortcomings. For example, PLA is compostable but not rapidly biodegradable in natural conditions. Consequently, more energy (such as elevated temperature) is required to facilitate its degradation, and it will not readily degrade if it escapes into the environment (such as via littering). As is the case with petroleum-based polymers, it is difficult to separate bio-based polymers from their paper substrates so that recycling still becomes a challenge.

While coating and lamination look similar (and their definitions vary depending on the targeted application), there are subtle differences between the two. In the case of lamination, two or more substrates (such as a paper and LDPE or PLA in the case of paper lamination) are first converted into film/layer/sheets and then combined together to form a multiple layer structure. On the other hand, coating involves the deposition of one or more materials onto the surface of a substrate, with paper being the substrate in the context of this discussion. In many cases, the coating is applied as a liquid, but solids and gases can also be applied onto substrates. In both the case of coating and lamination, the substrate is provided with a new property that it would not otherwise possess, such as liquid repellency and grease resistance in the case of paper.

Coatings that are frequently applied onto paper substrates include paraffin wax- and latex-based coatings. While both of these coatings can impart paper with water and oil-repellency, they also have certain shortcomings. For example, paraffin wax is a petroleum-based material that is nonrenewable as well as nonbiodegradable, and it is only suitable for use at room temperature or at low temperatures. Latex-based coatings tend to be expensive and they also pose environmental concerns, as they are difficult to dispose of and they can break down in the environment to form microplastics. As is the case with laminates mentioned above, it tends to be difficult to remove these coatings from their paper substrates, which impedes recycling efforts.

Fluorinated materials such as PFAS have been widely used to impart substrates such as paper with liquid repellency, and thus they are widely used on paper-based wrappers for fast food, and desserts as well as paperboard-based takeout containers. Their popularity stems from the fact that fluorinated substances have very low surface tension which gives them superior water and grease resistance. Although they offer excellent antiwetting properties, PFAS pose significant threats to the environment and to human health as they persist in the human body and environment due to their very stable nature [10]. The presence of PFAS in drinking water is a big concern and these chemicals have been found in drinking water in amounts larger than permissible levels [12]. Some companies have already begun this shift, and it is likely that regulations may emerge that will limit or restrict the use of these materials.

Despite the challenges encountered with coated or laminated paper, it is still a highly desirable material for packaging purposes. The EU has banned the use of disposable plastic plates as well as straws and is likely to ban plastic cups. Meanwhile, Canada is banning single-use plastic shopping bags. The move away from such plastics is likely to result in paper having an even greater role than it does now in the packaging and single-use food container sector. Therefore, it will become increasingly important to develop effective ways to enhance the antiwetting and liquid barrier properties of paper without sacrificing its recyclability.

While fluorine-based materials provide excellent antiwetting performance, there are alternatives that can still provide good liquid repellency while avoiding the issues that are associated with PFAS. One such alternative is silicone-containing materials such as polydimethylsiloxane (PDMS). Recently, Rabnawaz and coworkers employed chitosan-*graft*-PDMS copolymers as coatings for paper substrates, which could impart both water and oil repellency [13]. The paper substrates were immersed into an aqueous solution of this graft copolymer, with the chitosan units serving to anchor the copolymer onto the paper substrate, while the PDMS imparted liquid repellency. While these liquid-repellent coated papers were PFAS free, they were also biodegradable and repulpable, as the coating could be readily separated from the coating. Consequently, these types of papers could be recycled.

While PDMS imparts liquid repellency, it is not in itself a bio-based polymer, and thus copolymers or coating formulations employing PDMS or other siloxanes are not completely biobased. However, it has been demonstrated that it is also possible to impart paper with both water and oil resistance without using either PFAS or silicone-containing materials. This was achieved via a dual-layer approach, where biobased materials were employed for both layers of the coating [14]. In this approach, the paper is first coated by a base layer of chitosan, and subsequently by a second layer composed of zein. The underlying chitosan layer is oil-resistant, while the upper zein layer is hydrophobic (water-resistant). In another dual-layer strategy [15], starch was applied onto paper as a base layer, while zein was applied as the upper layer. Further efforts are needed in this area to ensure viable replacements for PFAS-coated and plastic-laminated paper.

Case Study: Nonwood Fibers as Paper Sources

There has been a significant shift toward the adoption of paper as a sustainable packaging material. A recent report has indicated that the global market for paper-based packaging was $265.8 billion USD in 2023, and is anticipated to grow to $420.6 billion USD by 2023 [16]

While paper is most often derived from wood pulp, there has been interest in the use of nonwood pulp as a source of paper to help expand the supply of source materials for paper and paperboard [17, 18]. The Confederation of European Paper Industries (CEPI), an organization representing the paper industry in Europe, had commissioned a study on the use of nonwood fibers for papermaking which was conducted by the nova-Institute [19, 20]. In their study, they found that there were at least 33 European companies utilizing nonwood fibers either for their products or in their product development activities. The study selected 16 of these companies for analysis as well as in-depth interviews. The materials employed as raw materials in that study included straw, textile waste, sugar cane bagasse, meadow grass, miscanthus, sugar beet as well as silphie. These nonwood materials were used as sources for various types of paper and paperboard products, such as technical paper, cigarette paper, tissue paper, security paper, specialty paper, in addition to paperboard and packaging.

The sources of nonwood fibers were categorized into two groups, where the first group was "residues, by-products, and co-flows," while the second group was known as "total fiber line" and includes desiccated feedstocks for fibers. The materials falling into the first group included textile waste, straw, flax tow, bagasse, and sugar beet fibers. The advantages of members of this group are that there are large volumes of these materials produced, they can be readily incorporated (in certain amounts or proportions) into existing pulp and paper processes, and they do not require additional or dedicated cultivation areas for their growth because they are by-products. Meanwhile, the drawbacks of these materials as pulp materials are that the quality of these fibers may be lower than that of wood and they also have applications in competing sectors. The study noted that flax tow is an exception in this category, as it has a superior quality compared to that of wood fibers but is produced at lower scales and has a higher cost.

Meanwhile, members of the second group of nonwood fiber sources included meadow grass, silphie, and miscanthus. On the positive side, these plants can offer high per-hectare yields as well as new marketing opportunities. Drawbacks of these plants are that they would require dedicated land to permit their cultivation, and they would be competing with agricultural cash crops such as corn or wheat. Another issue is that no value chains currently exist for miscanthus, meadow grass, or silphie. On the other hand, flax and hemp flax can grow without requiring much fertilizer and hemp could be incorporated into cash crop rotations. The growth and processing of hemp and flax can be costly, although they can offer higher quality pulp than that obtained from wood fibers [20].

Among the companies interviewed in the CEPI-commissioned study, the most common motivations from a sustainability perspective for using nonwood fibers included establishing regional circularity and local supply, finding ways to use biomass and side-streams for material use rather than as bioenergy or waste generation, and because these alternatives had a lower footprint than wood. From a business perspective, the most common motivations were found to be innovation marketing, declining availability, and increasing costs of virgin wood fibers or paper for recycling, a desire to diversify the range of raw materials as well as product diversification.

While these sources may have their advantages and disadvantages, the adoption of nonwood pulp fiber into paper production streams can offer significant environmental benefits such as lower carbon footprint and reduced deforestation [21] while also helping to diversify supply chains.

Paper Packaging Case Study: Metsä Paper Board and Dassault Systeme's 3DExperience Technology

With the advancement of computational power that has been seen in recent decades, it is apparent that computational simulations can also have an important role in helping to advance packaging sustainability. Based in Finland, Metsä Board is one of the Europe's leading manufacturers of paperboard and has a major role in the paper-based packaging industry, and it has a strong reputation for packaging sustainability.

To facilitate its expansion, Metsä Board has moved toward using software simulations to guide its development of new products, by employing the 3DEXPERIENCE program developed by Dassault Systèmes. This approach offered numerous benefits. They found that they could provide product recommendations 85% more quickly than was the case with producing physical prototypes. Due to the software-based nature of this modeling technology, new designs for packaging could also be shared securely over the cloud, thus facilitating collaboration.

This software was also used by Metsä Board to develop their own service for their clients, known as Metsä Board 360 services, which allows their customers to use virtual models of their packaging to compare their performance with new packaging formats under various conditions. Metsä Board has reported that they can use this software to recommend lighter board that offers the same performance in comparison with the paperboard that they are replacing, offering cost savings, and sustainability benefits [22].

This technology has allowed Metsä Board to develop solutions that are more suitable for e-commerce, as these types of packaged products tend to have more touch points along the supply chain than is the case for products sold in stores. They were able to use this technology to design packaging that was able to hold up well against this handling, while still being lightweight.

Dassault Système's 3DExperience platform also had a role in allowing Metsä to develop packaging solutions that helped to combat the COVID-19 pandemic. In particular, it was used to develop paperboard packaging for COVID-19 vaccine vials that could withstand extremely cold temperatures of −70 °C, all while being shipped across the world.

4.6 Concluding Remarks

In response to environmental concerns, there is growing pressure from society and governments to find sustainable solutions to packaging needs, and paper-based packaging will be an important tool in enabling these solutions. Paper is an inexpensive, lightweight, versatile, and environmentally friendly material that can provide a very effective packaging material and hence is widely considered an alternative for some of the plastics packaging. Awareness of the ways through which paper can degrade and how to address this issue while also enhancing its barrier properties will help ensure that it will have a very bright future and that it can become adopted for packaging applications and fulfill the role that is currently served by environmentally unfriendly materials.

Problem Set for Chapter 4

1 You have two fillers available: A and B. Filler B is mildly basic while A is acidic. Which filler will you use for the paper to obtain a more stable paper, considering all other functions are the same?
A Filler A.
B Filler B.
C Either one.

2 What strategies are used to prevent biological degradation in paper?
A Slimicide.
B Copper salt.
C Chlorinated phenols.
D All of the above.

3 How can the wet strength of paper be improved?

4 What are the advantages and disadvantages of using a PLA laminated/coated paper cup compared to an LDPE laminated/coated cup?

5 If the dry tensile strength of paper is 200 lbs/in and it decreases to 10 lbs/in when it becomes fully saturated with water, would this paper be suitable for wrapping wet materials like those used in butcher wrap? Why or why not?

6 What is the advantage of polyamide-polyamine-epichlorohydrin (PPE) resin over UF resin in paper wet strength applications?
   A Use of an acidic medium for curing, thus providing better stability in acidic medium.
   B Use of neutral or alkaline conditions for curing, thus providing better stability in acidic medium.
   C Both a and b.
   D None of the above.

7 Why do some paper substrates turn yellow faster than others?
   A Because they contain cellulose.
   B Because they contain lignin.
   C Because they contain hemi-cellulose.
   D Because they contain some inorganic salts.

8 Discuss the end-of-life (EoL) solutions for a wet-strength paper.

References

1 Bergh, N.O.2018). Binders for paper coating: starches, proteins and latices. *The Fundamentals of Papermaking Materials, Materials, Trans of the XIth Fund Res Symp*, Cambridge, 1997 (ed. C.F. Baker). pp. 193–207. Manchester: FRC.
2 Holland, A. (2012). *Papermaking Additives and Fillers*. George Weil & Sons Ltd.
3 Chauhan, V.S. and Bhardwaj, N.K. (2017). Efficacy of dispersion of magnesium silicate (talc) in papermaking. *Arab. J. Chem.* 10: S1059–S1066. https://www.sciencedirect.com/science/article/pii/S1878535213000245 (assessed on 15 November, 2024).
4 Gurnagul, N. and Zou, X. (1994). The effect of atmospheric pollutants on paper permanence: a literature review. *TAPPI J.* 77: 199–204.
5 Gould, I. and Hingston, J. (2002). Slimicides. In: *The Biocides Business* (ed. D.J. Knight and M. Cooke), 225–232. Wiley https://onlinelibrary.wiley.com/doi/abs/10.1002/352760197X.ch9.
6 Myers, R.G. (1916). Water absorption by vulcanized fiber. *J. Ind. Eng. Chem.* 8 (12): 1108–1109.
7 Fischer, S., Leipner, H., Thümmler, K. et al. (2003). Inorganic Molten salts as solvents for cellulose, https://doi.org/10.1023/A:1025128028462. *Cellulose* 10 (3): 227–236.

8 Gonçalves, M., Ribeiro, M., Paiva, N.T. et al. (2020). Resin and volatile content of melamine-impregnated paper assessed by near infrared spectroscopy, a simulation of the industrial process using a laboratory-scale gantry. *Eur. J. Wood Wood Prod.* 78 (6): 1279–1286.

9 Roberts, R.J. and Evans, P.D. (2005). Effects of manufacturing variables on surface quality and distribution of melamine formaldehyde resin in paper laminates. *Compos. Part A Appl. Sci. Manuf.* 36 (1): 95–104.

10 Keim, G.I. and Hercules Powder Co.: United States. (1957). Wet-strength paper and method of making same. U.S. patent 2,926,116, application date 5 September 1957 and patented 23 February 1960. https://patents.google.com/patent/US2926116A/en (accessed 15 November, 2024).

11 Crisp, M.T. and Riehle, R.J. (2018). Regulatory and sustainability initiatives lead to improved polyaminopolyamide-epichlorohydrin (PAE) wet-strength resins and paper products. *TAPPI J.* 17: 519–532. https://imisrise.tappi.org/TAPPI/Products/18/SEP/18SEP519.aspx (accessed on 15 November, 2024).

12 State of Michigan. (2021). Overview of Michigan's Screening Values & MCLs: Per- and polyfluoroalkyl substances (PFAS). State of Michigan. https://www.michigan.gov/-/media/Project/Websites/mdhhs/Folder4/Folder25/Folder3/Folder125/Folder2/Folder225/Folder1/Folder325/PFAS_-_Overview_of_Michigan_Values_FINAL.pdf?rev=f493083f6ee149a0b5fff9f10f5f47dc#:~:text=The%20Michigan%20PFAS%20Action%20Response%20Team's%20Science%20Advisory%20Board's%20report,high%20based%20on%20updated%20science (accessed 15 November, 2024).

13 Li, Z., Rabnawaz, M., Sarwar, M.G. et al. (2019). A closed-loop and sustainable approach for the fabrication of plastic-free oil- and water-resistant paper products. *Green Chem.* 21 (20): 5691–5700.

14 Kansal, D., Hamdani, S.S., Ping, R. et al. (2020). Food-safe chitosan–zein dual-layer coating for water- and oil-repellent paper substrates. *ACS Sustain. Chem. Eng.* 8 (17): 6887–6897.

15 Kansal, D., Hamdani, S.S., Ping, R., and Rabnawaz, M. (2020). Starch and zein biopolymers as a sustainable replacement for PFAS, silicone oil, and plastic-coated paper. *Ind. Eng. Chem. Res.* 59 (26): 12075–12084.

16 Research and Markets. (2024). Paper packaging materials - global strategic business report. https://www.researchandmarkets.com/reports/338765/paper_packaging_materials_global_strategic?utm_source=GNOM&utm_medium=PressRelease&utm_code=5n7rxg&utm_campaign=1648313+-+Global+Paper+Packaging+Materials+Market+Report+2022-2026+-+Growing+Environmental+Concerns+Regarding+Use+of+Plastics+in+Packaging+Application+Spurs+Demand+for+Paper&utm_exec=chdo54prd (accessed 26 March 2024).

17 Danielewicz, D. (2023). Industrial hemp as a potential nonwood source of fibres for European industrial-scale papermaking—a review. *Materials* 16: 6548. https://doi.org/10.3390/ma16196548.

18 Abd El-Sayed, E.S., El-Sakhawy, M., and El-Sakhawy, M.A.-M. (2020). Non-wood fibers as raw material for pulp and paper industry. *Nord. Pulp Pap. Res. J.* 35: 215–230. https://doi.org/10.1515/npprj-2019-0064.

19 Confederation of European Paper Industries (CEPI). (2023).Press release: Non-wood fibre is a new innovation frontier for Europe's paper and board sector, shows nova-Institute study. https://www.cepi.org/press-release-non-wood-fibre-is-a-new-innovation-frontier-for-europes-paper-and-board-sector-shows-nova-institute-study/ (accessed 26 March 2024).

20 Confederation of European Paper Industries (CEPI). (2024). Growing interest in non-wood pulp for paper, hygiene products and packaging. https://www.cepi.org/wp-content/uploads/2023/02/Cepi_Non-Wood-Fibre_Exec-Summary_130212022.pdf (accessed 26 March 2024).

21 Frechka, V. (2023). Non-wood fibre is a new innovation frontier for cellulose and paper sector. LinkedIn. https://www.linkedin.com/pulse/non-wood-fibre-new-innovation-frontier-cellulose-paper-frechka (accessed 26 March 2024).

22 Metsä Board. (2024). Dassault Systemes. https://www.3ds.com/insights/customer-stories/metsa-board-packaging?_gl=1*ryruy8*_ga*MzI1MjM2MjAwLjE3MTA5NDUxMTk.*_ga_DYJDKXYEZ4*MTcxMDk0NTExOS4xLjAuMTcxMDk0NTExOS42MC4wLjA (accessed 19 March 2024).

5
Glass

Glass has a vital role in our daily lives. For example, it is used in the windows of the homes we live in, cars that we drive, and is used as dinnerware, precision instruments (such as the lenses of cameras, microscopes, and telescopes), decorative containers (such as vases), as well as containers that are used to package certain medications, cosmetics, and food (such as jars and bottles). Although glass is sometimes produced naturally (as with volcanic glass), the first reported case of it being made by humans was 4,000 years ago in Mesopotamia [1].

5.1 Advantages of Glass

For packaging applications, glass has several desirable properties that set it apart from other materials. Glass is chemically inert, recyclable, and does not generate pollution (chemicals or microplastics) when it is disposed of (even if it escapes into the environment). Glass is a nonrenewable material but is produced from readily abundant and inexpensive raw materials (such as sand, etc.), it exhibits excellent optical properties, it has high intrinsic strength, and also has excellent gas and water vapor barrier properties [2]. On the one hand, glass is transparent and thus allows consumers to see the contents in a glass jar or bottle before purchasing an item. On the other hand, glass can be made into **different colors** (often green, brown, or blue) for situations where **protection against ultraviolet (UV) light is desired**.

Since glass has been used for such a long time and has generally been used more extensively in past decades than it is currently, items packaged in glass can instill a sense of nostalgia in consumers (for example, when “old fashioned” glass soda bottles are used instead of more “modern” plastic bottles or soda cans). Consequently, glass is often used to package luxury items such as expensive

Degradation, Stabilization, and Recycling of Packaging Materials, First Edition.
Muhammad Rabnawaz, Susan E.M. Selke, and Ian Wyman.

perfumes or cosmetics [3]. The attractive appearance of glass and its smooth texture further add to this premium appeal.

5.2 Disadvantages of Glass

While glass has numerous advantages, it also has **some shortcomings**. Glass is often quite brittle and can easily be broken during handling or transport. As a result, other materials or additional space may be needed to help protect glass containers, such as cardboard or paperboard boxes with compartments for glass bottles or straw or foam to help absorb impacts. In addition, glass is relatively heavy, which leads to higher fuel costs during transport [3]. Moreover, there have been concerns regarding the mining of silica sand [4]. Also, glass is typically not recommended for freezing applications below 32 °F or 0 °C. This is because glass has a very low elongation at break, meaning it cannot handle much deformation before breaking. When water or other beverages are frozen, they expand by approximately 9% by volume. This expansion can exert too much pressure on the glass, causing it to shatter.

5.3 Glass Chemistry

5.3.1 Composition

In contrast with plastics and paper, glass is not a carbon-based material (one exception is organic glass, which will be briefly discussed later in this chapter). Rather, it is a fully oxidized network structure typically comprised of elements such as oxygen, silicon, aluminum, and sodium and calcium ions. The composition of glass can vary depending on the type of glass in question or how it is made. In most cases, glass is made from three main components which include a **former, a flux, and a stabilizer** [5].

Formers are the main ingredients that form the glass structure. The most common former is silica (silicon dioxide). Silica is derived from sand, which forms a strong three-dimensional glass network. However, silica has a very high melting point, which makes the melt-processing treatment that is required to produce glass energy intensive.

A **flux** is blended with the former to lower the melting point of the glass former. The most commonly used flux is soda (sodium carbonate), which lowers the melting point of the silica, thus facilitating the glass production process and helping to minimize the energy requirements. Other common examples of fluxes include potash (potassium carbonate) and lithium carbonate.

Stabilizers are added to make the glass water-resistant and to improve its durability. The stabilizer serves to prevent the glass from being dissolved by water, such as when it is stored in humid environments. The most common stabilizer is lime (calcium carbonate), but others, such as alumina (aluminum oxide), may also be used.

Let's consider soda-lime-silica glass, which is one of the most common types of glass. This material is prepared from silica (the former), sodium carbonate (the flux), and calcium carbonate (the stabilizer). Soda-lime packaging glass also has a small amount of aluminum silicate added into the mix to get the best performance properties.

5.3.2 Glass Making Process

Formers, fluxes, and stabilizers are physically blended together; then the mix is typically smelted at high temperatures of 1,500 °C or higher, which leads to the oxidation of fluxes and formers such as sodium carbonate and calcium carbonate, thus yielding sodium oxide, potassium oxide, and calcium oxide, respectively, along with the release of carbon dioxide. This glass formation process is normally conducted in a furnace, where the introduced materials are melted into molten glass. There are two stages to this melting process, as the materials are first melted from the solid state into a liquid, and then the gasses that had formed during this melting process are removed to help ensure that there are no bubbles in the glass. Once the molten glass is both uniform and gas-free, it is then moved to the working end of the furnace and allowed to cool until a desired viscosity is reached [6]. When the appropriate temperature and viscosity are reached, the molten glass then undergoes a **forming process**, where it is made into glass jars or bottles. Depending on the type of container to be made, the molten glass will undergo forming processes such as blow and blow, wide mouth press and blow, or narrow neck press and blow [2, 6, 7]. Following the forming process, the glass jars or bottles are then subjected to an annealing process, which is used to remove stresses that may have been introduced to the glass containers during their forming and handling, particularly due to nonuniform cooling. This annealing process involves heating the glass bottles or jars in an oven at a certain temperature (typically slightly below the softening point of the glass in question) and then cooling them slowly to allow stresses to be removed [6].

In most cases, natural gas is used as the fuel for glass furnaces, and in some cases, natural gas is combined with oxygen [6]. It has recently been announced that a furnace powered by a combination of zero-carbon electricity and hydrogen will be opened in the UK [8, 9]. From an environmental perspective, it is undesirable to use air as a fuel for glass furnaces, as the nitrogen that is the predominant component of air will become oxidized to form nitrogen oxides such as nitric oxide and nitrogen dioxide, the former of which can attack the ozone layer [10].

While the focus of this chapter is on inorganic glasses such as those used for packaging such as jars, and bottles, it should be noted that from a technical standpoint, the term glass refers to a state of matter rather than to a specific material with a defined composition. In particular, glass is considered to be in a vitreous state, and it shares certain properties with both liquids and solids. Glasses will flow like liquids at higher temperatures, but they do not solidify at a defined temperature when they are cooled. Rather than solidifying, glass will become progressively more viscous as it is cooled until it effectively behaves as a solid in terms of its physical properties. A glass is generally classified as a supercooled liquid or as an amorphous solid. However, while most supercooled liquids are unstable and will crystallize if they are shaken or cooled, glass is much less prone to undergo crystallization. While glass has the physical properties of a solid, it lacks the ordered atomic structure of a typical solid. This situation occurs because as the liquid is cooled and becomes more viscous, the atoms eventually become unable to move and are essentially locked in place so that they cannot assume an ordered structure that would typically be observed among solids. In many cases, a glass is formed when a liquid is cooled relatively quickly so that its viscosity increases before the atoms can move sufficiently to assume an ordered arrangement [11].

5.3.2.1 Common Types of Glass and Modification Strategies

There are various types of glasses, with soda-lime glass being the most common. Some common types of glasses are summarized below.

- **Soda-lime glass (or soda-lime-silica glass):** This is a common type of oxide glass, which we have discussed earlier in this chapter. Soda-lime glass is obtained by adding Na_2O as a flux to SiO_2 and CaO as a stabilizer. Flux lowers the glass transition temperature so that it softens at lower temperatures, and thus, lower temperatures can be used to produce the glass. Specifically, the fluxes act by interfering with the stable Si–O–Si linkages, as the Na^+ cations break some of these linkages due to the cations interacting with terminal $Si–O^-$ units [12]. In other words, the flux disrupts the Si–O–Si network and makes it less stable so that it requires less heat energy to break apart. The result is a lower melting point for the mixture of silica and flux than would be the case for pure silica. Soda-lime glass is commonly used in packaging (i.e., jars and bottles) as well as windows.
- **Borosilicate glass:** Borosilicate glass is obtained by using silica (SiO_2) (former in 70–80%), boric oxide (B_2O_3 7–13%), alumina (Al_2O_3 2–3%), soda (Na_2O) and/or potash (K_2O) 4–8% and lime (CaO) <1%. This type of glass has higher heat resistance than soda-lime glass as it has a lower thermal expansion coefficient.

It is often used as glass pans for cooking in ovens or for laboratory glassware [12]. In addition, borosilicate glasses are often used instead of soda-lime glass for the packaging of certain pharmaceutical products or medications such as vaccines due to its higher chemical inertness.

- **Aluminosilicate glass:** Aluminosilicate glasses can exhibit high electrical resistivity and good moisture resistance. In addition, aluminosilicate glass fibers have robust mechanical strength and are used as reinforcing agents for polymer composites [2]. *Aluminosilicate* glass is obtained by using silica (SiO_2) (former in 57–72%), alumina (Al_2O_3 16–25%), soda (Na_2O) and/or potash (K_2O) 5–15% and lime (CaO) <0–10%.
- **Organic glasses:** One thing that glass and organic glasses (glassy polymers) have in common is their fully amorphous structures (lacking any crystallinity). In glassy polymers, the polymer chains are arranged in a random manner just like glass where there is no crystallization. Consequently, glass and glassy polymers are transparent because their disordered structure allows light to pass through these materials without being scattered by crystalline regions. Examples of glassy polymers include polystyrene, polymethyl methacrylate (PMMA), and polycarbonate, among others. These polymers are used in a wide range of applications, including packaging, coatings, lenses, and more, due to their unique properties. Organic glasses are polymers such as thermoplastics and thermosets. In the case of thermoplastics, the polymer chains are interacting with one another rather weakly via van der Waals interactions and/or via dipole-dipole interactions, while in the case of thermosets, the polymer chains are crosslinked and are interacting with one another primarily via covalent bonds [2]. Currently, most lenses used in eyeglasses are actually comprised of organic glasses rather than traditional soda-lime-glass [2] thanks to their lightweight nature, excellent impact resistance, UV protection, and low cost. As organic glasses are comprised of polymers, they fall within the category of plastics, which were discussed in further detail in Chapter 2.

5.3.2.2 Stability of Glass

While glass is brittle and easily broken due to mechanical stress or physical impacts, on the other hand, it tends to be more resistant to corrosion than metals and more resistant to chemical deterioration than plastics. In general, glass tends to have better heat resistance than wood, paper, or plastics. If you have noted that glass is widely perceived as being infinitely recyclable, it is because glass has fully oxidized Si–O–Si linkages. In contrast, plastics, paper, and wood have C—H and C—OH bonds that are oxidizable and hence such materials are susceptible to degradation/oxidation during processing, use, and reprocessing.

However, it should be noted that glass is vulnerable to thermal shock, which arises due to sudden changes in temperature. In general, the thermal shock that is encountered during cooling tends to cause more damage to glass than that encountered during heating. During rapid cooling, compressive stresses are on the outer surface, and tensile stresses are on the inner surface of a filled glass container. On the other hand, during rapid heating, tensile stress is on the outer surface, while compressive stress is on the inner surface. Since the outer surface is more exposed to damage that greatly reduces tensile strength, and because glass generally fails under tension and not compression, rapid cooling is more likely to lead to failure. The stresses encountered during rapid heating or cooling dissipate when the final temperature is reached [6, 7]. Glass is toughened against thermal shock, chemical corrosion, and mechanical stress by a variety of approaches, as discussed below.

5.4 Chemical Corrosion

Although glass is generally inert and less susceptible to chemical degradation than some other common types of packaging materials, **it is not completely inert** and is more vulnerable to degradation by some chemicals than others. In general, glass is more susceptible to chemical degradation when it is exposed to alkaline media than it is to neutral or acidic media. However, one acid that glass is particularly vulnerable to is **hydrofluoric acid**, as the fluoride ions of this acid can form strong bonds with the silicon atoms that are present in glass [13]. In fact, hydrofluoric acid is often used to give glass a "frosted" opaque appearance [14]. This may be desirable for decorative purposes, to allow writing to be applied or etched onto glass surfaces in a templated manner, or for situations where natural light is desired while still providing privacy, such as glass walls for office cubicles or washroom windows. Glass can undergo three main types of corrosion, including leaching, etching, and weathering [6].

5.4.1 Leaching

In the case of alkali-containing glass (such as soda-lime glass), an ion exchange reaction can take place between the $-Si-O^- X^+$ units (where X is an alkali/alkaline cation such as sodium, potassium, calcium, or lithium ions, albeit in most cases it is sodium) and H_3O^+ ions, which are present in acidic media. This reaction is shown below:

$$\mathrm{Si-O^{-}X^{+} + H_2O \rightarrow \mathbf{Si-OH} + XOH}$$

The reaction leads to the formation of silanol (Si–OH) groups as well as the leaching of alkali ions from the glass. In addition, water can become incorporated into the glass through this ion exchange reaction.

5.4.2 Etching

The etching of glass occurs in alkaline media, and in these situations, hydroxide ions react with siloxane bonds, thereby leading to a gradual breakdown of the network structure of the glass. This reaction is shown below [2]:

$$Si-O-Si+OH^{-} \rightarrow {}^{-}O-Si+Si-OH$$

While an alkaline environment may be encountered when a base is directly added to the glass, an alkaline environment may also be obtained due to the leaching of sodium ions from the glass network, especially in highly humid environments. This latter situation leads to a degradation process known as weathering, which is discussed in further detail below. As mentioned earlier, it should be noted that the term etching is also often also used to describe the treatment of glass with certain acids, most notably hydrofluoric acid.

5.4.3 Weathering

Glass is prone to weathering when it is exposed to warm and humid environments. In these cases, the moisture in the air can condense on the surface of the glass and form an aqueous layer. As sodium ions leach out from the glass, this film becomes alkaline and causes the glass to corrode. This situation occurs when a pH value of approximately 9 is reached, at which point hydroxide ions can react with the siloxane bonds of the glass, thereby cleaving some of the glass's siloxane bonds [2]. This weathering phenomenon often leads to a white crust or residue on the surface of the glass. Interestingly, this moisture-induced weathering phenomenon is generally not encountered when glass is submerged in water, as in that case, the escaping sodium ions are dissipated into the surrounding water rather than forming an alkaline microenvironment on the glass's surface. As much of the corrosion described above stems from the presence of alkali ions (primarily sodium ions) in glass, one strategy to mitigate this issue is to reduce the amount of alkali ions that are present, such as through the use of **borosilicate or aluminosilicate glass** that is enriched with boron trioxide or alumina, respectively, at the expense of sodium ions [2].

Several types of treatment can be used to improve the chemical durability of glass. One simple method is to modify the chemical composition of the glass (reducing the sodium content, for example). Glass can also be treated with acid to remove and neutralize alkaline substances at its surface. Treatment with sulfur dioxide (SO_2) can also be used to form sodium sulfate through the reaction with the available sodium ions. The sulfur dioxide may even occur naturally in the combustion gases used in the annealing furnace.

5.5 Physical Stability and Strength of Glass

In addition to chemical corrosion and weathering, glass is also susceptible to physical damage due to impacts and other physical forces. The physical strength of glass is an important consideration when this material is used in various applications, such as windows for buildings and vehicles (where it must offer good visibility while also providing protection to occupants or passengers), packaging (where it must be able to withstand impacts that may be encountered during transport and storage), as well as glassware (where it must hold up against day-to-day use). Unfortunately, glass is rather brittle and can easily become broken if it is subjected to physical impacts. When it is used as a food packaging material, care must be taken to prevent or minimize the likelihood of the glass breaking down into small glass fragments that could become mixed with food and accidentally ingested [7].

One factor that can affect the strength of glass is the presence of surface defects, as stress becomes concentrated at these defects so that the strength of the glass will diminish. In general, glass will break when it is subjected to a certain amount of tensile stress, and this strength of the glass surface is referred to as the ultimate tensile strength. In general, the glass will break when a certain level of tensile stress is applied, and if defects are present, the glass will break at these defects as the stress concentrates at these surface defects or cracks [6].

There are a number of ways to strengthen glass to help it withstand physical impacts. One of these strategies is to ensure that the glass does not have physical defects. As mentioned above, stress is focused on defect sites, which weakens the strength of the glass. One way to achieve this is to handle the glass very carefully during its production, such as the forming and annealing stages, to ensure defect-free, mechanically robust glass [6]. Another strategy is to use thicker glass to help ensure it has the desired physical strength. Although using thicker glass can be effective, it also has drawbacks due to the increased material costs associated with producing the thicker glass and potentially higher transportation costs if the glass is used for packaging applications due to its higher weight. Alternatively, the physical strength of glass can be enhanced by altering the chemical composition of the glass surface, such as by reducing the sodium content. When the glass surface has a reduced sodium ion content, and when this glass cools, the bulk of the glass with its higher coefficient of thermal expansion results in a greater degree of contraction than would be "natural" for this reduced-alkali content glass – giving a compressed surface layer. This means that in order for the glass to break, an applied tensile stress must not only be large enough to cause the glass to fail, but it must first overcome the existing compressive stress. Unfortunately, this approach is only effective if the layer of compressive stress reaches deeper into the glass than any of the surface defects.

As noted above, surface defects greatly weaken the strength of glass. However, weathering can have a surprising effect on glass defects. For example, if a new scratch is applied onto a glass surface and then the glass is submerged in water for a given amount of time (such as a day), the glass will actually become stronger. Likewise, the glass will also become stronger if a scratch is made and then the glass is thermally annealed, and its strength will be enhanced by a similar degree as is achieved after immersion in water. On the other hand, the glass will not become strengthened if a crack is made, and then the glass is annealed and allowed to age in water. This strengthening phenomenon can be understood through close examination of the new crack before and after the aging treatment, as the size of the crack will become larger during this process due to hydrolysis reactions of the silicate bonds at the damaged area. Apparently, this higher strength is related to interactions between the crack's lateral and radial components. In effect, the tips of the crack become blunted as the stressed bonds are removed during the aging process. Consequently, the stress becomes focused on a larger area so that it is less concentrated than would be the case if the scratch had not received the aging treatment.

The fatigue resistance of glass can be evaluated by subjecting it to continuous stress. In general, the fatigue resistance of glass will degrade if it is placed in a humid environment. If glass is simultaneously subjected to stress and exposed to water, the Si—O—Si bonds will be strained and undergo hydrolysis, thus leading to the loss of glass in the strained area such as the tip of a crack. If this stress reaches a critical level, it will produce a microfracture, thus sharpening the tip and generating more stressed bonds, which will, in turn, undergo hydrolysis. Consequently, while aged scratched glass is stronger than its nonaged counterparts, it is also more susceptible to fatigue. On the one hand, these cracks are initially larger so that the stress is distributed over a larger area, and consequently, the aged glass can withstand a higher initial stress. On the other hand, the aged glass will fail more quickly than the nonaged glass if the stress is continuously applied.

5.5.1 Strategies to Modify or Strengthen Glass

Besides the approaches mentioned thus far, there are a variety of strategies that can be employed to strengthen or modify glass. Some common examples include chemical modification, chemical strengthening, thermal strengthening or toughening (sometimes referred to as tempering), as well as employing a fused salt mixture. These strategies will be described below.

5.6 Chemical Modification and/or Strengthening

- **Treatment with sulfur dioxide:** Glass can be treated with sulfur dioxide at an elevated temperature to help protect it against chemical degradation. The sulfur dioxide will react with alkali ions (typically sodium ions) to form sodium sulfate, thereby helping to remove free alkali ions from the glass surface and preventing them from reacting with the glass surface. As mentioned earlier in this chapter, it is noteworthy that sulfur dioxide may be present in the furnace used for the annealing treatment of glass during production. While helping to protect the glass against chemical degradation, this treatment also serves to strengthen the glass, as lowering the alkali content in this way helps to lower the thermal expansion coefficient of that glass. As the glass cools following this treatment, the bulk of the glass (the interior) will still contract at a greater rate than would be favored by the surface layer, which has a lower alkali content (due to its lower thermal expansion coefficient), thereby introducing a compressed surface layer to the glass.

- **Treatment with an acid:** Treatment with acid is somewhat similar to the sulfur dioxide treatment strategy described above in that it serves to reduce the alkali ion content near the glass's surface. This helps to prevent or mitigate the formation of an alkaline environment if the glass is placed in a humid environment. In this case, the alkali ions are neutralized by the acid so that they can then be removed [16].
- **Ion exchange:** Ion exchange processes can also be used to remove alkali ions. For example, sodium cations can be exchanged for hydrogen cations, and subsequently, the hydrogen cations can be removed via treatment with oxygen at elevated temperatures.
- **Surface coatings:** Another strategy to enhance the durability of glass substrates has been to apply thin coatings such as those comprised of silicon oxide, metal oxides, or metal fluorides. However, a limitation of these coatings is that they are often susceptible to the formation of pinhole defects. Some success has been achieved by treating glass surfaces with silicone, which imparts hydrophobicity to the glass and thus makes it less susceptible to hydrolysis.

Two common types of coatings include hot-end coatings and cold-end coatings [7]. Among these two, the former is applied to hot bottles prior to annealing. For example, tin chloride is applied in this manner and then becomes oxidized to produce tin oxide or organotin. These coatings can help with the adhesion of the subsequent cold-end coating and can introduce roughness onto the surface, thereby filling cracks while hardening and compressing the surface. Cold-end coatings are applied onto glass containers after they have been cooled. Their primary function is typically to provide lubrication, thereby reducing friction and consequently reducing damage. In addition, coatings are sometimes applied onto glass containers to provide compatibility with adhesive labels [7].

5.6.1 Strengthening via the Fused Salt Mixture Approach

Sodium-containing glass can be thermally strengthened via immersion in a fused salt mixture at a temperature that is below the glass's annealing point. The smaller sodium ions on the surface of the glass will be replaced by larger potassium ions via base exchange reactions. Under ideal conditions, the glass would expand to provide space for the larger potassium ions, but this expansion is not possible because the temperature falls below the annealing point. In particular, the internal layers of the glass (which have not undergone this exchange reaction) prevent this expansion from taking place, thereby producing compressive stress in the surface layers. This is the fundamental process behind what is known as chemically strengthened glass. A well-known example of such glass is Gorilla Glass [15], which is used in many smartphones and other touchscreen devices. It provides improved resistance to scratches and breakage compared to regular glass.

5.6.2 Thermal Strengthening or Toughening

Thermal strengthening or toughening is another approach to strengthen glass. This process involves heating glass and subsequently cooling its surface rapidly with air jets so that a compression layer is introduced at the glass's surface. In addition, a large amount of stored energy is introduced into the glass, which affects its behavior if the glass does eventually fail. In the event that the glass does fail, its stored energy is liberated very quickly and is translated into the surface energy of the resultant glass fragments. Consequently, the glass fails in such a way that the resulting glass fragments have a large overall surface area (when they are taken together). A given amount of material will have a larger surface area if it exists in the form of many smaller fragments than as fewer larger fragments. As a result, this glass will shatter into numerous small fragments. These smaller fragments can do less harm than would be the case with larger glass fragments, thus enhancing the safety of the glass if it is used as windows for buildings or automobiles.

Many types of automobile glass, particularly windshields, are constructed as laminated glass, which includes a layer of polyvinyl butyral (PVB) that is sandwiched between two layers of glass. The purpose of the PVB layer is primarily safety oriented. If the glass is impacted and shatters, the PVB layer can hold the broken pieces of glass together.

5.6.2.1 Recyclability of Glass

One of the advantages of glass is that it is highly recyclable, and it can be repeatedly recycled [3]. While glass is highly recyclable, glass jars or bottles used in the packaging sector are often accompanied by other components that will need to be sorted from the glass itself, such as metal or plastic lids, as well as paper- or plastic-based adhesive labels.

In glass production, the term **cullet** refers to recovered glass that has either been recovered from post-consumer glass packaging (recovered from glass bottles, jars, etc., after they have been disposed of by consumers) or alternatively has been recovered internally from within a factory (such as faulty batches of glass bottles, etc.) [6]. The use of this recovered glass or cullet has its advantages and disadvantages. On the one hand, cullet may require sorting from other components of the former glass products, such as metal or ceramics, as well as labels. In addition, cullet may come in different colors (green, amber, etc.) due to the color of the former glass product. Depending on the type or color of glass being made in a given process, only a certain amount of a given color of cullet can be accommodated while still achieving the desired color of the final glass product. One of the shortcomings of glass recycling is that a remelting process is still required to prepare containers from recycled glass. On the other hand, the use of cullet can lead to energy savings, as cullet melts more readily than its pristine counterparts. This could be attributed to the fact the cullet (recycled glass) already has distorted Si–O–Si networks due to the flux from the previous processing cycle. Meanwhile, the Si–O–Si network of virgin glass must undergo partial distortion with the aid of a flux. In addition, less dust is produced when cullet is used than is the case when glass is produced from pristine materials.

5.7 Conclusions

Overall, glass is a highly versatile material that can be produced in numerous formats, in a wide range of colors, and which can be treated or modified to enhance its suitability for a given application. It is extensively used in the transportation and construction sectors. Although it does not dominate the packaging sector to the same extent that it has in the past, it remains a widely used material, particularly for premium items or for products that require chemical inertness. While it can be expensive to transport and requires high temperatures to produce, it still offers a number of environmental benefits, as it can be repeatedly recycled and it is environmentally benign if it enters the environment, without causing harmful contaminants such as persistent microplastics or toxic chemicals [4]. Taken together, it is likely that glass will continue to have an essential role in our everyday lives in both the short-term and long-term future.

Problem Set

1 Which of the following packaging materials cannot be further oxidized?

A Glass.

B Metal.

C Plastics.

D None of the above.

2 Which glass is more stable against chemical corrosion?
   A Soda lime.
   B Borosilicate.
   C Potash glass.
   D None of the above.

3 What is the major function of using **flux** in glass?
   A To lower the melting point of the silica used to make glass and thus minimize the energy requirements.
   B To increase the melting point of the silica in making the glass.
   C To improve glass water resistance and durability.
   D All of the above.

4 Describe two methods that are used to improve the strength of glass.

References

1 Corning Museum of Glass. (2023). Origins of glassmaking. Corning Museum of Glass, Corning, NY. https://whatson.cmog.org/exhibitions-galleries/origins-glassmaking (accessed 6 June 2023).
2 Le Bourhis, E. (2008). *Glass: Mechanics and Technology*. Weinheim: Wiley-VCH GmbH & Co.
3 Rabnawaz, M. *Sustainable Packaging Materials: Winning Solutions*. Carl Hanser Verlag GmbH Co KG.
4 Lee, C. (2023). Glass or plastic: which is better for the environment? BBC. https://www.bbc.com/future/article/20230427-glass-or-plastic-which-is-better-for-the-environment (accessed 5 June 2023).
5 Corning Museum of Glass. (2023). Chemistry of glass. Corning Museum of Glass, Corning, NY. https://www.cmog.org/article/chemistry-glass (accessed 7 June 2023).
6 Robertson, G.L. (2012). *Food Packaging: Principles and Practice*, 3e. Taylor & Francis Group.
7 Saha, N.C., Ghosh, A.K., Garg, M., and Sadhu, S.D. (2022). *Food Packaging: Materials, Techniques and Environmental Issues*. Springer Nature Singapore Pte Ltd.
8 Vending International. (2023). Hydrogen powered furnace to change the face of UK glass manufacturing industry., Vending International. https://vendinginternational-online.com/hydrogen-powered-furnace-to-change-the-face-of-uk-glass-manufacturing-industry/ (accessed 7 June 2023).
9 Diageo. (2022). Encirc and Diageo announce hydrogen powered furnace to change the face of UK glass manufacturing industry https://www.diageo.com/en/news-and-media/press-releases/2022/encirc-and-diageo-announce-hydrogen-powered-furnace (accessed 21 November 2024).

10 European Environment Agency. Nitrogen oxides, NOx. https://www.eea.europa.eu/help/glossary/eper-chemicals-glossary/nitrogen-oxides-nox (accessed 7 June 2023).

11 Laidler, K.M. and John, H. (1982). *Physical Chemistry*. Menlo Park, CA: The Benjamin-Cummings Publishing Company, Inc.

12 Shriver, D. and Atkins, P. (1999). *Inorganic Chemistry*, 3e. New York: W. H. Freeman and Company.

13 Guidelines for the safe use of hydrofluoric acid. Department of Chemistry and Chemical Biology, Harvard University. (2013). https://chemistry.harvard.edu/files/chemistry/files/safe_use_of_hf_0.pdf (accessed 7 June 2023).

14 Spencer, C.D. and Ott, L. (1927). The frosting of glass by mixtures containing hydrofluoric acid and alkali fluorides. *J. Am. Ceram. Soc.* 10: 402–410.

15 Corning. The illustrated history and science of Corning® Gorilla® Glass. https://www.corning.com/gorillaglass/worldwide/en/illustrated-history-of-gorilla-glass.html (accessed 8 June 2023).

6

Degradation and Stabilization of Metals

Metals used in packaging were inspired by glass packaging. The process for forming tin cans was invented by Philippe de Girard and patented by merchant Peter Durand in 1810. The patent was sold to two Englishmen, Bryan Donkin and John Hall, in 1812. The first commercial canning factory was established in London in 1813; it produced cans for the Royal Navy. By 1820, tin cans were used for gunpowder, seeds, and turpentine. Early cans, which were sealed with a tin-lead alloy, posed a risk of lead poisoning. The American Can Company, founded in 1901, once produced 90% of U.S. tin cans. Technological advancements in the 1920s lowered the cost of cans. The first canned beer, sold in 1935, became an instant sales success. In 1935, most synthetic plastics were not commercialized. Today, metal packaging materials such as aluminum and steel are still widely used (Figure 6.1).

6.1 Benefits of Metals as Packaging

Barrier properties: Metals offer excellent protection against light, moisture, and air, thereby maintaining product quality. They are suitable for barrier applications and are, therefore, seen as viable alternatives to some types of multilayer plastic packaging.

Durability: Metals are strong, robust, and can withstand physical impacts.

No microplastics: Metals do not create microplastics.

Recyclability: Metals are highly recyclable without losing quality, contributing to sustainability.

Temperature tolerance: Metals can endure high-heat processing for sterilization.

Long shelf life: Metals provide extended shelf life for perishable goods.

Versatility: Metals are suitable for various products, from food to hazardous materials.

Degradation, Stabilization, and Recycling of Packaging Materials, First Edition.
Muhammad Rabnawaz, Susan E.M. Selke, and Ian Wyman.

Figure 6.1 Photograph of a typical can used in food packaging applications. *Source:* Image reused from Wikimedia Commons [1].

6.2 Disadvantages of Metals as Packaging

- **Weight:** Metals are heavier than plastic or paper, increasing transportation costs.
- **Cost:** Metals are generally more expensive as a raw material than plastics or paper.
- **Corrosion:** Metals are susceptible to rust or corrosion when exposed to certain conditions.
- **Limited flexibility:** Metals are less flexible than materials like plastic, limiting design options.
- **Energy-intensive:** Metals are often more energy-intensive in both production and recycling processes. Note that recycled aluminum is much more energy efficient than extraction of aluminum from its ore.
- **Resource intensive:** Metals are resource-intensive, as the extraction of metals can be environmentally damaging.

6.3 Basic Aspects of Metal Corrosion

All packaging materials except glass are prone to oxidation. Glass still degrades, but via corrosion rather than oxidation. Interestingly, metals also oxidize, often referred to as rust formation. Corrosion is usually defined as the formation of rust on a metal surface or, more broadly, as oxidation of a metal surface. (Rust, by definition, is iron oxide so that the first definition is limited to iron and steel.)

A better definition of corrosion does not limit the chemical change to oxidation. This definition says corrosion is a destructive attack on a metal through a chemical or electrochemical reaction with the environment [2]. This definition is reflected in the following chemical equation:

$$M \rightarrow M^{+z} + Ze^{-}$$

where M is the metal before oxidation (loss of electrons), M^{+Z} is the oxidized metal, and z is the number of electrons lost from the metal atom during this change.

In packaging applications, the metals of most interest are **steel and aluminum**. Steel is highly susceptible to corrosion, particularly the formation of rust. Aluminum is less susceptible, not because it does not oxidize, but rather because the oxide tends to adhere tightly to the aluminum, thus forming a protective coating that acts as a barrier to further reaction. In contrast, iron oxide does not adhere well to steel (or iron), instead readily flaking off and thus not deterring further attack. Photographs of metal containers which had undergone corrosion are shown in Figure 6.2.

Protection of metals from corrosion is often a requirement for their successful use in packaging.

For many years, tin coating has been used to protect steel against attack. In the last several decades, **tin-free steel (TFS) has emerged**, where steel is coated with chromium/chromium oxide. Steel can also be coated with other metals, such as zinc and aluminum, to reduce corrosion, but these techniques do not have any significant use in packaging applications. **Organic coatings** (plastics) are

Figure 6.2 Photographs of various metal containers that have undergone corrosion. The can at the left had originally packaged pickled cucumbers, while that at the right had packaged shellfish. *Source:* The photograph at the left is reprinted from Ref. [3] and at the right is reprinted from Ref. [4] Wikimedia Commons.

also used to impede corrosion, and they are **applied to substrates such as aluminum, tinplate or TFS.** However, these coatings do not provide sufficient protection to be used on unmodified steel.

6.4 Elements Required for Corrosion

Corrosion occurs in metals when the following elements are present:

1) An anode (negative electrode) and a cathode (positive electrode);
2) An electrolyte (a corrosive environment); and
3) A circuit connecting the anode to the cathode (Figure 6.3).

6.4.1 Anodes and Cathodes

The anode and cathode may be two different metals or may be two areas of the same metal that differ in chemical potential due to slight differences in chemical composition, slight differences in substances adsorbed onto the metal surfaces, differences in stress within the metal, etc. All else being equal, the greater the difference between the potential of the anode and cathode, the greater will be the rate of corrosion. The metal with the lowest (most negative) potential forms the anode, and that with the highest (most positive) potential forms the cathode. The metal at the anode dissolves (becomes oxidized), while that at the cathode does not dissolve.

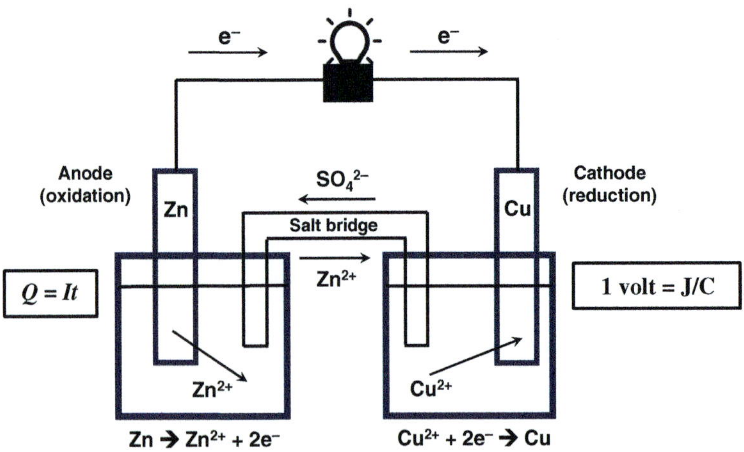

Figure 6.3 Figure illustrating the components of a galvanic cell. In the above illustration, copper sulfate ($CuSO_4$) is used as the electrolyte.

The reaction at the anode is oxidation of the metal, e.g.,:

$$Sn \rightarrow Sn^{++} + 2e^{-}$$
$$Al \rightarrow Al^{+++} + 3e^{-}$$
$$Fe \rightarrow Fe^{+++} + 3e^{-}$$
$$Fe \rightarrow Fe^{++} + 2e^{-}$$

The reaction at the cathode is reduction, e.g.,:

$$2H^{+} + 2e^{-} \rightarrow H_2$$
$$O_2 + 2H_2O + 4e^{-} \rightarrow 4OH^{-}$$
$$\text{metal ion} + \text{electrons} \rightarrow \text{metal}$$
$$\text{electron acceptor} + \text{electrons} \rightarrow \text{products}$$

6.4.2 Electrolyte (the Internal Circuit)

In electrochemical reactions, electrolytes are substances that dissociate into ions when they are dissolved or melted, thereby producing an electrically conductive medium. As electrons are "given off" at the anode and "taken up" at the cathode, there must be a flow of electrons from the anode to the cathode for the reaction to continue. Similarly, there must be a continuous supply of the species being reduced at the cathode. The electrolyte provides for this movement and is also generally important in supplying the ion that is being reduced since it is (with rare and insignificant exceptions) not the same as the ion that is produced by the initial oxidation reaction. The movement of ions in the electrolyte completes the internal circuit, allowing the overall electrochemical process to proceed. Electrons flow through the external circuit while ions flow through the electrolyte. The electrolyte provides the medium where the reduction (gain of electrons) occurs at the cathode and oxidation (loss of electrons) occurs at the anode. The electrolyte serves as the interface between the anode and cathode, where ions can easily migrate, making it essential for electrochemical cells.

For example, in a simple copper-zinc galvanic cell, the electrolyte might be a copper sulfate solution. Copper ions (Cu^{2+}) are reduced at the cathode, gaining electrons to form solid copper, while zinc is oxidized at the anode, losing electrons to form zinc ions (Zn^{2+}). The sulfate ions in the electrolyte move toward the anode to maintain electrical neutrality, completing the internal circuit.

6.4.3 Circuit

This consists of the path taken by electrons outside the battery as they migrate from the anode to the cathode. This could include wires, a load (like a light bulb or a motor), and a switch. The external pathway allows the flow of electrons from the anode to the cathode. When the circuit is closed, electrons can move freely, allowing for the completion of the electrochemical reaction. If the circuit is disconnected, the flow of electrons between the anode and cathode will cease. This interruption means that the electrochemical reaction responsible for transferring ions and electrons will be halted. In practical terms, this means reduced corrosion as the absence of an electrical circuit slows down the corrosion process, particularly for galvanic corrosion, since the electron flow is essential for the reaction to continue. Interruption means no current flow and thus no electrochemical reaction.

6.5 Role of Liquid Water in Corrosion

In most situations where we encounter corrosion, liquid water plays an important part, serving various roles:

1) Carrier for ions that are needed to maintain electrical neutrality.
2) Carrier for corrosive species such as dissolved oxygen or hydrogen ions. Here the hydrogen ions stem from the product.
3) Corrosive species (water as a source of H^+ and ^-OH ions and free radicals)
4) Generator of hydrogen ions when water hydrates metal ions, e.g.,

$$Sn + H_2O \rightarrow SnO + 2H^+$$

As mentioned earlier, whether a metal (or a part of a metal) is an anode or cathode depends on its electrochemical potential. The galvanic series (Figure 6.4) is an ordering of various metals from most negative to most positive in terms of their relative potential, at standard conditions, measured against a standard hydrogen electrode:

The galvanic series gives us an idea of whether a particular metal will likely be the anode or cathode when it is in a "corrosion cell" (see Figure 6.5). Metals with lower relative potential behave as anodes in electrochemical cells. The higher the potential difference between the two metals, the faster the anode will dissolve. For example, in an electrochemical cell composed of magnesium and silver electrodes, magnesium would act as the anode and would be oxidized, while silver would act as the cathode and would be reduced, since the standard electrode potential for the Mg/Mg^{2+} couple is −2.37 V, and for the Ag/Ag^+ couple is +0.80 V.

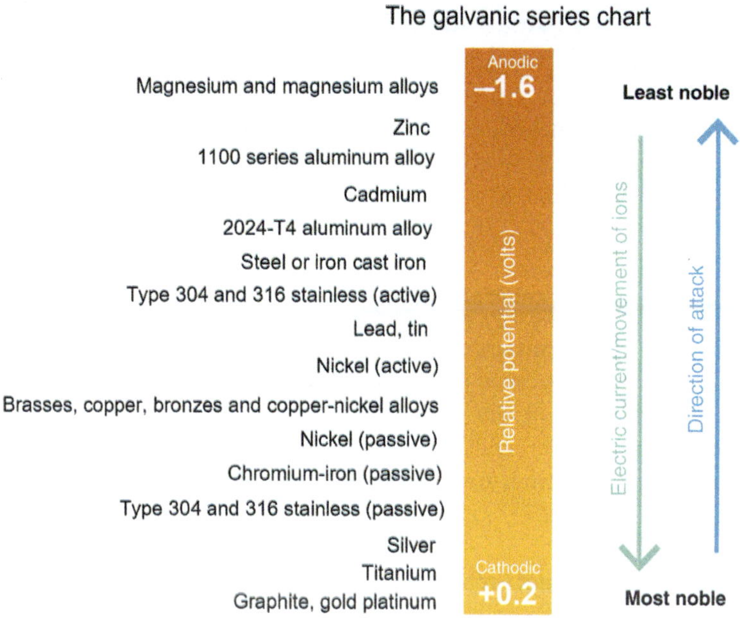

Figure 6.4 Schematic depiction of the galvanic series. *Source:* Image used with permission from Daniel Overbey [5].

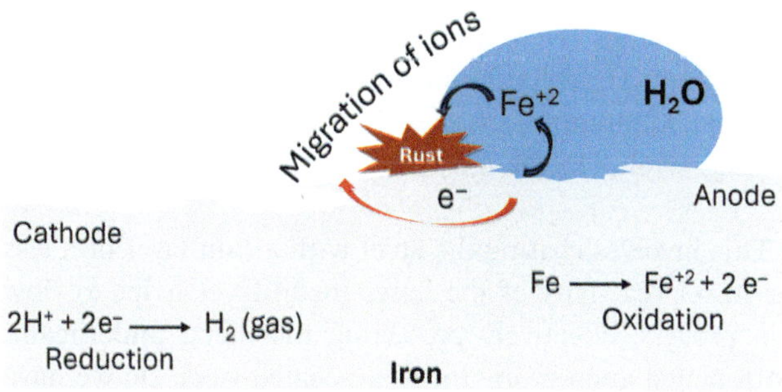

Figure 6.5 Schematic depiction of a typical corrosion cell, showing corrosion of iron occurring at the anode, while the metal at the cathode is protected from corrosion.

A more negative standard electrode potential means a greater tendency to lose electrons (i.e., to be oxidized). However, the potential is affected by the conditions in a particular situation. As we will discuss further, this means that assumptions about the direction of the reaction based solely on its relative position in the galvanic series are likely to be unreliable.

6.6 Methods for Protecting Metals Used in Packaging from Corrosion

To ensure the long life of metals used in packaging, it is crucial to minimize their susceptibility to corrosion. Various techniques are employed to either prevent or slow down the corrosion process.

6.6.1 External Environment Protection

Minimizing electrolyte contact: One approach to mitigating external corrosion is to reduce the contact between the metal and any electrolyte. This is why it is advisable to keep metal cans dry.

Barrier materials: Another strategy involves inserting a barrier between the metal surface and an electrolyte or a liquid that could serve as one.

6.6.2 Internal Environment Protection

In scenarios where the contents themselves are not conducive to corrosion, such as water-free products like powdered milk, the risk of corrosion is inherently low. However, for items like canned beverages and vegetables, where water is an integral component, additional measures are needed.

Organic coatings: These coatings act as a physical barrier between the can's metal and its contents (product), serving as an electrolyte.

Inorganic oxide coatings: Aluminum cans benefit from a naturally occurring aluminum oxide coating, while TFS often uses chromium or chromium oxide coatings.

Anodic protection: This involves coating the steel with a thin layer of a less reactive metal. The lower reactivity of the outer metal layer helps to slow down the corrosion process, effectively protecting the metal underneath. Examples include tin-plated and chromium oxide-coated steel. Do we have both an anode and a cathode if steel is coated with tin? Yes – even if the tin layer completely covers the surface of the can (leaving no steel exposed), areas on the tin would be sufficiently dissimilar in electrochemical potential that we would find that some areas will act as anodes while others behave as cathodes. In practice, some steel is usually exposed, so we will usually have one metal acting as the anode and the other as the cathode. Therefore, it is "normal" to have some corrosion occurring in any uncoated tinplate can containing a moist or wet product. It is also possible to have "abnormal" corrosion, leading to premature failure.

In "normal" tinplate corrosion, there is slow dissolution of the tin coating, which is referred to as even etching of the tin. The reduction reaction produces hydrogen gas, which then diffuses through the steel at a slow rate. After many years, when nearly all the tin has dissolved, the buildup of hydrogen inside the can will be sufficient to cause the can to swell, and this swelling marks a failure point.

Premature failure can be caused by an increased rate of reaction. In rapid detinning, the tin layer dissolves at an accelerated rate so that hydrogen is evolved at an accelerated rate, and the can fails by hydrogen swelling well before the time it "should" fail. Another type of premature failure involves a different corrosion mechanism – **pitting**. In this case, we have dissolution of steel either alone or along with tin dissolution. The steel dissolution is concentrated in small areas (pits) [6], and eventually results in failure by hydrogen swelling or through perforation of the can.

There are also abnormal types of corrosion that do not lead to early failure. In headspace detinning, rather than dissolving evenly, the tin layer preferentially dissolves at the interface between the product and the headspace, usually within a week or so after processing. Normally, this process changes to even etching within a short time, but the rate of etching is faster than under normal circumstances. A similar situation is called milk plate staining because it was once a common phenomenon when evaporated milk was packaged in soldered tinplate cans. This process consisted of localized detinning adjacent to the soldered side seam and the ends. Tin sulfide staining also involves the appearance of discoloration in the can, usually first appearing immediately after processing, intensifying during the first few months, and then leveling off. Asparagus staining involves discoloration not of the can but of the broth containing the asparagus, resulting from the formation of an iron salt with a component of the asparagus. It is considered corrosion because the iron originates from the metal of the can.

If the tinplate can has a lacquer coating (organic coating), the expectation is that there will be much less electrochemical reaction. Detinning can still occur, through the enamel film. Even etching of the tin is considered normal and is acceptable with some products. If the detinning is localized, however, it is considered abnormal and can lead to enamel lifting and flaking, then rapid detinning, and premature failure due to hydrogen swelling. Lacquered cans can also fail through pitting corrosion. Staining can also occur, especially sulfide black corrosion, which can result in black deposits on the can as soon as 24 hours after processing.

A number of things that may be present in the contents of a can contribute to acceleration of corrosion. These influences include oxygen, sulfur, copper, nitrates, fungicides, pesticides, and colorants, among others. As little as 1 ppm of sulfur, in some cases, can cause a 50% reduction in shelf life. Less than 0.5 ppm of copper can initiate electrochemical pitting. So, as packaging engineers, we must ensure that we know how the nature of a product or impurity can affect metal packaging.

To understand what goes on during corrosion, we need to consider both which metal is acting as the anode and which behaves as the cathode (the direction of the reaction), and also the overall rate of reaction. The electrochemical potential determines the direction of the reaction. As mentioned earlier, the anode metal will dissolve during the oxidation reaction, and the cathode metal will be protected from dissolution, since it is involved in the reduction reaction. The rate of the reaction is controlled by the amount of current that is flowing. Faraday's law states that the passage of 96,500 coulombs will cause the dissolution of a gram equivalent weight of the metal concerned (where the gram equivalent weight is the atomic weight divided by the valence).

6.7 Structure of Tinplate and Tin-Free Steel

To understand the mechanism of **corrosion in steel**, it is useful to review the composition of packaging steel. One factor influencing the amount of corrosion is the quality of the base steel. Three types of steel are commonly used in can manufacturing, as given in Table 6.1.

The can thickness is normally between 0.20 mm (for small cans) and 0.29 mm (for large institutional cans) (Figure 6.6). The steel is subjected to **pickling**, involving a dilute sulfuric acid treatment along with an electric current. Prior to pickling it may be cleaned in dilute sodium hydroxide, also using an electric current. For a metal, pickling is a chemical process that involves the use of an acidic solution to remove impurities, such as stains, inorganic contaminants, rust, or scale from ferrous metals, copper, and aluminum alloys. This treatment prepares the metal surface for further processes such as galvanizing, enameling, plating, and painting.

Table 6.1 Types of steel typically used in cans.

L	Low and controlled content of residuals, suitable for highly corrosive products
MC	Rephosphorized steel, high strength, low corrosion resistance
MR	Less restriction on residuals than L, most used for moderately corrosive products

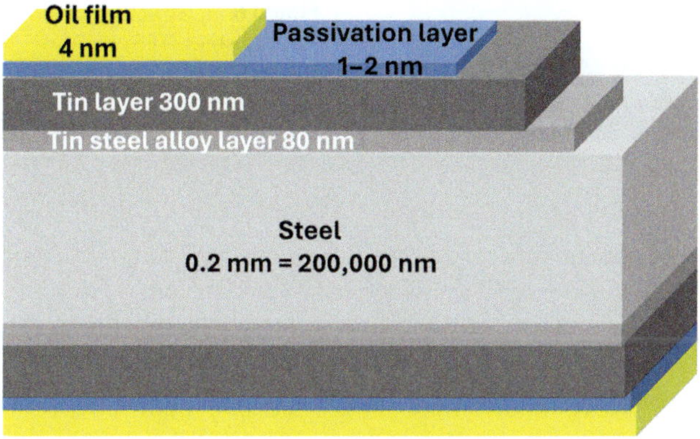

Figure 6.6 Structure of tinplate – not to scale! Note that all layers combined are just a fraction of that of the steel's thickness.

To prepare tinplate steel, a tin layer (typically about 0.4 μm in thickness) is then applied through an electrolytic process. The thickness of the tin layer varies, depending on the application and on whether this layer is applied onto the inside or outside surface of the can. Usually the tin layer inside the can is thicker than that outside. In the plating operation, a negative charge is given to the steel, and direct current flows through the electroplating bath, causing Sn^+ ions to be reduced to metallic tin on the steel surface.

Subsequently, a flow brightening (or flow melting) process is performed. During this process, the coated sheet is heated to about 450 °F, just below the melting point of the tin. During this process, there is some diffusion of tin into the steel sheet, forming a tin/steel alloy layer with a thickness of about 0.08 μm (80 nm). For best corrosion protection, a dense, closely packed alloy layer is formed. The tinplate is then immediately quenched with water.

The next step is passivation, aimed at preventing uncontrolled oxidation of the tin, which can lead to adhesion failure of a subsequently applied organic coating. There are three main passivation techniques:

1) cathodic sodium dichromate (CDC);
2) sodium dichromate dip (SDCD); and
3) cathodic sodium carbonate (CSC).

These three methods are listed in decreasing effectiveness in preventing corrosion. In all these processes, a colorless surface film, a few nanometers thick, of nonstoichiometric tin oxide is formed. In the case of the CDC and SDCD processes, this layer also contains chromium.

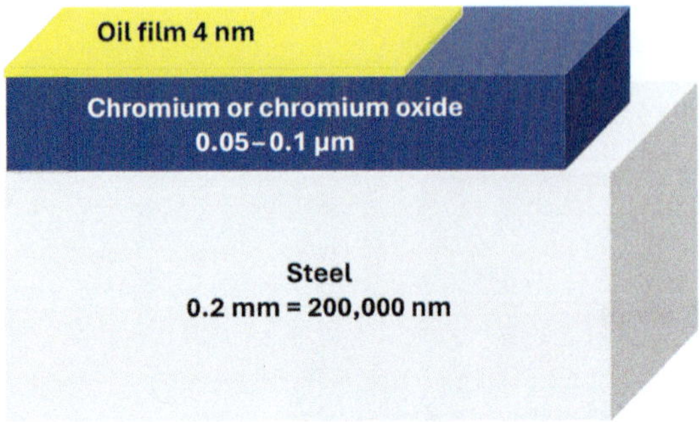

Figure 6.7 Structure of tin-free steel (TFS).

Finally, an oil film, typically about 4 nm thick, is applied. Either natural or synthetic oils can be used to produce this film. The film is approximately monomolecular (one molecule) in thickness, and functions to minimize abrasion damage. Natural oils used in this role are often cottonseed or palm oil. The most commonly used synthetic oils to prepare these oil films are dioctyl sebacate, acetyl tributyl citrate (ATBC), and butyl stearate.

A close inspection of tinplate steel suggests that packaging engineers are using tin as a protective layer for steel, which is in turn is protected by chromium, and finally, the chromium layer is coated with an organic material.

For TFS, chromium metal or more frequently a chromium/chrome oxide layer, is electrochemically deposited onto the base steel as shown in Figure 6.7. The total thickness is usually in the range of 0.05–0.1 μm—much thinner than the tin layer in the tinplate. An oil film is also used to minimize abrasion damage.

In spite of these efforts to avoid surface damage, we find that in a can, there will be some degree of exposure of the steel through defects in either the tin coating or in the chromium coating as well as at any locations where the edge of the coated sheet is exposed. In general, the thicker the coating, the smaller the amount of steel exposed through coating defects, but such exposure cannot be completely avoided.

6.8 Corrosion in Plain (Uncoated) Tin Cans

If we examine the **outside** of a tinplate can that is not protected by a lacquer coating, we find that the exposed steel is anodic with respect to the exposed tin, as we would expect from the galvanic series. Therefore, on exposure to an electrolyte, the steel will corrode (dissolve), resulting in the production of rust on the outer surface of a steel can.

Figure 6.8 In tinplate steel can, the corrosion chemistry involves several steps: (a) and (b) occur in the presence of oxygen and generate minimal hydrogen gas, and (c) takes place in an oxygen-free environment and primarily produces H_2 gas.

(a) **Aerobic corrosion (occurs with oxygen)**

- $4Fe + 3O_2 \rightarrow 2Fe_2O_3$
- $4Fe + 6H_2O + 3O_2 \rightarrow 4Fe(OH)_3$

(b) **Oxide film reduction**

- $4Fe_2O_3 + Fe \rightarrow 3Fe3O_4$

(c) **Anaerobic corrosion (occurs without oxygen)**

- $3Fe + 4H_2O \rightarrow Fe_3O_4 + 4H_2$
- $Fe + 2H_2O \rightarrow Fe(OH)_2 + H_2$
- $3Fe(OH)_2 \rightarrow Fe_3O_4 + 2H_2O + H_2$

On the inside of the can, we might expect the same phenomenon to occur. However, as has been mentioned earlier, the polarity of an electrode (and thus its position as either more or less active than a metal to which it is electrochemically coupled) can be affected by its surroundings. When an acidic product is packaged in an uncoated can (without organic coating of the tinplate), once any residual oxygen in the can is used up, this change in polarity is usually sufficient to render the tin anodic to the steel. Consequently, inside the can the normal occurrence is that tin will dissolve (corrode, oxidize), while the steel will be protected (Figure 6.8).

This "**reversal of polarity**" comes about due to the combined effect of several different types of factors, none of which alone are sufficient to cause it to occur.

One factor that causes a change in the polarity of an electrode is the current density (current per unit area). Metals differ in how they are affected by the amount of current that is flowing. Also, in the case of tinplate, there is a large difference in the amount of exposed metal – and it is only on the metal exposed to the electrolyte that the reactions take place (the current flows). Since the area of exposed tin is much larger than the area of exposed steel, the current density is much greater on the steel than on the tin. Consequently, the polarity of the steel is changed more than the polarity of the tin.

Another effect that occurs is the influence of a buildup of reaction products around the electrode. As corrosion occurs, cations build up around the anode and anions build up around the cathode. This shifts the electrode potential, through a process that is known as concentration polarization [7]. Related to this is resistance polarization, which occurs because current flows through a resistance related to a buildup of reaction products on the electrode, or a change in availability of reactants caused by depletion of reacting species, etc. These factors are also related to the current density. Since the current density is much lower on the tin than the steel, these factors will, in general, also cause greater change in the polarity of the steel than that of the tin.

Another factor is polarization related to the presence of specific reactions or species. In particular, the hydrogen overpotential[1] is associated with the reaction:

$$2\,H^{+} + 2\,e^{-} \rightarrow H_2$$

Tin tends to become polarized much more than steel by the presence of hydrogen. While the details of these effects and why they occur are beyond the scope of this book, we do need to understand what the result is. In particular, the end result is that when a tinplate can containing an acid product is first sealed and there is still some residual oxygen present in the can, the steel will act as the anode and we get iron dissolution. As the oxygen is removed, the electrode polarity normally reverses, and then tin becomes the anode. The key factors in causing this reversal of polarity are the significantly higher current density on the steel due to its much smaller exposed area; the reduction in the amount of exposed steel that occurs as the reaction proceeds and stannous hydroxide (which has very low solubility) precipitates on the steel, thereby reducing its exposed area; the higher hydrogen overpotential of tin compared to steel; and, in some cases, a decrease in the stannous ion concentration due to the formation of soluble complexes with some organic ligands in the product. These factors combine to make the tin much more active than it normally is, causing it to become less noble than steel.

The effect of oxygen is primarily manifested through its relationship to the hydrogen overpotential. During "normal" corrosion, hydrogen gas is generated, as has been discussed earlier in this chapter. In the presence of oxygen, this will not occur, as the end product of combination of the hydrogen and oxygen is water. Therefore, when oxygen is present, even though the other factors can still operate, there is not a sufficient change in polarity to render the tin anodic (so on the outside of the can the steel will remain anodic). Even in the absence of oxygen, some factors can cause this reversal of polarity to fail to occur or to be incomplete. Therefore, our next task is to go back to the discussion of the types of corrosion that we have described as occurring in tinplate, and explain them.

6.8.1 Normal Detinning

Tin is anodic to steel, so there is complete cathodic protection of the steel. The rate of corrosion (the couple current) is controlled by the current density at the cathode. Therefore, the exposed area of the steel, by determining the cathodic current density for a given amount of corrosion, controls the couple current and thus determines the amount of corrosion that can be supported.

---

1 **High hydrogen overpotential** refers to the relatively large voltage required to drive the electrochemical reduction of hydrogen ions to form hydrogen gas at an electrode.

This is the situation that is normally encountered with citrus products, peaches, and apricots, and other low pH products that are stored in plain tin cans.

6.8.2 Rapid Detinning

In this situation, the tin is sufficiently anodic to protect the steel, but the couple current is high. This can be due to exposure of large steel areas (which lowers the current density), the use of faster corroding steel (increasing the difference in the polarity of the electrodes), or from the presence of strong complexing agents that remove Sn^{2+} ions from solution and thereby render the tin more anodic.

This type of corrosion is characteristic of aggressive citrus products such as lemon juice and tomato products.

6.8.3 Partial Detinning and Pitting

In this situation, the tin is generally anodic to the steel, but local areas on the steel are more anodic than the tin. In other words, reversal of polarity is not complete. Some localized areas of the steel function as cathodes and some as anodes. Tin will dissolve from the anodic tin, but steel will dissolve in those areas of exposed steel that are anodic. In these areas, corrosion will, by iron dissolution, eat into the steel, thus producing pitting.

This situation is characteristic of the use of inferior steel. It also tends to occur with certain particularly problematic products, such as prunes or pear nectar.

6.8.4 Pitting Only

In this situation, the steel is anodic to the tin. The polarity has not reversed in this case. Therefore, there is no tin dissolution. Steel dissolution, as discussed earlier, leads to pitting.

This type of corrosion is quite rare, and is generally limited to highly corrosive products such as pickles.

We will return to some general discussion of the effects of the product being stored on corrosion later in this chapter.

6.9 Corrosion in Lacquered Cans

As has been discussed earlier in this chapter, the idea behind the use of can coatings (lacquers) is to interpose a high resistance in the electrolytic pathway, thereby minimizing corrosion. Again, **if the protection were complete**, we would have no corrosion, but in practice, there are always defects in the coating, resulting from pores or scratches that expose both tin and steel.

Recall that one of the factors influencing the "normal" reversal of polarity inside a plain tinplate is the disparity in the area of exposed steel and exposed tin. When we

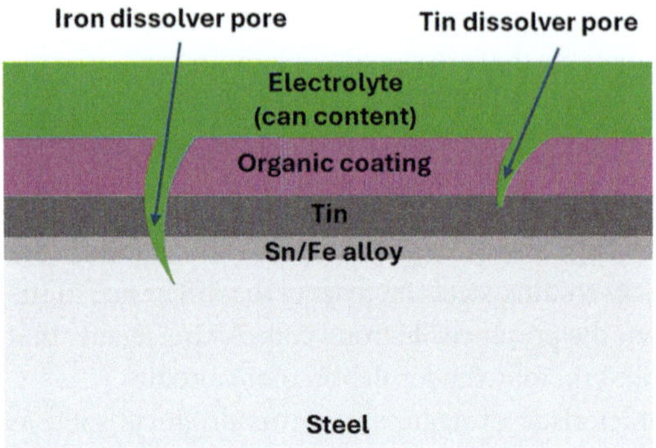

Figure 6.9 Corrosion in a lacquered can, showing tin-dissolving (left) and iron-dissolving (right) pores in the lacquer.

add a coating, we significantly reduce this disparity. Consequently, the difference in current density between the tin and steel electrodes will be much smaller. Of course, for a given corrosion rate, the overall current density on the tin would be much higher. Therefore, we would expect the overall rate of corrosion to be considerably lower in a lacquered can than in an unlacquered can. However, corrosion in the lacquered can is, on average, considerably more likely to result in iron dissolution rather than tin dissolution than corrosion in an unlacquered can. The coating minimizes the difference between the exposed surfaces of steel and tin, leaving steel as the anode and tin the cathode, as per the galvanic series. It should be noted that lacquered cans may be either tin dissolvers if steel is not exposed or there is too little exposure, making the tin anodic (see Figure 6.9 Left) or iron dissolvers if iron is anodic (see Figure 6.9 Right).

If tin is anodic, the rate of corrosion increases as more iron is exposed, so areas of exposed iron should be kept as small as possible. If iron is anodic, the rate of corrosion increases as more tin is exposed, so the areas of exposed tin should be minimized.

The number of pores, and therefore the amount of exposed metal, decreases as the coating thickness increases, according to the general relationship shown in Figure 6.10. Coating thicknesses generally is in the range 1.5–10 μm depending on the needs of the product [8].

In plain (uncoated tinplate cans), failure is usually due to hydrogen swelling or to pitting. Failure in lacquered cans is often due, instead, to reduction in the bond between the lacquer and the metal, which can eventually cause lifting of the lacquer film from the metal substrate (delamination of the coating). While this failure can sometimes be due to stresses during fabrication, it is more often due either to a subsequent breakdown in adhesion or to corrosion. The following detachment mechanisms are common:

1) **Water displacement:**
   There is always at least a monomolecular layer of water between the oxide film on the metal surface and the organic coating. The oxide layer attracts water

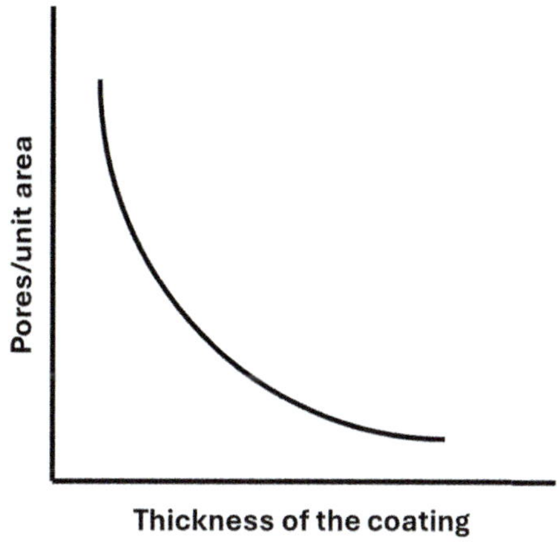

Figure 6.10 General relationship between the thickness of a lacquer coating and the number of pores per unit surface area. The coating thickness inside metal cans is typically between 1 and 10 μm.

because of its high polarity. If the water layer is thicker than this, it generally results in fewer bonds between the metal and the coating. The chemical composition of the oxide layer affects the amount of absorbed water. Chromium oxide forms a stable bond with lacquers and coatings, minimizing delamination. Occasionally, polyacrylic acid is used as a primer to enhance the adhesion between metals and organic coatings.

2) **Cathodic detachment:**

   Reduction of residual oxygen in the can or other reactions leading to hydroxyl ion (OH^-) formation on the cathode can cause the development of a high pH and lead to a water displacement type of loss of adhesion. See the chemical equation leading to OH^- ions formation.

   $$O_2 + 2\,H_2O + 4\,e^- \rightarrow 4\,OH^-$$

3) **Anodic undermining:**

   Corrosion of the surface layer of the metal beneath the organic coating leads to the formation of crevices between the metal and the coating, with consequent loss of adhesion. This type of corrosion tends to be promoted by the presence of chlorides and sulfates. For example, in lacquer-coated tinplate steel, tin dissolves nonuniformly beneath the coating. This Sn^{+2} diffuses from its location, leaving behind crevices that weakens surrounding coating-metal interactions.

   $$Sn \rightarrow Sn^{2+} + 2\,e^-$$

6.10 Effects of Products, Processing, and Storage

6.10.1 Effects of Products

At several points, we have mentioned that certain products tend to be associated with certain types of corrosion issues. Several types of components that occur in some products are known to act as **corrosion accelerators.** These include:

- Oxygen,
- Sulfur compounds,
- Nitrates,
- Trimethylamines,
- Anthocyanins (found in fruit),
- Dihydroascorbic acid, and
- others.

We have already discussed the role of oxygen.

Sulfur compounds, often found in meat and vegetables, have a number of modes of action. Such compounds can react with exposed steel to form FeS on the Fe layer in which the iron is anodic even in an anaerobic media if the pH is in the range of 3.6–4.6. This results in localized corrosion of the steel. Due to the high reactivity of iron (Fe) and sulfur (S) with each other; they readily combine when in proximity, especially in an environment devoid of oxygen. Once the first layer of FeS forms on the iron surface, it can act as a nucleation site, speeding up the formation of additional FeS layers. In anaerobic conditions (absence of oxygen), like those found inside cans, FeS formation is often favored.

A combination of **sulfur compounds** with fungicidal residues can cause inactivation of the tin coating through the formation of a protective sulfide film on the tin. The coated tin fails to provide electrochemical protection for the steel, again leading to steel dissolution. Furthermore, residual SO_2 affects the polarity of the electrodes. A concentration as small as 1 ppm can significantly accelerate corrosion.

In contrast to what might be expected, there is no direct proportionality between acidity or pH and corrosion. In acidic conditions, protons (H^+) facilitate the oxidation of metals like tin or iron, leading to corrosion. However, the type of acid and its concentration also matter. Some acids may form protective oxide layers on the surface, reducing corrosion. Meanwhile, alkaline products react with tin or organic coatings but do not generally corrode base steel. The relationship between corrosion and pH is complex and not solely determined by the acidity or alkalinity of the environment. Various factors, such as electrochemical potential, ion concentration, and the presence of specific ions, contribute to the corrosion process.

Nitrates, which are often found in plants, water supplies, and heavily fertilized soils, tend to be especially common in foods such as green beans, spinach, turnips, lettuce, beets, and radishes. During storage, the nitrates act as corrosion accelerators and are reduced to ammonia. While there are differing mechanisms, one effect of nitrates is to act as a new pathway for electron transfer, thus increasing tin dissolution through the following sequence of reactions:

$$4\,Sn \rightarrow 4\,Sn^{2+} + 8\,e^-$$

$$NO_3^- + 2\,e^- + 2\,H^+ \rightarrow NO_2^- + H_2O\,(\text{slow})$$

$$NO_2^- + 6\,e^- + 8\,H^+ \rightarrow NH_4^+ + 2\,H_2O\,(\text{fast})$$

The reactions involving nitrates are catalyzed by the presence of ferrous ions. Tin nitrite ($Sn(NO_2)_2$) forms when Sn^{+2} and NO_2^- combines. $Sn(NO_2)_2$ is generally more soluble in water than tin hydroxide ($Sn(OH)_2$). The latter is virtually insoluble and forms a gel-like precipitate, making it useful as a flocculant in water treatment. In contrast, tin nitrite has better solubility due to the stable complex it forms with tin ions. As a result, $Sn(OH)_2$ creates a protective layer on tin, while $Sn(NO_2)_2$ dissolves, leaving the tin surface vulnerable to further corrosion.

When products are aggressive – tend to promote rapid corrosion – the solution is generally to use a lacquered can. However, for certain products, there are actually benefits to a limited amount of corrosion. These benefits are usually associated with the production of a reducing atmosphere (**H_2**) within the can that acts to limit oxidative changes in the product. For example, a reducing atmosphere can prevent discoloration (purpling) of some red fruits, such as cherries, plums, and strawberries. It also slows nonenzymatic browning reactions, and ascorbic acid loss. In these cases, it can be beneficial to use an uncoated can so that an adequately reducing atmosphere is quickly formed.

As was already discussed, oxygen can affect the corrosion mechanism, participating in oxidation processes and acting as a depolarizer:

$$O_2 + 4\,H^+ + 4\,e^- \rightarrow 2\,H_2O$$

The oxygen may be present in the headspace, dissolved and/or absorbed in the product, or simply entrapped by particulates. Oxygen promotes attack on steel.

Oily or fatty products present no particular corrosion problems.

6.10.2 Effects of Processing and Storage Conditions

Processing and storage conditions can also affect corrosion. Usually, since the time required for processing is short relative to the total storage time, **corrosion during this period is not significant**. However, the mode of corrosion during processing may be different than the later mode of corrosion during storage. A can may be a tin dissolver under one set of conditions and an iron dissolver under another. In addition, correct cooling after thermal processing is important both to avoid paneling of the can and because incomplete drying can lead to external corrosion.

During storage, increasing the temperature generally accelerates the rate of corrosion, but once again a change of mechanism can occur, leading to apparent paradoxes. **For example,** canned fish stored at 40 °C may exhibit no significant corrosion problems, while rapid pitting may be encountered when the same product is stored at 20 °C, due to a change in mechanism from tin-dissolving at the higher temperature to iron-dissolving at the lower temperature.

There can be both safety and toxicity concerns associated with the presence of dissolved metals in products such as food. Lead presents a problem even at very low levels. Even though lead is normally cathodic to both tin and steel in all foods, there can be lead dissolution in some cases in lacquered cans where lead is exposed in the side seam but there is little tin exposure. More commonly, when lead dissolution has occurred, it has originated from solder splashes or dust that adhered to the surface of the lacquer, where it was not electrically coupled to other metals in the can and therefore was not protected. **In the United States, the use of lead solder is not permitted for food cans.** Soldered cans are metal cans sealed using solder, a fusible metal alloy. Once the solder cools and solidifies, it forms a strong bond that effectively seals the can, making it airtight and leak-proof. For a time, companies using soldered cans for food were using mostly tin solder (which was more expensive than lead). However, there is no longer any food packaging in soldered cans of any type in United States. For a number of years several major U.S. states have prohibited the use of lead (along with cadmium, mercury, and hexavalent chromium) in packaging of any type (with very rare exceptions), and consequently the use of lead solder in cans, along with soldered cans in general, has disappeared. These laws pertain to imported cans as well as those produced domestically, so they had a worldwide effect even before many other countries passed similar legislation.

Iron and tin are relatively nontoxic, under most circumstances. A 50 ppm limit on dissolved iron is recommended, mostly because of concern about off-flavor. For tin, salts have low toxicity, although some alkyl derivatives are highly toxic. A 200 ppm limit on tin salts is the general guideline in the United States, again mostly due to taste considerations. In general, tin has less of an influence on the flavor of food products than is the case with iron.

6.11 VCI Packaging Materials

In addition to concern about the corrosion of metal packages themselves, a corrosion-related packaging concern is the prevention of **corrosion of metal products**, especially those that are exposed to high humidity for extended periods of time during shipment or storage.

One approach to limiting these corrosion problems is to store the product in a high-barrier package, perhaps with a desiccant included to maintain low relative humidity. In essence, this approach is to prevent corrosion by avoiding the presence of an electrolyte.

Another approach is to coat the metal product with oil or grease, thus interposing a non-conductive barrier between the electrolyte and the metal.

A different approach is to use packaging that has been treated with additives that act to protect nearby metals from corrosion. The action of these materials requires that the inhibitory compounds volatilize, so they are known as VCI materials (**volatile corrosion inhibitors or vapor corrosion inhibitors** – there is no consensus on which of these words the V stands for) [9]. These materials were first discovered in 1943, and first came into large-scale use for protection of rifles during the Korean War. Despite this reasonably long history, the mechanism by which VCI materials act is not really understood. Also, there are various VCI materials that may act in different ways. The three most common **THEORIES** are that:

1) **VCI materials** prevent oxygen from reaching the metal surface by combining with it in the air;
2) **VCI materials** work like a neutralizer, removing H^+ ions; and
3) **VCI vapor** is absorbed onto the metal surface and forms a protective layer that helps prevent corrosion.

The last of these appears to be the most popular theory.

Precise VCI formulations in commercial use are proprietary. Known formulations include diisopropyl amine nitrate and dicyclohexylamine nitrate; dicyclohexylammonium nitrite; sodium nitrite and urea; sodium nitrite, urea, and monoethanol amine benzoate; amine salts; diisopropyl amine nitrite; and dicyclohexylamine nitrite. Commercial formulations generally contain a combination of ingredients.

VCI materials exist in various forms. One of the most common approaches is to apply the VCI as a coating on kraft paper. The object to be protected may be wrapped in the paper or the paper may simply be included with the object in the container. A VCI may also be placed in the container as a crystal or tablet. Rather than using a separate paper, VCI coatings can be applied onto corrugated board. More recently, VCI materials have been incorporated into plastic films and foams.

VCI materials are sometimes used in direct contact with the product being protected. In other cases, direct contact is avoided and may even accelerate corrosion; products work strictly in the vapor phase.

Currently, there is no standardized test for VCI performance. Each manufacturer tends to have its own set of test conditions. The U.S. military has a set of test procedures, but they are rather complicated to use, and not necessarily any more representative of real-life performance than the simpler tests that are used by manufacturers.

User specifications sometimes call only for a certain loading of VCI in the paper that is meant to be used for metals, rather than specifying actual performance. It is common practice to measure coating weight by the weight loss of the paper after washing and drying, without regard to whether it is active or inert ingredients that are removed, or even whether some of the measured weight loss is due to loss of paper components that are unrelated to the coating. While the coating weight may be useful to VCI manufacturers as a process control measurement, it is likely to be of limited if any use to end users, who may be evaluating performance of materials from different manufacturers that have quite different formulations.

Figure 6.11 illustrates the ability of VCI materials to reduce metal corrosion. The top pictures are the controls, which were stored at elevated temperature and relative humidity. Meanwhile, the bottom pictures are samples that were stored under identical conditions, but wrapped in VCI paper. On the left is copper, in the middle is steel, and on the right is galvanized steel (steel coated with a layer of zinc). Note the reddish discoloration on the copper, and the obvious rust on the steel, in the top pictures. The galvanized steel has also corroded. In contrast, the metals protected with VCI are much less affected by the exposure. These pictures were taken during tests performed in the School of Packaging at Michigan State University a number of years ago.

Galvanized steel has a zinc layer which is more reactive than steel, so when it is exposed to corrosive elements like oxygen and water, it oxidizes before the steel does. This "sacrificial" action of zinc provides two mechanisms of protection:

Barrier protection: The zinc layer physically separates the steel from the corrosive environment, reducing the rate at which the moisture and oxygen can reach the steel surface.

Galvanic protection: In areas where the zinc coating might be damaged or scratched away, the zinc still provides protection by corroding first due to its higher electrochemical potential (negative potential) compared to steel.

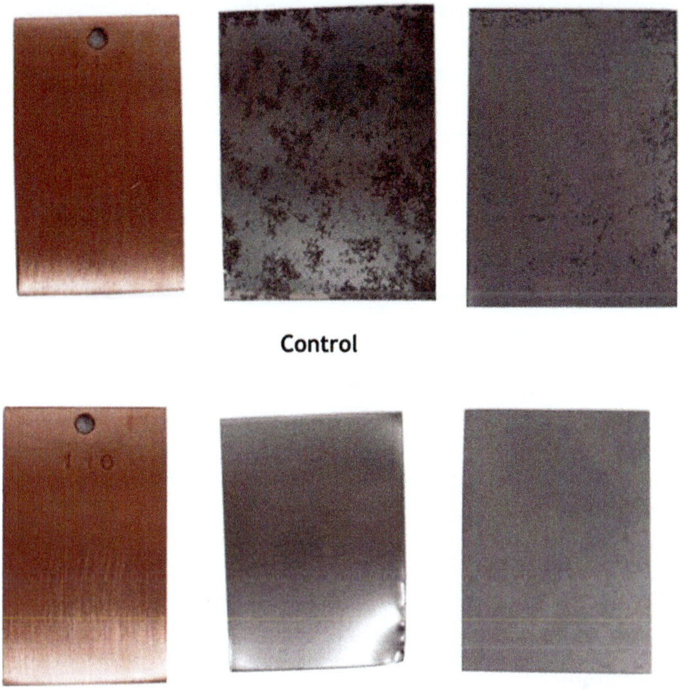

Figure 6.11 Metal plaques (copper, steel, and galvanized steel) stored at 95 °F and 100% RH at Michigan State University's School of Packaging. The samples shown in the top row were wrapped in kraft paper. The samples shown in the bottom row were wrapped in VCI-coated kraft paper.

6.12 Corrosion of Aluminum

Aluminum is much more resistant to corrosion than steel. It is not that aluminum does not oxidize, as in fact it oxidizes quite rapidly. Rather, oxidation of aluminum results in the creation of a thin film of aluminum oxide that tightly adheres to the underlying metal and protects it from further reaction.

A freshly exposed surface of aluminum metal will acquire, within about 5 minutes, a layer of aluminum oxide (with associated water molecules, the composition is $Al_2O_3{\cdot}3H_2O$) that is about 8 Angstroms (or 0.8 nm) thick. Within a few days, the thickness of the oxide film will increase to 32–38 Angstroms, and then stop increasing. **This coating is slightly porous, so the underlying metal can still be attacked by aggressive formulations of contents.** If this is a concern, the aluminum can be oxidized anodically in an acid bath through a process called

Table 6.2 Relationship between the purity of aluminum and the corrosion rate.

Purity of aluminum (%)	Relative corrosion rate
99.998	1
99.970	1,000
99.80	3,000
99.70	5,000
99.60	7,500

Source: Based on M. Johnson, "The technology of aluminum can linings," Spray Technology & Marketing, September 1999, pp. 24–36.

anodizing. This treatment increases the thickness of the oxide layer, and can provide a 2–10-fold increase in resistance to corrosion.

The rate of corrosion for aluminum is very much affected by the purity of the metal. Aluminum that is 99.99999% pure is reported to be able to hold concentrated hydrochloric acid [10]. The relative corrosion rates for aluminum of differing purity are given in Table 6.2, where the basis of comparison is the corrosion rate of 99.998% aluminum.

It should be noted that this is to some extent a simplification, as the corrosion rate depends not only on the percentage of other ingredients but also on what those ingredients are. Exposure conditions also play a role, of course. In particular, copper, iron, nickel, and cobalt promote corrosion, while manganese improves corrosion resistance (primarily by reducing the effect of iron in accelerating corrosion).

Aluminum alloys are identified by a **4-digit number**, with the first digit indicating the principal alloying element. If the first digit is **1**, the alloy is at least 99% aluminum by weight. A first digit of **3** indicates that manganese is the principal ingredient other than aluminum, and **5** indicates magnesium.

The most common aluminum alloys used for aerosol cans are AA-**1**060 and AA-1070 alloys, which are guaranteed to be at least 99.60% and 99.70% aluminum, respectively, and usually actually contain about 99.73% and 99.81% aluminum. Beverage cans commonly use AA**3**004 for the body (1.0–1.5% Mn, 0.8–1.3% Mg), AA5182 for the can lids (4.0–5.0% Mg, 0.20–0.50% Mn), and AA5052 for the can tab (2.2–2.8% Mg, 0.15–0.35% Cr). Aluminum foil commonly uses AA1100 (99.0–99.3% Al), AA1145 (99.45–99.60% Al), AA3003 (1.0–1.5% Mn), or AA5052.

Corrosion in aluminum cans most often is due to localized pitting. Crevice corrosion, at defect areas such as lining blisters or cracks and under can curls, is also common. When galvanic corrosion occurs, aluminum is usually anodic, so it is dissolved. It should be noted that pH plays an important role, with aluminum usually well protected by the oxide film for pH ranging between about **4.0 to about 8.5,** at room temperature, but the aluminum is subject to attack when the pH is higher or lower than that range.

As with steel, the product can act as a promoter or inhibitor of corrosion. In particular, products containing over 0.1 ppm of copper or mercury, or 0.5 ppm of iron ions can result in plating of these metals onto an uncoated aluminum container, where they become cathodes and result in pitting of the aluminum container. Cl^-, OH^-, and H^+ ions also promote corrosion, as do some soaps such as sodium lauryl sulfate. Hydrogen peroxide and potassium dichromate, on the other hand, act to strengthen the oxide film and thereby inhibit corrosion. Sodium benzoate and sodium nitrite also inhibit corrosion.

In contrast to steel cans, the presence of oxygen in an aluminum can generally improve its corrosion resistance.

Despite the inherent corrosion resistance of aluminum, most aluminum cans are coated. The coating both improves the corrosion resistance and helps with manufacturing by increasing slip.

6.13 Lacquer Coatings for Cans

Various formulations of coatings are commonly used for both steel and aluminum cans. On the inside of the can, the coatings serve a protective function, while outside they can be either protective or decorative (printed) or both. In most cases, the coating is applied prior to can manufacture, when the material is in a flat sheet. An exception is coating for draw and ironed Al beverage cans, where the coating is applied after the can body has been formed. Also, for 3-piece cans, coating of the side seam is done after the body is formed.

Coatings for the inside of steel cans (linings) fall into the following major groups:

- Acrylic coatings;
- Epoxyamine coatings;
- Epoxy-ester coatings;
- Epoxy-phenolic coatings;
- Oleoresin coatings;
- Phenolic coatings;
- Polybutadiene coatings; and
- Vinyl coatings.

Epoxyphenolic resins are the most widely used, and are suitable for canned meat, fish, vegetables, and fruits. They have largely replaced the oleoresins that once were very common.

White vinyl lacquers are often used where the white appearance of the inside of the can provides a marketing advantage or where their lack of any taste or odor is an asset.

For aluminum cans, epoxy-phenolics are the most common can linings, followed by epoxyphenolic-vinyl resins, which are also called organosols. A newer category, polyamide imide (frequently abbreviated as PAM of PAI) is growing in use. Amino resins are also used to some extent.

Coatings for the outer surfaces of cans are often polyesters (alkyd resins) and acrylics.

Various methods are available for testing the performance of can liners and coatings. Many of these are electrochemical in nature, measuring current flow or resistivity as an indicator for the degree of protection against corrosion. While these provide fast information, they sometimes do not accurately represent long-term behavior, so static storage tests are often also necessary.

6.13.1 Bisphenol A (BPA)

A growing concern for users and manufacturers of coated cans is the presence of bisphenol A (BPA). It is widely used as an ingredient in coating formulations, but various groups have expressed concern about its toxic effects, particularly for infants and fetuses, even at very low exposure levels. While most regulatory agencies have determined its use is safe, particularly in the very low levels found in coatings, not everyone is convinced. Consequently, some product companies are actively seeking to use BPA-free alternatives. Currently, the U.S. government is re-evaluating its policies regarding BPA.

The pro-BPA-safety viewpoint can be found, among other places, at: https://www.factsaboutbpa.org/

Meanwhile, the anti-BPA-safety viewpoint can be found, among other places, at Ref. [11]:

The FDA's current advice is found at Ref. [12]:

It is important to note that in the past, most of the concern (as it relates to packaging) was directed toward polycarbonate, rather than can linings, where potential exposure is generally much smaller. However, can linings are increasingly being mentioned as a concern.

On the polycarbonate side, the American Chemistry Council, which represents chemical manufacturers, actually called on the FDA to change the law to ban the use of BPA in baby bottles and children's sippy cups, and the FDA did so [13].

Companies like Sherwin-Williams and PPG aim for a new molecule they created to serve as a universal can liner, offering the benefits of BPA without its drawbacks [14]. Recently, PPG won the IP rights for can coating against Sherwin Williams [15].

Case Study: Kingston Aluminum Technology's Aluminum Bottles

While aluminum is widely used to manufacture beverage cans, it can also be used to produce bottles as well. From a sustainability perspective, aluminum is an intriguing rival to plastic bottles. While aluminum is more expensive than plastic, it is also lightweight and can be recycled for an infinite number of cycles, while in contrast, the recyclability of plastic has its limitations, especially when it is used for food contact applications [16].

It is also noteworthy that aluminum cans have the highest recycling rate among beverage containers [17]. Another advantage that aluminum has is that it is nonbreakable, in contrast with glass bottles. Aluminum packaging has been very popular in recent years, with over 70% of new beverage products being marketed in aluminum cans, with a shift toward aluminum and away from plastic bottles and other formats occurring in response to environmental considerations [17]. While soda and beer have traditionally been the main beverages offered in aluminum cans, other beverages are increasingly being offered in aluminum packaging, such as wine and even water. Aluminum has excellent oxygen barrier properties and offers protection against light, both of which can otherwise be detrimental to beverage shelf lives [17].

While one of the challenges with aluminum is that it has been more difficult to design aluminum packaging in unique shapes as readily as is the case with plastic [17, 18], Kingston Aluminum Technology has developed an approach that allows them to design custom-shaped aluminum bottles. These bottles, marketed under the name Alumishape™ can be produced in various sizes (185–355 mL) [19] and can be manufactured in both small and large production runs. These bottles can also be produced with roll-on pilfer-proof (ROPP) or crown closures [19], offering an additional level of flexibility to beverage producers. The ability for production to be viable at small runs provides a significant degree of flexibility, allowing them to accommodate both large and small companies as clients, including local or startup companies who may not have the budget for larger-scale production. A photograph showing various Alumishape™ bottles manufactured by Kingston Aluminum Technology is provided in Figure 6.12.

Kingston Aluminum Technology's patented blow molding process enables them to use a 66% smaller production line than is used for conventional aluminum bottle production, while also enabling them to produce a lighter bottle. During their production, 80% less volatile organic compounds are produced, and 60% less energy is consumed for the production of each bottle.

For their innovation, Kingston Aluminum Technology was awarded the "Own The Change" Special Award at SIAL Canada, North America's largest food innovation trade show, in 2023 [20].

Figure 6.12 Examples of Alumishape™ bottles manufactured by Kingston Aluminum Technology [20]. *Source:* Image reused with permission from Kingston Aluminum Technology.

6.14 Concluding Remarks

Given their versatility, metal containers are crucial in the packaging sector. They have excellent mechanical properties as well as barrier properties and can be readily formed into useful shapes. They are also relatively lightweight (especially in the case of aluminum) when compared to glass and are highly recyclable. In addition, they do not create microplastics. As noted in this chapter, they can be susceptible to corrosion, and a number of strategies have been developed to mitigate this issue. Careful consideration is often needed with regard to the potential environmental or health implications of these strategies. It is likely that more strategies will continue to develop as technology progresses and metal containers will continue to offer important tools for the packaging sector [21].

Problem Set for Chapter 6

1 What is the purpose of coating tin on steel?
- **A** To minimize/prevent steel dissolution.
- **B** To minimize/prevent tin dissolution.
- **C** Tin plating has no real function.
- **D** None of the above.

2 What is the role of water in corrosion?
  A Carrier for corrosive species such as dissolved oxygen or hydrogen ions.
  B Acts as a corrosive species/electrolyte (source of H^+ and OH^- ions and free radicals).
  C It generates nitrogen gas.
  D Both a and b.

3 Why does aluminum (Al) corrode slower than steel despite Al reacting faster with oxygen than is the case with steel?
  A The formation of a protective Al_2O_3 layer.
  B The formation of a protective Fe_2O_3 layer.
  C The formation of a protective FeS layer.
  D All of the above.

4 A polymer upon pyrolysis gives 90% of its parent monomer; which of the following answers are correct about this polymer?
  A It degrades primarily by random scission.
  B It degrades primarily by depolymerization (unzipping).
  C It degrades equally by both random scission and depolymerization.
  D None of the above.

5 If you need to pack a product with aggressive components, such as nitrate or sulfur, into a steel can, how would you approach this problem?

References

1 Zenz, R. https://commons.wikimedia.org/wiki/File:Konservendose-1.jpg. (This file is licensed under the Creative Commons Attribution-Share Alike 3.0 Unported license. https://creativecommons.org/licenses/by-sa/3.0/deed.en). (accessed 18 November 2024).
2 El Ibrahimi, B., Nardeli, J.V., and Guo, L. (2021). *An Overview of Corrosion*, 1–19. Washington, DC: American Chemical Society https://pubs.acs.org/doi/10.1021/bk-2021-1403.ch001.
3 Nagel, N. https://commons.wikimedia.org/wiki/File:Gew%C3%BCrzgurken_Auslese,_Casa_de_los_Winter_-_El_Cofete_-_Jandia_-_Fuerteventura.jpg. (This file is licensed under the Creative Commons Attribution-Share Alike 3.0 Unported license. https://creativecommons.org/licenses/by-sa/3.0/deed.en). (accessed 18 November 2024).
4 Auckland Museum. https://commons.wikimedia.org/wiki/File:Can,_food_(AM_2007.41.1-2).jpg. This image has been released as "CCBY" by Auckland Museum. (This file is licensed under the Creative Commons Attribution 4.0 International license. https://creativecommons.org/licenses/by/4.0/deed.en). (accessed 18 November 2024)

5 Overbey, D. Beware of Galvanic Action (It's a Thing). BE: Building Enclosure. https://www.buildingenclosureonline.com/blogs/14-the-be-blog/post/87725-beware-of-galvanic-action-its-a-thing (accessed 26 January 2024).

6 Guo, P., La Plante, E.C., Wang, B. et al. (2018). Direct observation of pitting corrosion evolutions on carbon steel surfaces at the nano-to-micro-scales. *Sci. Rep.* 8 (1): 7990.

7 Acar, E.T., Hinkle, P., and Siwy, Z.S. (2018). Concentration-polarization-induced precipitation and ionic current oscillations with tunable frequency. *J. Phys. Chem. C* 122 (6): 3648–3654.

8 Guide to Tinplate, ITRA Ltd. https://www.tinplategroup.com/wp-content/uploads/2019/08/Guide-toTinplate.pdf (accessed 26 January 2024).

9 Baker, H.R. (1954). Volatile rust inhibitors. *Ind. Eng. Chem.* 46 (12): 2592–2598.

10 Matsusaka, K., Osono, H., and Kino, T. (2000). Corrosion of ultra-high purity aluminum in hydrochloric acid. *Mater. Trans. JIM* 41 (1): 203–209.

11 Bisphenol A – Toxic plastics chemical in canned food, Environmental Working Group (2007). https://www.ewg.org/research/bisphenol-toxic-plastics-chemical-canned-food (accessed November 18 2024).

12 Bisphenol, A. (2024). Use in Food Contact Application. U.S. Food & Drug Administration. http://www.fda.gov/NewsEvents/PublicHealthFocus/ucm064437.htm (accessed 24 January 2024).

13 Tavernise, S. (2012). FDA makes it official: BPA can't be used in baby bottles and cups. *New York Times* https://www.nytimes.com/2012/07/18/science/fda-bans-bpa-from-baby-bottles-and-sippy-cups.html (accessed November 18 2024).

14 Cornwall, W. (2020). To replace controversial plastic additive BPA, a chemical company teams up with unlikely allies. *Science* 367 (6476): 380–383. https://doi.org/10.1126/science.367.6476.380.

15 Sherwin-Williams seeks new trial after beverage can IP loss, Law360. (2022). https://www.law360.com/articles/1481379 (accessed 24 January 2024).

16 Wiseman, R. (2020). Design for sustainability: making a case for metal cans, Inside Packaging, Issue 55. https://inside-packaging.nridigital.com/packaging_nov20/metal_cans_sustainability (accessed 26 March 2024).

17 Sabetta, L. (2022). Sustainable attributes drive adoption of aluminum cans from beverage-makers, Beverage Industry. https://www.bevindustry.com/articles/94799-sustainable-attributes-drive-adoption-of-aluminum-cans-from-beverage-makers (accessed 26 March 2024).

18 Metal Cans vs Plastic Bottles for Sustainable Packaging, Levapack. (2023). https://www.levapack.com/metal-cans-vs-plastic-bottles/ (accessed 26 March 2024).

19 Creative, Innovative and Sustainable, Kingston Aluminum Technology. https://katinc.ca/#sustainability (accessed 26 March 2024).

20 SIAL (2023) Innovation Winners, Nutrition for NON-Nutritionists (2023). https://nutritionfornonnutritionists.com/2023/06/sial-2023-innovation-winners/. (accessed 18 November 2024)

21 Jadhav, N. and Gelling, V.J. (2019). Review—The use of localized electrochemical techniques for corrosion studies. *J. Electrochem. Soc.* 166 (11): C3461–C3476.

Further Reading

Here are some links for additional reading about corrosion:

Nimmo, B. and Hinds, G. *Corrosion Basics*. Windsor, CA, http://resource.npl.co.uk/docs/science_technology/materials/life_management_of_materials/publications/online_guides/pdf/beginners_guide_to_corrosion.pdf: Testech (accessed 18 November 2024).

Jacobson, G. (2003. NACE International, https://www.nace.org/Corrosion-Basics/). Corrosion - A natural but controllable process. *AMPTIAC Q.* 7 (4): 39–45. (accessed 18 November 2024).

Corrosion Fundamentals. eFunda. http://www.efunda.com/materials/corrosion/corrosion_basics.cfm (accessed November 18 2024)

There are a number of other sources, as well.

BPA substitutes may be just as bad as the popular consumer plastic. https://www.science.org/content/article/bpa-substitutes-may-be-just-bad-popular-consumer-plastic (accessed November 18 2024)

7

Solid Waste Issues

7.1 Overview of Packaging Waste in U.S. Municipal Solid Waste

In this chapter, we will turn our attention to the end-of-life (EoL) issues associated with packaging and the environment. In other words, what happens to all the packaging material when we no longer need it to contain and protect products and do the other jobs associated with packaging functions?

To begin, we will take a broad look at municipal solid waste (MSW), with a more detailed look at the packaging component of the MSW stream. The U.S. EPA divides solid waste into three main categories:

1) Municipal solid waste,
2) Hazardous waste, and
3) Nonhazardous industrial waste.

7.1.1 Municipal Solid Waste

MSW includes residential and commercial waste, including office waste. Of particular importance for our purposes, it includes all waste packaging, regardless of how the packaging was used [1]. So, for example, the packaging used for parts delivered to an auto assembly plant is classified as MSW at the end of its lifetime. MSW also includes tires and car batteries, but not the cars themselves.

One of the major sources of information on U.S. MSW is the EPA, through a series of annual reports prepared for many years by Franklin Associates, a consulting firm. One of the most recent reports is titled *Advancing Sustainable Materials Management: 2018 Fact Sheet*, which was published in July 2020 [2]. It includes information on Municipal Solid Waste generation, recycling,

Degradation, Stabilization, and Recycling of Packaging Materials, First Edition.
Muhammad Rabnawaz, Susan E.M. Selke, and Ian Wyman.

composting, other food management, combustion with energy recovery, and landfilling [2]. Unfortunately, it appears that the EPA is no longer producing updates to these reports.

Another good source of information historically were the reports in ***BioCycle* magazine** titled *The State of Garbage in America* which reported on data from their every-other-year survey of state solid waste personnel. The most recent survey was published in October 2010, and contained data for 2008 [3]. One thing this survey provided that the EPA data did not was a look at state-to-state differences. Unfortunately, this survey has been discontinued, but the EPA does now report some state and regional (partial state) data. However, no discussion of state-to-state differences is included. *BioCycle* provides data which can be found at the following address. https://www.biocycle.net/category/biocycle-surveys/

The EPA breaks municipal solid waste down into the following categories:

1) Durable goods,
2) Nondurable goods,
3) Containers and packaging, and
4) Other wastes (food waste, yard trimmings, and miscellaneous inorganic wastes).

Durable goods are products that have an appreciably long lifetime, such as furniture and appliances. The EPA defines durable goods as products with a lifetime of three years or more, although there are some exceptions. In this EPA analysis, the durable goods category includes large and small appliances, furniture and furnishings, carpets and rugs, rubber tires, lead-acid automotive batteries, consumer electronics, and other miscellaneous durable goods such as luggage, sporting goods, and household goods.

Nondurable goods are products with a relatively short lifetime (less than three years), such as newspapers, magazines, batteries, and so forth. The **Department of Commerce** defines nondurable goods as products with a lifetime of less than three years [4]. The EPA followed this definition to the greatest extent possible for this analysis.

Containers and packaging: The EPA defines containers and packaging as products that are assumed to be discarded during the same year that the products they contain are purchased [5].

Other wastes. This category covers food waste, yard trimmings, and miscellaneous inorganic waste.

Some other notation used in EPA data: "generated" means, essentially, what becomes discarded into the waste stream and includes those things that become recycled. This category then can be divided into the "recovered" and "disposed" categories. The "recovered" category includes both materials that are recycled and materials that are composted (recycling + composting).

Tables 7.1 and 7.2 present details regarding the containers and packaging segment of MSW, which amounted to about one third of the total MSW stream for many decades but has been slowly declining (and in 2018 was **about 28%** of the total). The first table shows tonnage and the second shows the percentage of the total.

Table 7.1 Products generated in U.S. MSW with details regarding packaging, in thousands of tons [5].

	Thousands of tons									
	1960	1970	1980	1990	2000	2005	2010	2015	2017	2018
Durable goods	**9,920**	**14,660**	**21,800**	**29,810**	**38,870**	**45,060**	**49,350**	**53,940**	**56,870**	**57,100**
(Detail in Table 14 of Reference [6])										
Nondurable goods	17,330	25,060	34,420	52,170	64,010	63,650	53,250	51,810	50,700	50,440
(Detail in Table 18 of Reference [6])										
Containers and packaging										
Glass packaging										
Beer and soft drink bottles**	1,400	5,580	6,740	5,640	5,710	6,540	5,670	5,320	4,830	4,650
Wine and liquor bottles	1,080	1,900	2,450	2,030	1,910	1,630	1,700	1,810	1,800	1,810
Other bottles and jars	3,710	4,440	4,780	4,160	3,420	2,290	1,990	1,990	3,220	3,330
Total glass packaging	6,190	11,920	13,970	11,830	11,040	10,460	9,360	9,120	9,850	9,790
Steel packaging										
Beer and soft drink cans	640	1,570	520	150	Neg.	Neg.	Neg.	Neg.	Neg.	Neg.
Cans	3,760	3,540	2,850	2,540	2,630	2,130	2,300	1,740	1,480	1,580
Other steel packaging	260	270	240	200	240	240	440	480	530	630
Total steel packaging	4,660	5,380	3,610	2,890	2,870	2,370	2,740	2,220	2,010	2,210

(Continued)

Table 7.1 (Continued)

	Thousands of tons									
	1960	1970	1980	1990	2000	2005	2010	2015	2017	2018
Durable goods	**9,920**	**14,660**	**21,800**	**29,810**	**38,870**	**45,060**	**49,350**	**53,940**	**56,870**	**57,100**
Aluminum packaging										
Beer and soft drink cans	Neg.	100	850	1,550	1,520	1,450	1,370	1,220	1,330	1,330
Other cans	Neg.	60	40	20	50	80	70	130	50	80
Foil and closures	170	410	380	330	380	400	460	490	500	510
Total aluminum packaging	170	570	1,270	1,900	1,950	1,930	1,900	1,840	1,880	1,920
Paper and paperboard packaging										
Corrugated boxes	7,330	12,760	17,080	24,010	30,210	30,930	29,050	31,330	32,540	33,260
Other paper and paperboard packaging										
Gable top/aseptic cartons†			790	510	550	500	540	590	590	630
Folding cartons			3,820	4,300	5,820	5,530	5,470	5,380	5,330	5,370
Other paperboard packaging	3,840	4,830	230	290	200	160	90	70	50	50
Bags and sacks			3,380	2,440	1,490	1,120	1,040	930	1,100	1,090
Wrapping papers			200	110	Neg.	Neg.	Neg.	Neg.	Neg.	Neg.
Other paper packaging	2,940	3,810	850	1,020	1,670	1,400	1,490	1,620	1,450	1,500
Subtotal other paper and *paperboard packaging*	*6,780*	*8,640*	*9,270*	*8,670*	*9,730*	*8,710*	*8,630*	*8,590*	8,520	8,640
Total paper and board packaging	14,110	21,400	26,350	32,680	39,940	39,640	37,680	39,920	41,060	41,900

Plastics packaging										
PET bottles and jars			260	430	1,720	2,540	2,670	2,980	2,960	3,130
HDPE natural bottles			230	530	690	800	800	760	770	750
Other containers	60	910	890	1,430	1,740	1,420	1,830	1,940	1,990	1,990
Bags and sacks			390	940	1,650	1,640	770	—	—	—
Wraps			840	1,530	2,550	2,810	3,160	—	—	—
Subtotal bags, sacks and wraps			*1,230*	*2,470*	*4,200*	*4,450*	*3,930*	*4,130*	4,140	4,200
Other plastics packaging	60	1,180	790	2,040	2,840	3,210	4,450	4,870	4,630	4,460
Total plastics packaging	120	2,090	3,400	6,900	11,190	12,420	13,680	14,680	14,490	14,530
Other packaging										
Wood packaging	2,000	2,070	3,940	8,180	8,610	9,230	9,770	9,770	11,560	11,530
Other misc. packaging	120	130	130	150	240	280	340	370	350	340
Total containers and packaging	27,370	43,560	52,670	64,530	75,840	76,330	75,470	77,920	81,200	82,220
Total product wastes	54,620	83,280	108,890	146,510	178,720	185,040	178,070	183,670	188,770	189,760

(*Continued*)

Table 7.1 (Continued)

	Thousands of tons									
	1960	**1970**	**1980**	**1990**	**2000**	**2005**	**2010**	**2015**	**2017**	**2018**
Durable goods	**9,920**	**14,660**	**21,800**	**29,810**	**38,870**	**45,060**	**49,350**	**53,940**	**56,870**	**57,100**
Other wastes										
Food‡	12,200	12,800	13,000	23,860	30,700	32,930	35,740	39,730	40,670	63,130
Yard trimmings	20,000	23,200	27,500	35,000	30,530	32,070	33,400	34,720	35,180	35,400
Miscellaneous inorganic wastes	1,300	1,780	2,250	2,900	3,500	3,690	3,840	3,990	4,040	4,070
Total other wastes	33,500	37,780	42,750	61,760	64,730	68,690	72,980	78,440	79,890	102,600
Total MSW generated - weight	88,120	121,060	151,640	208,270	243,450	253,730	251,050	262,110	268,660	292,360

* Generation before materials are recycled, composted, managed by other food pathways, combusted with energy recovery or landfilled. Details may not add to totals due to rounding.

** Includes carbonated drinks and noncarbonated water, teas, flavored drinks and ready-to-drink alcoholic coolers and cocktails.

† Includes milk, juice and other products packaged in gable top cartons and liquid food aseptic cartons.

‡ In 2018, the food waste measurement methodology was expanded to include additional sources of generation and management pathways beyond composting, combustion with energy recovery and landfilling. Please see https://www.epa.gov/facts-and-figures-about-materials-waste-andrecycling/food-material-specific-data.

Neg., less than 5,000 tons or 0.05%; NA, not available;—, detailed data not available.

Text in **bold** font represents categories and totals within the table.

Text in *italics* font highlight the subcategories and subtotals within the table.

Source: [5] / U.S. Environmental Protection Agency / Public Domain.

Table 7.2 Products generated in U.S. MSW with details on packaging, in percent of total [7].

Products	Percent of total generation									
	1960	1970	1980	1990	2000	2005	2010	2015	2017	2018
Durable goods (%)	**11.3**	**12.1**	**14.4**	**14.3**	**16.0**	**17.8**	**19.7**	**20.6**	**21.2**	**19.5**
(Detail in Table 14 of Ref. [6])										
Nondurable goods (%)	**19.7**	**20.7**	**22.7**	**25.0**	**26.3**	**25.1**	**21.2**	**19.8**	**18.9**	**17.3**
(Detail in Table 18 of Ref. [6])										
Containers and packaging										
Glass packaging										
Beer and soft drink bottles** (%)	1.6	4.6	4.4	2.7	2.3	2.6	2.3	2.0	1.8	1.6
Wine and liquor bottles (%)	1.2	1.6	1.6	1.0	0.8	0.6	0.7	0.7	0.8	0.6
Other bottles and jars (%)	4.2	3.7	3.2	2.0	1.4	0.9	0.8	0.8	1.2	1.1
Total glass packaging (%)	**7.0**	**9.8**	**9.2**	**5.7**	**4.5**	**4.1**	**3.7**	**3.5**	**3.8**	**3.3**
Steel packaging										
Beer and soft drink cans (%)	0.7	1.3	0.3	0.1	Neg.	Neg.	Neg.	Neg.	Neg.	Neg.
Cans (%)	4.3	2.9	1.9	1.2	1.1	0.8	0.9	0.7	0.6	0.5
Other steel packaging (%)	0.3	0.2	0.2	0.1	0.1	0.1	0.2	0.2	0.2	0.2
Total steel packaging (%)	**5.3**	**4.4**	**2.4**	**1.4**	**1.2**	**0.9**	**1.1**	**0.9**	**0.7**	**0.8**
Aluminum packaging										
Beer and soft drink cans (%)	Neg.	0.1	0.6	0.7	0.6	0.6	0.5	0.5	0.5	0.5
Other cans (%)	Neg.	Neg.	Neg.	Neg.	Neg.	Neg.	0.03	0.04	0.02	0.03
Foil and closures (%)	0.2	0.3	0.3	0.2	0.2	0.2	0.2	0.2	0.2	0.2
Total aluminum packaging (%)	**0.2**	**0.5**	**0.8**	**0.9**	**0.8**	**0.8**	**0.8**	**0.7**	**0.8**	**0.7**

(Continued)

Table 7.2 (Continued)

Products	Percent of total generation									
	1960	1970	1980	1990	2000	2005	2010	2015	2017	2018
Paper and paperboard packaging										
Corrugated boxes (%)	8.3	10.5	11.3	11.5	12.4	12.2	11.6	12.0	12.1	11.4
Other paper and paperboard packaging										
Gable top/aseptic cartons[†] (%)			0.5	0.2	0.2	0.2	0.2	0.2	0.2	0.2
Folding cartons (%)			2.5	2.1	2.4	2.2	2.2	2.1	2.0	1.8
Other paperboard packaging (%)	4.4	4.0	0.2	0.1	0.1	0.1	0.0	Neg.	Neg.	Neg.
Bags and sacks (%)			2.2	1.2	0.6	0.4	0.4	0.4	0.4	0.4
Wrapping papers (%)			0.1	0.1	Neg.	Neg.	Neg.	Neg.	Neg.	Neg.
Other paper packaging (%)	3.3	3.1	0.6	0.5	0.7	0.6	0.6	0.6	0.5	0.5
Subtotal other paper and paperboard packaging (%)							*3.4*	*3.3*	*3.1*	*3.0*
Total paper and **board Packaging (%)**	**16.0**	**17.7**	**17.4**	**15.7**	**16.4**	**15.6**	**15.0**	**15.3**	**15.3**	**14.3**
Plastics packaging										
PET bottles and jars (%)			0.2	0.2	0.7	1.0	1.1	1.1	1.1	1.1
HDPE natural bottles (%)			0.2	0.3	0.3	0.3	0.3	0.3	0.3	0.3
Other containers (%)	0.1	0.8	0.6	0.7	0.7	0.6	0.7	0.7	0.7	0.7
Bags and sacks (%)			0.3	0.5	0.7	0.6	0.3	—	—	—
Wraps (%)			0.6	0.7	1.0	1.1	1.3	—	—	—

Subtotal bags, sacks and wraps (%)			*0.8*	*1.2*	*1.7*	*1.8*	*1.6*	*1.6*	*1.5*	1.4
Other plastics packaging (%)	0.1	1.0	0.5	1.0	1.2	1.3	1.8	1.8	1.7	1.5
Total plastics packaging (%)	**0.1**	**1.7**	**2.2**	**3.3**	**4.6**	**4.9**	**5.4**	**5.5**	**5.3**	**5.0**
Other packaging										
Wood packaging (%)	2.3	1.7	2.6	3.9	3.5	3.6	3.9	3.7	4.3	3.9
Other misc. packaging (%)	0.1	0.1	0.1	0.1	0.1	0.1	0.1	0.1	0.1	0.1
Total containers and packaging (%)	**31.1**	**36.0**	**34.7**	**31.0**	**31.2**	**30.1**	**30.1**	**29.7**	**30.2**	**28.1**
Total product wastes (%)	**62.0**	**68.8**	**71.8**	**70.3**	**73.4**	**72.9**	**70.9**	**70.1**	**70.3**	**64.9**
Other wastes										
Food[‡] (%)	13.8	10.6	8.6	11.5	12.6	13.0	14.2	15.2	15.1	21.6
Yard trimmings (%)	22.7	19.2	18.1	16.8	12.5	12.6	13.3	13.2	13.1	12.1
Miscellaneous inorganic wastes (%)	1.5	1.5	1.5	1.4	1.4	1.5	1.5	1.5	1.5	1.4
Total other wastes (%)	**38.0**	**31.2**	**28.2**	**29.7**	**26.6**	**27.1**	**29.1**	**29.9**	**29.7**	**35.1**
Total MSW generated (%)	**100.0**	**100.0**	**100.0**	**100.0**	**100.0**	**100.0**	**100.0**	**100.0**	**100.0**	**100.0**

* Generation before materials are recycled, composted, managed by other food pathways, combusted with energy recovery or landfilled. Details may not add to totals due to rounding.
** Includes carbonated drinks and noncarbonated water, teas, flavored drinks and ready-to-drink alcoholic coolers and cocktails.
[†] Includes milk, juice, and other products packaged in gable top cartons and liquid food aseptic cartons.
[‡] In 2018, the food waste measurement methodology was expanded to include additional sources of generation and management pathways beyond composting, combustion with energy recovery and landfilling. Please see https://www.epa.gov/facts-and-figures-about-materials-waste-andrecycling/food-material-specific-data
Neg., less than 5,000 tons or 0.05%; NA, not available; —, detailed data not available.
Text in **bold** font represents categories and totals within the table.
Text in *italics* font represents subcategories and subtotals within the table.
Source: [6] / U.S. Environmental Protection Agency / Public Domain.

You can see in Table 7.1 that the total amount of MSW generated in the United States in 2018 was 292,360 thousand tons (**292 million tons**). Of this, containers and packaging accounted for 82.22 million tons. (Note: it is not uncommon for data for previous years to be revised slightly when the next report comes out.)

Figures 7.1 and 7.2 show the 2018 breakdown by product category and by material. (Note that all data in this and other graphs is from this EPA report, unless indicated otherwise.

7.1.2 Products Generated in MSW, 2018

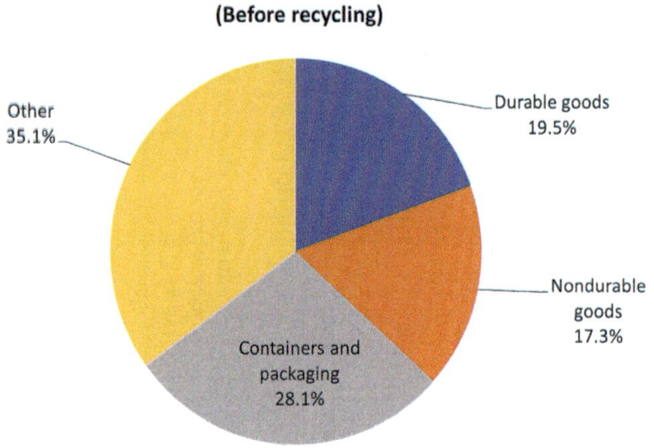

Figure 7.1 Products generated in MSW in the United States in 2018 U.S. Environmental Protection Agency / Public Domain.

7.1.3 Materials Generated in MSW, 2018

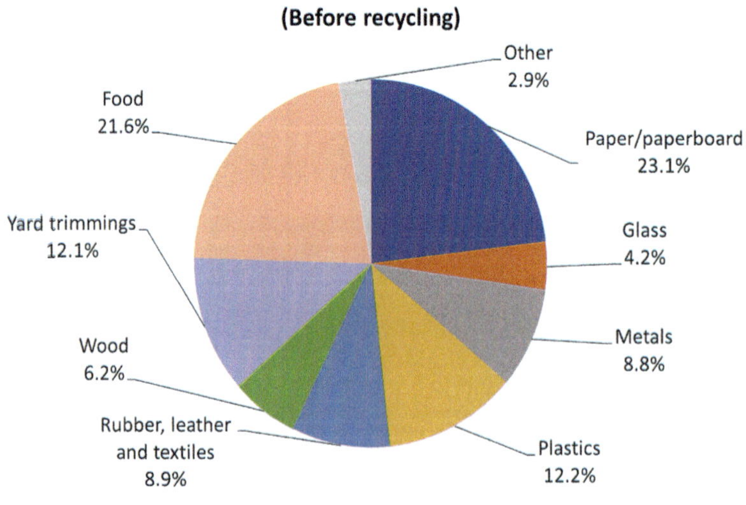

Figure 7.2 Materials generated in MSW in 2018 [10] / U.S. Environmental Protection Agency / Public Domain.

It is generally accepted that there needs to be an integrated approach to the management of MSW – that there is not a single "best" solution. This has become known as **integrated solid waste management** [8, 9]. Within this concept, it is also generally agreed that there is a hierarchy of solutions, implying that those at the top of the hierarchy have inherent benefits over those lower in the hierarchy. This is the EPA's *hierarchy of integrated solid waste management* and has been expressed in various ways at various times. The first generation was laid out like this:

- Reduction/Reuse
- Recycling
- Incineration
- Landfill

Reuse was sometimes rated slightly lower than reduction. Composting was added later on two lines, with backyard composting on a par with reduction and reuse, and organized composting (involving collection) on a par with recycling. Incineration always implicitly means incineration with energy recovery.

Later the EPA modified this approach to a three-category split between waste reduction (or prevention), recycling, and disposal. In the 2005 report, the hierarchy was modified again, to reflect energy recovery through combustion as being of higher value than landfilling or combustion without energy recovery. That version is stated as follows:

"EPA's integrated waste management hierarchy includes the following four components, listed in order of preference:

- Source reduction (or waste prevention), including reuse of products and on-site (or backyard) composting of yard trimmings
- Recycling, including off-site (or community) composting
- Combustion with energy recovery
- Disposal through landfilling or combustion without energy recovery."

In 2007, the phrase "or combustion without energy recovery" was removed from the hierarchy altogether. To date, subsequent reports have maintained this language, and the 2011 and 2013 reports express the idea graphically as shown in Figure 7.3, which displays a hierarchy of waste management options, ranging from most preferred to least preferred. Source reduction and reuse are the top strategies to prevent waste generation. Subsequent priorities, such as recycling, composting, energy recovery, and disposal, address waste after it has been generated. If both composting and recycling options exist for a particular solid waste, then it is better to recycle that item rather than to compost it, as recycling allows the materials to remain in the loop for the next cycle of usage.

Table 7.3 presents US waste management data from 1960 to 2018. It shows the amount of waste generated, recycled, composted, combusted with energy recovery, and sent to landfills. In 2018, a new category was added of "Other Food Management" including animal feed, donation, bio-based materials, and other

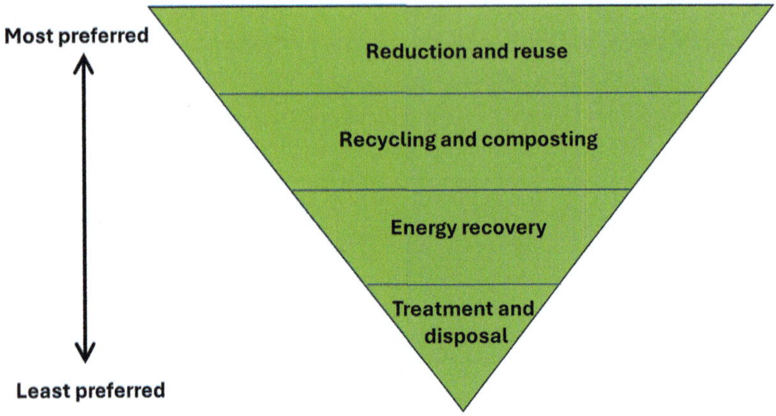

Figure 7.3 Solid waste management hierarchy, with better options shown at the top of this inverted triangle, and less desirable options at the bottom of this triangle [10].

options. Over the years, total waste generation has increased from 88.1 million tons in 1960 to 292.4 million tons in 2018. The recycling percentages have grown significantly, from 6.4% in 1960 to 23.6% in 2018. The data also suggests that landfilling has been significantly reduced over the past several decades.

Figure 7.4 suggests that total waste generation is continuously growing due to the increase in population, but per capita waste generation has remained fairly stable. This is because the United States is a developed nation and perhaps the standard of living has reached a plateau. For emerging economies, one should expect that their per capita waste generation is still growing. Note that per capita energy consumption is different and is still increasing, thanks to the popularity and widespread adoption of electronic devices.

About half of all MSW generated is disposed of by landfilling, though the percentage has declined, as shown in Figure 7.5. As shown in Figure 7.6, 50.0% of MSW was landfilled, 23.6% was recycled, 8.5% was composted, and 11.8% was incinerated for energy recovery in 2018. One should not confuse the recycling and composting of MSW with that of plastic because plastic recycling rates are much lower than those of MSW recycling.

Note that the amount of landfilled waste has held fairly steady since 1985, although the total amount of MSW has risen substantially. However, the number of landfills fell sharply between 1988 and 2002, as shown in Figure 7.7, from the 2010 EPA report. Numbers were relatively stable for a time, with a small increase occurring sometime between 2005 and 2009. The EPA did not report a number for 2011 or for 2013, instead repeating the 1,908 number from *BioCycle's* 2010 report, which provided data for 2008. In its 2015 report, the EPA said there were 1,738 active landfills, a further decline. In the 2018 report, EPA listed only 1,269 active

Table 7.3 Generation, recycling, composting, combustion with energy recovery, and disposal of MSW (in millions of tons).

Activity	1960	1970	1980	1990	2000	2005	2010	2015	2017	2018
Generation	88.1	121.1	151.6	208.3	243.5	253.7	251.1	262.1	268.7	292.4
Recycling	5.6	8.0	14.5	29.0	53.0	59.2	65.3	67.6	67.0	69.1
Composting*	neg.	neg.	neg.	4.2	16.5	20.6	20.2	23.4	27.0	24.9
Other food management**	—	—	—	—	—	—	—	—	—	17.7
Combustion with energy recovery†	0.0	0.5	2.8	29.8	33.7	31.7	29.3	33.5	34.2	34.6
Landfilling and other disposal‡	82.5	112.6	134.3	145.3	140.3	142.2	136.3	137.6	140.5	146.1

* Composting of yard trimmings, food and other MSW organic material. Does not include backyard composting.

** Other food management pathways include animal feed, bio-based materials/biochemical processing, codigestion/anaerobic digestion, donation, land application, and sewer/wastewater treatment. Details might not add to totals due to rounding.

† Includes combustion of MSW in mass burn or refuse-derived fuel form, and combustion with energy recovery of source separated materials in MSW (e.g., wood pallets, tire-derived fuel).

‡ Landfilling is what remains after recycling, composting, other food management and combustion with energy recovery are accounted for. Landfilling includes other disposal methods such as combustion without energy recovery.

neg. (negligible) = less than 5,000 tons or 0.05%. — A dash in the table means that data are not available.

Source: U.S. Environmental Protection Agency / Public Domain.

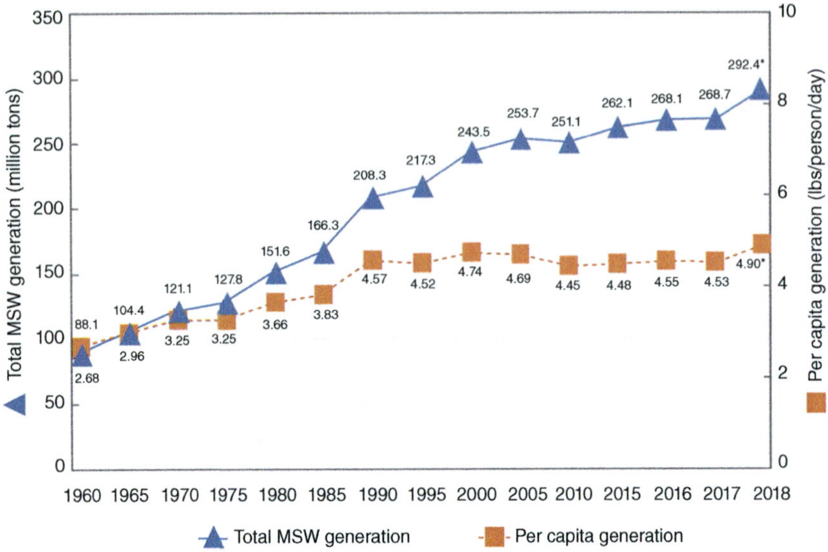

Figure 7.4 Total and per capita MSW generation rates. *The rise in MSW generation from 2017 to 2018 was mainly due to the fact that the EPA enhanced its food measurement methodology to provide a fuller account for all the ways by which wasted food is managed throughout the food system. [11] / U.S. Environmental Protection Agency / Public Domain.

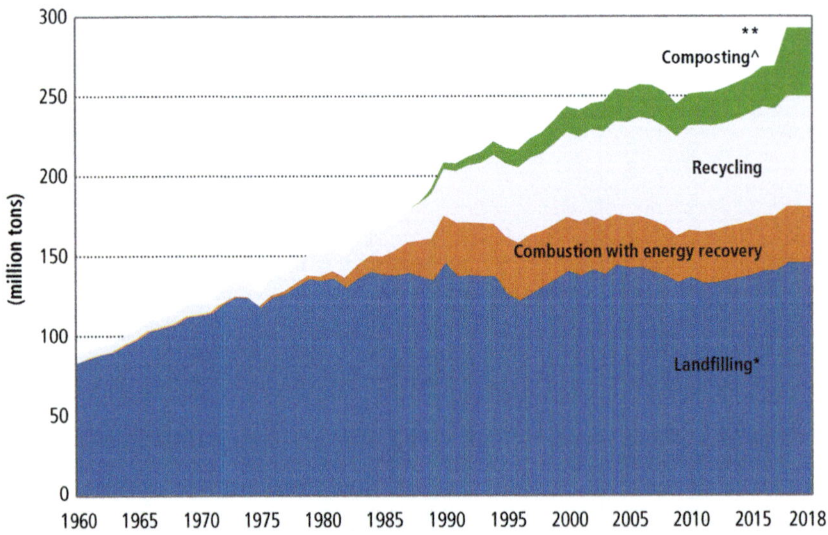

^ In this figure composting and other food management pathways are combined.

* Landfilling after composting, food waste management, recycling and combustion with energy recovery. Includes combustion without energy recovery. The top line measures generation, because generation = recycling + composting + combustion with energy recovery + landfilling.

** MSW generation rose considerably from 2017 to 2018 mainly because EPA enhanced its food measurement methodology to more fully account for all the ways wasted food is managed throughout the food system.

Figure 7.5 Recycling, composting, combustion, and landfilling of MSW from 1960 to 2010. U.S. Environmental Protection Agency / Public Domain.

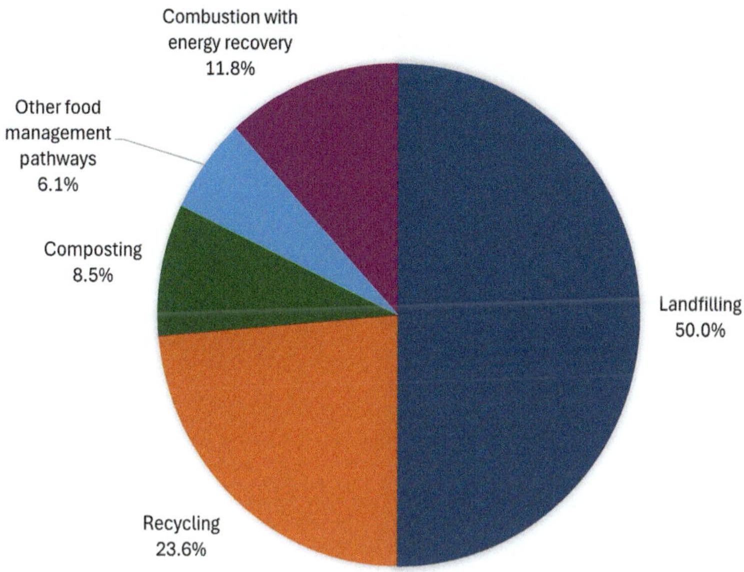

Figure 7.6 MSW disposition in 2018. [12] / U.S. Environmental Protection Agency / Public Domain.

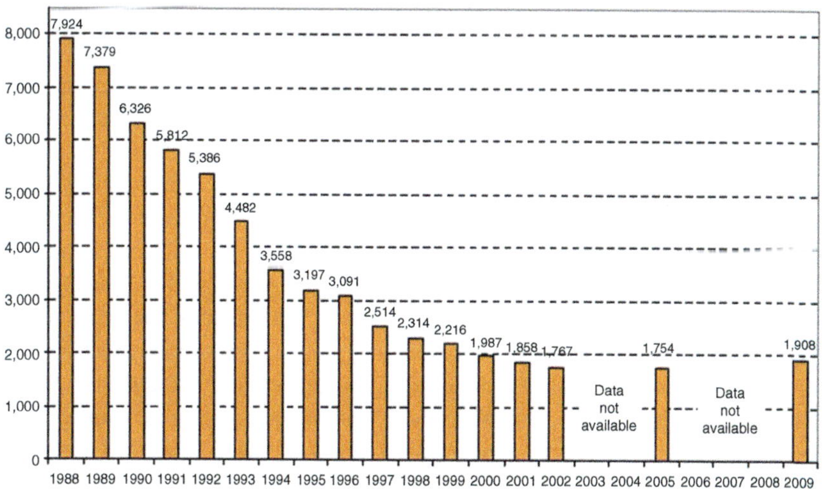

Figure 7.7 Number of landfills in the United States from 1988 to 2009. [13] / U.S. Environmental Protection Agency / Public Domain.

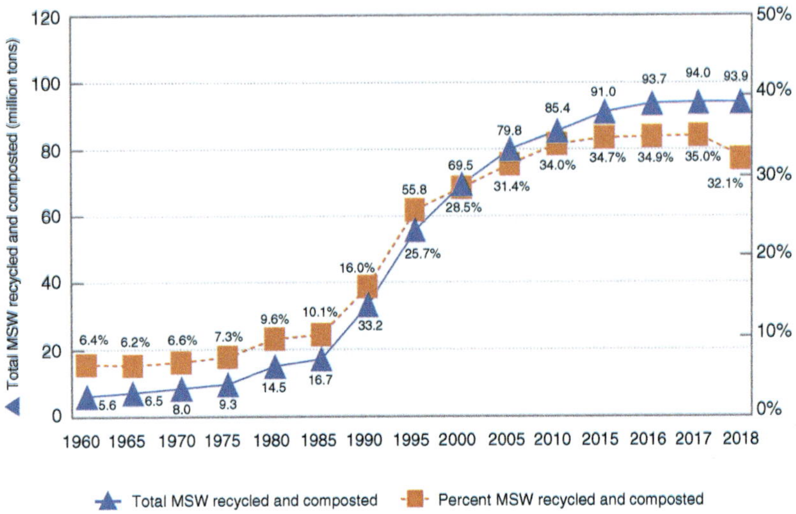

Figure 7.8 Recycling and composting of U.S. MSW between 1960 and 2018. U.S. Environmental Protection Agency / Public Domain.

landfills. Obviously, the **average size of current landfills** must be much larger than used to be the case, to accommodate roughly the same volume of waste.

Figure 7.8 shows the increase in recycling tonnage and recycling rates between 1960 and 2015.

Figure 7.9 displays a comparative analysis between the United States, European Union 27 (that is, the 27 countries which comprise the EU) and some selected countries within the EU such as Sweden, the Netherlands, Germany, Belgium, and Austria [14, 15]. If you are looking at the comparative analysis between the EU 27 and the United States, one of the major differences is in landfilling. The United States relies heavily on landfilling at 69%, while landfilling only accounted for 38% of end-of-life outcomes in the EU 27. When it comes to recycling and composting, Europe surpasses the United States with recycling and composting accounting for 40% of the end-of-life outcomes in the EU 27, compared to 24% in the United States. If you examine waste-to-energy conversion as a method of municipal solid waste treatment, you will see that Europe incinerates 22% of their waste compared to 7% in the United States. Reasons for these differences between the EU and the United States include: 1) The United States has more landfilling opportunities compared to the EU, so the EU relies more on incineration. 2) The EU has an extended producer responsibility (EPR) system [15] in place that requires companies to pay for the handling of their municipal solid waste. This is not the case in the United States. As a result, there are more than 30,000 material recovery facilities (MRFs) in the

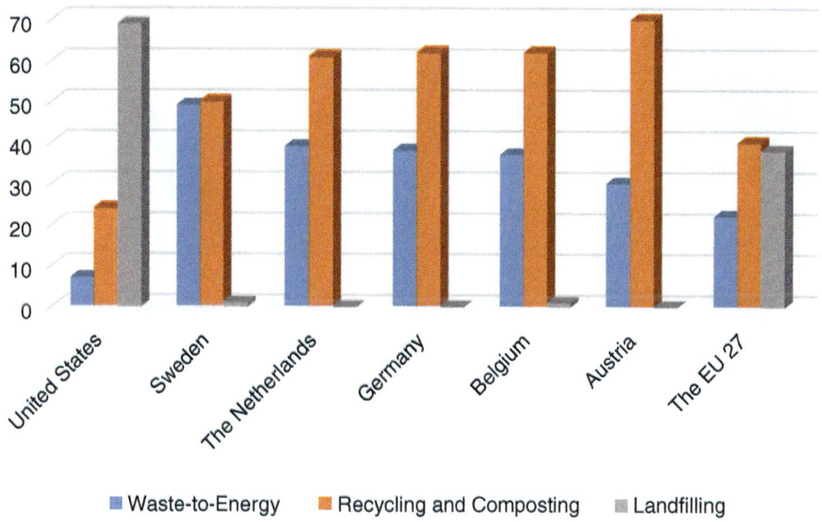

Figure 7.9 A comparative analysis of MSW EoL outcomes (by percentage) in the EU, United States, and some selected countries [14]. **Note:** *BioCycle* consistently reported somewhat higher landfill disposal rates than the EPA (e.g., a rate of 69.3% in 2008, compared to the EPA's figure of 54.3% for that year). It is likely that the data for the United States in the graph is taken from *BioCycle*.

EU compared to about 532 in the United States. The outcome is evident: the EU has much higher recycling rates compared to the United States.

7.2 Disposal of Packaging Materials

7.2.1 Landfills

As we have seen, landfills remain the most common way of handling MSW in the United States. However, the proportion of MSW being handled by landfilling has decreased over the last two decades or so, as incineration and recycling increased. Different states within the United States differ appreciably in the proportion of MSW that is landfilled, as is shown in Figure 7.10 [16].

While the above information is rather dated, as it is based on 2008 data, the pattern of state-to-state has not changed significantly. As mentioned in the last section, a good source on state and regional disposal and recycling information was *BioCycle's* biennial *State of Garbage in America* report, which has been discontinued. As noted earlier in this chapter, ***BioCycle* has consistently reported somewhat higher landfill disposal rates than the EPA,** due to differences in their methodologies.

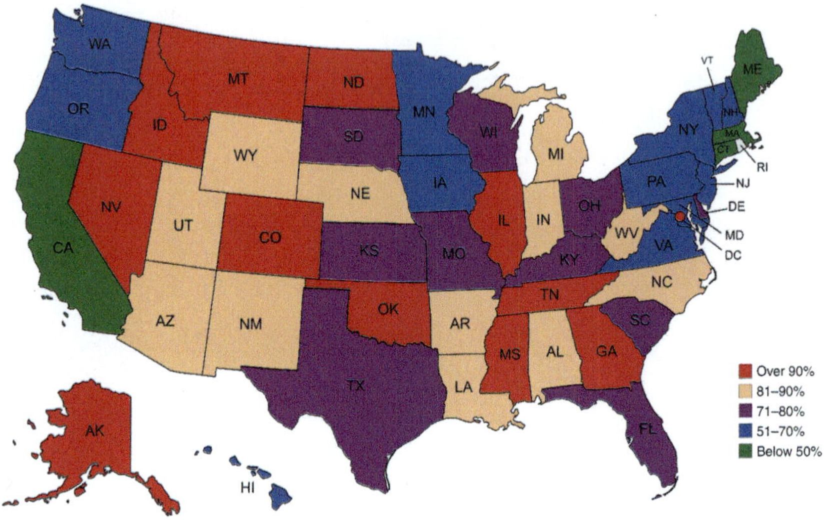

Figure 7.10 Landfill disposal rates in the United States. Based on 2008 data from the 2010 BioCycle "The State of Garbage in America" report. Map created using MapChart [Maps created with MapChart are licensed under a Creative Commons Attribution-ShareAlike 4.0 International License (CC BY-SA 4.0)].

The total number of landfills in the United States has decreased steadily between 1988 and about 2000 and is now down to 1,269, as we saw in the last section. However, most concern about overall landfill capacity has abated.

In general, five years is seen as the minimum acceptable capacity, as it takes at least that long to plan and construct new capacity. The reason that the landfill capacity did not decline substantially even though the number of landfills decreased is that the average size of the new landfills that are opening is much larger than the average size of the landfills that are closing.

An indirect measure of the decline in concern over landfill space availability can be found in the 2018 EPA report, which includes data on national average landfill tipping fees, adjusted to 2018 dollars, as shown in Figure 7.11. A fee paid by anyone disposing of waste at a landfill is referred to as the Landfill Tipping Fee. We can see that fees rose dramatically between 1986 and 1995, dropped between 1995 and about 2004, and have risen somewhat since then, but only to about the same rate as in 1995. The Environmental Research & Education Foundation (EREF) reported that the **national** unweighted tip[1] fee rose by 8% in 2021 [17].

1 The unweighted average is the average of all fees, with the fees of each reporting landfill being equally weighted. The weighted average is the average of all fees based on the relative amount of MSW managed at each reporting landfill.

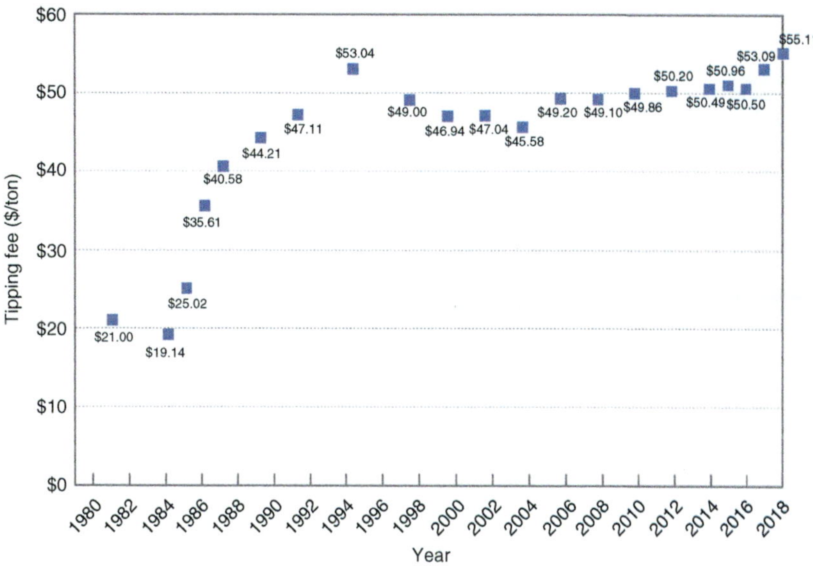

Figure 7.11 Average landfill tipping fees in the United States from 1980 to 2018.

More recently, EREF reported that the unweighted average tipping fee was $58.47 in 2023, which was 3% lower than the 2022 fee of $58.47 [18]. The landfill tipping fees reported in 2023 nationwide for the United States and different regions are shown in Figure 7.12.

Structure of a Modern Landfill: If you are interested, a description of the structure of a landfill can be found at the following link: http://people.howstuffworks.com/landfill6.htm

"Sanitary landfills" were a major improvement over the open dumps that they replaced. The structure of landfills has evolved over time, as regulators and landfill operators became more knowledgeable about the hazards presented by landfills and developed ways to reduce those hazards.

Modern landfill design incorporates two bottom liners; usually one is synthetic and one is comprised of clay, but sometimes both are synthetic, usually made from high-density polyethylene [19, 20]. There is also a moisture barrier layer at the top of the landfill, once the cell is complete and it is capped. Note that there are two different uses of the term **"cell."** A typical structure of a modern landfill is shown in Figure 7.13.

A landfill cell may be a day's accumulation of garbage, which is covered with a layer of dirt and then more garbage is eventually piled adjacent to it, and/or on top of it. **Another common meaning**, and the way we used the term in the first sentence of this paragraph, is a section of a large landfill facility that is relatively

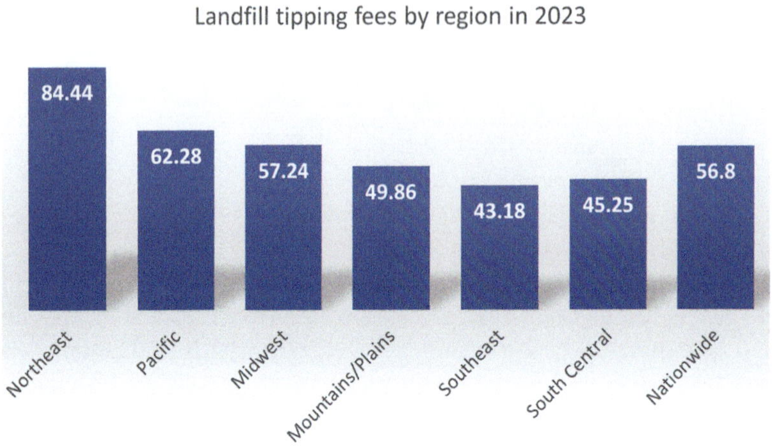

Figure 7.12 Average cost to landfill municipal solid waste in the United States in 2023 (in U.S. dollars per ton) [18].

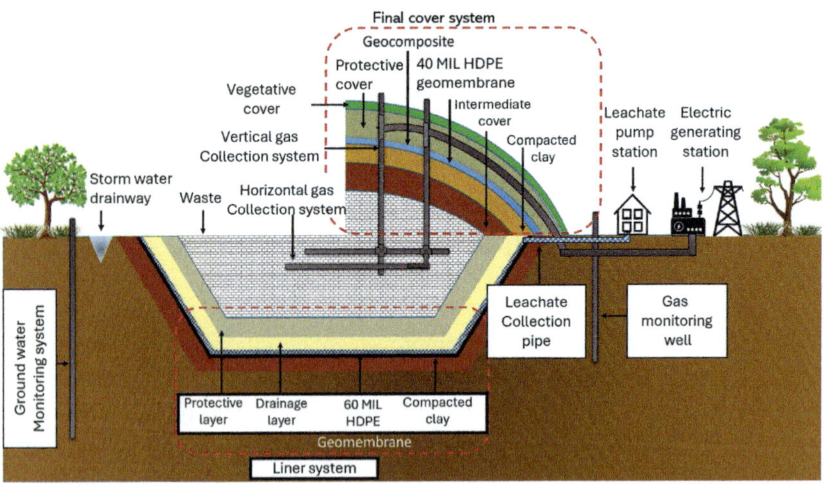

Figure 7.13 Illustration showing the structure of a modern landfill [20]. Reproduced from [20] / with permission of Elsevier.

self-contained, and that contains a large number of the small daily cells of garbage. This large cell usually contains a year's worth of garbage. At the end of the year, a new cell will become active, and the cell used the previous year will be capped.

A landfill cap includes a barrier layer to help keep water out, a layer of soil (usually on top of a drainage layer), and then vegetation. **Vegetation** has an important role as it stabilizes the soil cover and prevents erosion, while also improving the aesthetics of the area. There are limits on the kind of vegetation that is considered appropriate for this purpose. **Generally, anything with deep roots that might penetrate the cap is considered unsuitable for use as part of a barrier layer.**

Landfills must provide a system for the collection, removal, and treatment of liquids (water) that percolate into the landfill or are contained in the wastes when they are placed in the landfill. This liquid is called **leachate** [19, 20]. Landfills must also provide control over the methane that is generated in the landfill once it becomes anaerobic (without oxygen). Systems for **leachate and gas collection** are incorporated during the design and construction of modern landfills; in older landfills, these systems may need to be added afterwards. It is much more economical to design them into the landfill than to retrofit. Landfills are also required to provide periodic testing of the groundwater around the landfill, as an early detection system for pollutants.

By law, landfill operators must provide for the "postclosure" care of the landfill. The length of time varies with state law but is usually around **20 years.** During this time period, the owner of the landfill must maintain the land and continue the **leachate collection and methane collection or flaring** as well as **monitoring the groundwater near the landfill.**

In the past, it was not uncommon to build factories, or even residences on top of closed landfills. Now, with modern landfill design and a greater understanding of what does and does not occur inside landfills, this is usually not acceptable. It is desirable to get some kind of use out of the land, but buildings should be avoided. **Golf courses or ski hills are less problematic uses**. Even here, however, it is important to realize that there will be changes in the land over time. In particular, there will be subsidence (sinking), usually in an uneven manner, as decay processes proceed [21, 22].

In some areas, mostly in Florida but also to some extent in New York and elsewhere, **landfills have been "mined."** When a landfill is mined, it is dug up, the material it contains sorted out, and fresh MSW placed in the landfill [23]. The major goal of this task is usually to recover the valuable space occupied by the old, full landfill. When the waste is dug up, much of it has generally decayed and become a humus-like material, which can be used as daily cover for the fresh garbage. Usually there is some recovery of materials from the landfill, especially **ferrous metals** (which can readily be separated using an electromagnet) and aluminum. In some cases, materials such as paper and plastic are incinerated. **Landfill mining is still quite rare in the United States, as it is not cost-effective in most cases.**

You can read about landfill gas-to-energy systems, if you wish, at https://www.epa.gov/lmop/basic-information-about-landfill-gas. The map at https://www.epa.gov/lmop/project-and-landfill-data-state shows current and potential landfill gas-to-energy sites (as of July, 2023).

7.2.2 Incineration

There are significant differences in the reliance on MSW incineration in different states. What they all have in common, as mentioned earlier, is that these facilities are waste-to-energy facilities. Figure 7.14 shows combustion rates for MSW by state, in 1999. **A number of states have no MSW incineration facilities.** Incineration is heavily concentrated in the New England (comprised of Connecticut, Maine, Massachusetts, New Hampshire, Rhode Island, and Vermont) and Mid-Atlantic regions, which have high population densities. On the other hand, these facilities are almost nonexistent west of the Mississippi (except in Hawaii), where landfill space is more abundant. For similar reasons, **incineration is much more prevalent in Western Europe** than it is in the United States.

The number of incineration facilities in the United States has fallen substantially from **102 in 2000 to only 80 in 2013, and further to 75 in 2018.** The EPA reports 37 operational facilities in the Northeast, 19 in the South, 14 in the Midwest, and only 5 in the West. In terms of capacity per population, the Northeast is by far the highest, at nearly **800 tons/day per million persons**; the South had roughly 280, the Midwest about 175, and the West about 100, according to the 2013 report. Subsequent reports did not update those values but provided total

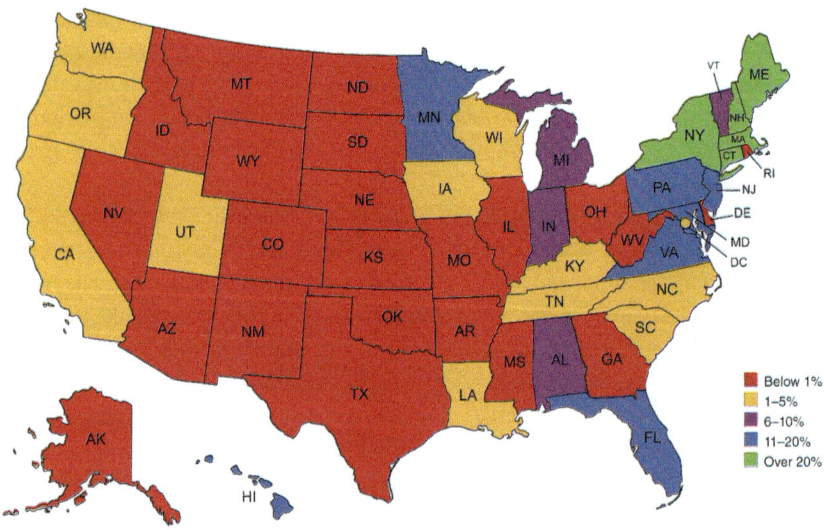

Figure 7.14 Waste to energy incinerations in the United States. Based on 2008 data from the 2010 BioCycle "The State of Garbage in America" report. Map created using MapChart (mapchart.net, Maps created with MapChart are licensed under a Creative Commons Attribution-ShareAlike 4.0 International License (CC BY-SA 4.0).

design capacity values of 44,807 tons per day for the Northeast, 32,194 for the South, 11,524 for the Midwest, and 6,530 for the West in 2018.

7.2.2.1 So, Which Is More Expensive, Landfilling or Incineration?

Generally, incineration is significantly more expensive to operate than landfilling. Residues from incineration occupy much less volume than un-combusted waste, but there are residues that must be disposed of. In many cases, the incinerator ash will contain a greater concentration of **heavy metals** which are not permitted in ordinary non-hazardous waste facilities. The recommended practice is to dispose of incinerator ash in "**ash monofills**" – landfill cells that are designed to contain only incinerator ash. One of the motivations behind the **Model Toxics Law**, which we will discuss in the next section, was to reduce the heavy metals going into MSW incinerators, as the only realistic way to decrease the heavy metal content in the ash. (For understandable reasons, the alternative, not controlling emissions and allowing the heavy metals in the fly-ash to be spewed out all over the neighborhood, was not seen as an acceptable way to reduce their presence in the ash!)

Incinerator ash can be categorized as bottom ash, which is the unburned or noncombustible residue from the incineration process, and **fly-ash**, which is the residue removed from the air stream by the emission control equipment [24, 25]. **Most of the heavy metals are in the fly-ash, rather than the bottom ash.** If fly-ash and bottom ash are kept separate, rather than combined as is the usual practice, it is sometimes possible to find markets for the bottom ash. These are typically low-value markets associated with building and construction, such as using ash as part of the aggregate in manufacture of cement or asphalt [26].

It should be noted that it is reasonably common practice to recover ferrous metals from incinerators, using electromagnets to pull out these magnetic materials. This can be done either before or after combustion. In some incineration facilities, there is also recovery of some other materials, often through hand-picking of selected loads to pull out large items such as appliances, or special materials such as wood and metal from building and construction debris or corrugated boxes from MSW originating in retail-heavy sectors such as shopping malls.

One of the newer developments in waste-to-energy incineration is the growing (though still small) **amount of combustion of source-separated MSW**. The most common is use of rubber tires as fuel in cement kilns, pulp and paper mills, industrial boilers of various types, and even dedicated tire-to-energy facilities. Wood waste and other wastes are also used in boilers that burn other types of solid fuel. Michigan State University (MSU) has used wastes along with coal as fuels for its campus power plant (although this practice has been discontinued since the use of coal on campus has been phased out).

7.2.3 Heavy Metals in Packaging – CONEG Model Toxics Law

7.2.3.1 Heavy Metals

One of the major concerns about MSW incineration is the **emissions** that result. In particular, the presence of **heavy metals in the ash** can lead to it being classified as hazardous waste, requiring much more expensive disposal. In an effort to reduce the amounts of heavy metals in incinerators, and as a by-product in landfills, regulators have then directed their attention toward the sources of these heavy metals so that efforts could be made to remove these items from the waste stream. **There are four heavy metals that were identified as of particular concern:**

1) Lead,
2) Cadmium,
3) Mercury, and
4) Hexavalent chromium (chromium with a +6 ionic charge).

One obvious question that must be addressed, in any effort to eliminate these metals from incinerator ash, is: What materials in MSW contain these metals?

7.2.3.2 Lead

In the 1980s when concerns about heavy metals in incinerator ash first surfaced, by far the largest source of **lead in MSW was automobile batteries.** As a consequence of this and other concerns about lead exposure, most states have banned the disposal of such batteries. In most states, all automobile batteries carry a hefty deposit fee as a way to ensure that they do not enter the MSW stream. **Other sources of lead included pigments, inks, and stabilizers in some plastics**. It is noteworthy that electric car batteries, especially those used in modern electric vehicles (EVs), rely on nonlead battery technologies. The most common type of battery used in electric cars is the lithium-ion (Li-ion) battery due to its high energy density, long cycle life, and declining cost [27]. There are companies in the United States that are recycling Li-Ion batteries.

7.2.3.3 Cadmium

A major use of cadmium is in nickel-cadmium (NiCd) rechargeable batteries. Like lead, cadmium was also found in pigments and inks.

7.2.3.4 Mercury

One major source of mercury was alkaline batteries, as mercury was used to prevent hydrogen buildup and explosion. The use of mercury in these batteries was discontinued in 1996, in response to concerns about mercury exposure, in part because of MSW incineration. The use of mercury as a preservative in outdoor paints has also been discontinued. As you have probably heard, mercury is also being phased out of thermometers and similar devices [28, 29]. Mercury has also been used in some types of switches, for example in automobiles. Most of these uses are disappearing, as well. There was never any significant use of mercury in packaging applications.

7.2.3.5 Hexavalent Chromium

Chromium is quite toxic in its +6 oxidation state, but not a significant concern as an un-ionized metal or as Cr^{3+}. Hexavalent chromium has also been used in **inks and pigments** as well as in plastics additives.

Figures 7.15–7.17 show sources of lead, cadmium, and mercury in U.S. MSW as of 2000.

7.2.3.6 Heavy Metals in Packaging

In addition to efforts to reduce some of the major sources of heavy metals in MSW, regulators and environmental groups have realized packaging was a significant, although certainly smaller source of heavy metals in MSW. It was widely perceived that, for the most part, there was no real need for the use of these toxic heavy metals in packaging. In the 1980s, **the Coalition of Northeastern Governors (CONEG)** put together a Packaging Task Force comprised of representatives of the legislative/regulatory community, of industry, and of environmental/citizens' groups. The **Task Force recommended that the use of these heavy metals in packaging be banned, except for unusual circumstances (lead shielding for shipments of radioactive material, for**

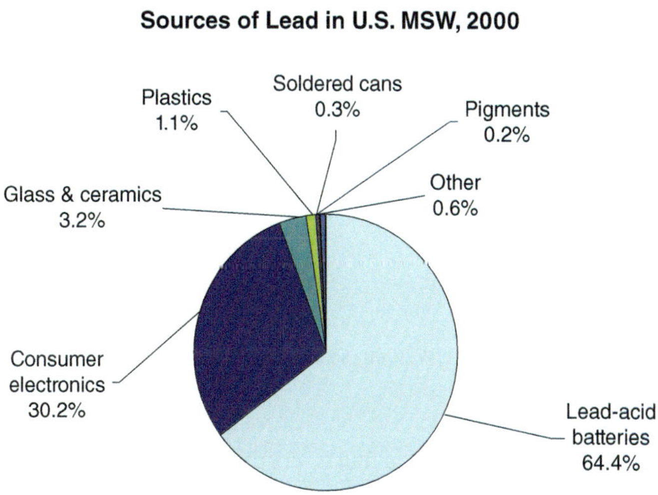

Figure 7.15 Sources of lead in U.S. MSW, 2000 (*Source:* U.S. Environmental Protection Agency / Public Domain).

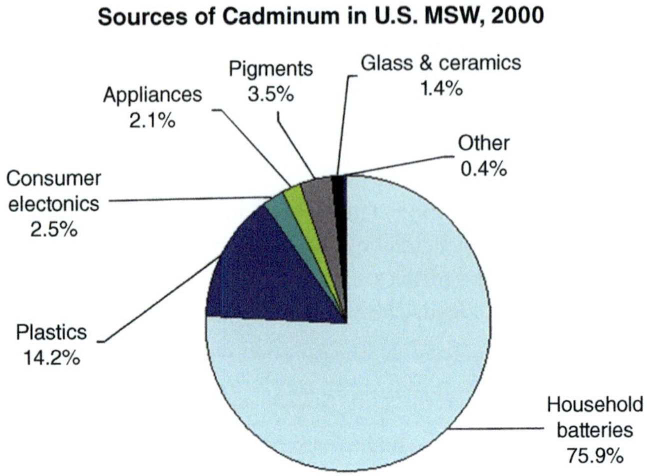

Figure 7.16 Sources of cadmium in U.S. MSW, 2000 (*Source:* U.S. Environmental Protection Agency / Public Domain).

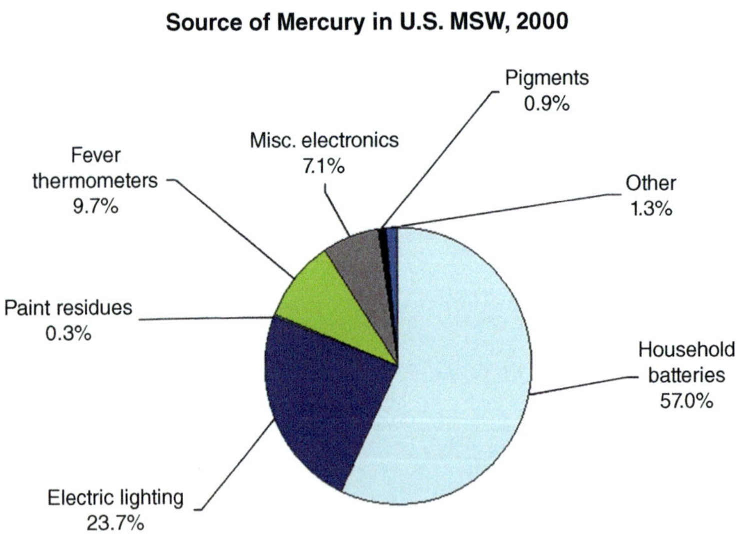

Figure 7.17 Sources of mercury in U.S. MSW, 2000 (*Source:* U.S. Environmental Protection Agency / Public Domain).

example). Recognizing that industry would find it very difficult to deal with multiple state laws that might conflict in important areas, the Task Force created what became known as the "**Model Toxics" law**. It consisted of a fill-in-the-blank template, designed to be easy for interested legislators in various states to use, simply requiring them to fill in the appropriate state-specific information such as the state agency in charge, the effective date, etc. That way, industry could have essentially one law with which to comply. **The 2012 revision of the Model Toxics law is available at** http://toxicsinpackaging.org/model-legislation/model/.

A fact sheet about Toxics in Packaging legislation can be found here: http://toxicsinpackaging.org/model-legislation/fact-sheet/.

At the time of this writing, **19 states have now ratified the Model Toxics law**; this number has not changed in several years. These states represent a significant fraction of the U.S. marketplace. Since it is rarely effective for manufacturers to package goods differently for sale in different states, and since the law applies to all packaged products sold in the state, rather than to those manufactured there, this quickly became de facto U.S. policy. In fact, since it applies equally to goods imported from outside the United States, **it has had a profound impact on packaging decisions in many areas around the world**. A number of other countries have now adopted, or are considering adoption, of very similar legislation. For example, the **EU used it as the basis for its limitations on heavy metals in packaging** [30].

Note that the Model Toxics laws ban all deliberate introduction of the target heavy metals ("No intentional introduction of any amount of the four metals is allowed."). **The 100 ppm by weight in the "Incidental Presence Concentration Limits"** section pertains only to the inadvertent presence of these metals [31]. In other words, if you would like to use cadmium-containing ink on the label for your package, because that is the only way you can get the precise color you want, it doesn't matter how little cadmium you use – any such use is totally prohibited.

These laws have resulted in widespread reformulation of inks and pigments. For the most part, these changes were invisible to consumers – but not to the manufacturers, as the replacements were usually more expensive.

Another change would probably have occurred even without these laws. Lead used to be a major ingredient in the solder used to produce the side seams in cans. **Because lead-based solder had been recognized as sometimes leading to small amounts of lead contamination in food, the FDA was already working toward the total elimination of lead in food packaging;** the Model Toxics law speeded that up and immediately expanded it to all cans [32]. The immediate solution was to substitute tin-based solder for lead. **The current situation is that soldered cans have been essentially entirely eliminated from the U.S. market, being replaced by welded cans, and by two-piece cans that have no side seams.**

Table 7.4 lists the results of advancements in Waste to Energy (WTE) pollution control technology that have significantly reduced the release of common and hazardous toxins from trash burners [33, 34]. In addition to

Table 7.4 Emission pollution control during WTE between 1990 and 2005 (Before and after the introduction of the Maximum Achievable Control Technology Standards).

Emission	1990 Emissions (tons/year)	2005 Emissions (tons/year)	Percent reduction (%)
Mercury	57	2.3	96
Cadmium	9.6	0.4	96
Lead	170	5.5	97
Particulate matter	18.600	780	96
Hydrogen chloride	57,400	3,200	94
Sulfur dioxide	38,300	4,600	88
Nitrogen oxides	64,900	49,500	24

Source: [33] / U.S. Environmental Protection Agency / Public Domain.

growing developments over time, it is also noteworthy that Maximum Achievable Control Technology (MACT) standards for MSW combustion facilities as part of the Clean Air Act.

7.3 Recovery

In the United States, the term recovery refers to both the **composting and recycling** of MSW. We will discuss composting in this section, and recycling in the next section. Please note that in **Europe and some other places,** the term "recovery" includes the recovery of energy from waste-to-energy incineration [35]. We will not use the word in that way, in this book.

7.3.1 Composting

When the EPA reports on composting, they generally include only "organized" composting and do not count "**back yard composting**" in which individuals divert certain materials from entering the waste stream.

By far, most composting of MSW is limited to **yard waste**. Some of these programs **accept leaves only; others include leaves, grass, tree trimmings, and other yard waste**. Figure 7.18 shows the number of yard waste composting programs in various parts of the United States in 2018 (The most recent data in that report was from 2017).

One of the prime movers behind the institution of yard waste composting programs was the decision in many states (25 in 2008, according to *BioCycle*)

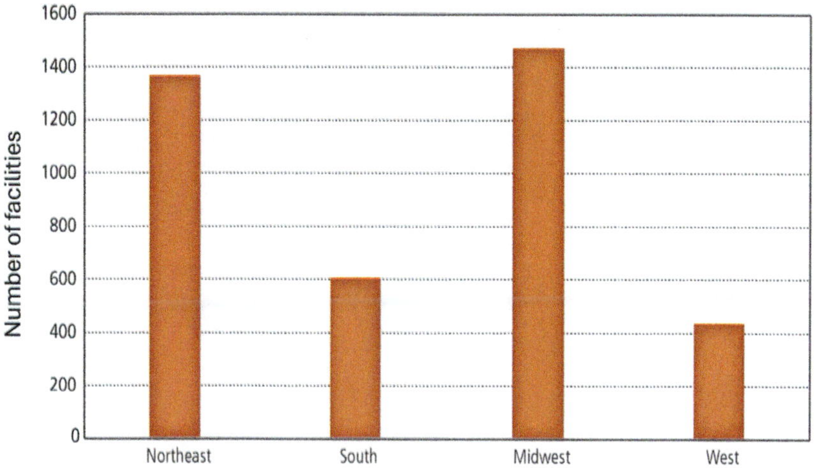

Source: Goldstein, N. "The State of Organics" BioCycle, October 2017. United States composting facilities data reported for 2015-2017. Facilities composting yard trimmings, yard trimmings and food waste, and mixed organics. Excludes 740 facilities composting manure, biosolids, mixed MSW, or not defined.

Northeast: Connecticut, Maine, Massachusetts, New Hampshire, New Jersey, New York, Pennsylvania, Rhode Island, Vermont

South: Alabama, Arkansas, Delaware, District of Columbia, Florida, Georgia, Kentucky, Louisiana, Maryland, Mississippi, North Carolina, Oklahoma, South Carolina, Tennessee, Texas, Virginia, West Virginia

Midwest: Illinois, Indiana, Iowa, Kansas, Michigan, Minnesota, Missouri, Nebraska, North Dakota, Ohio, South Dakota, Wisconsin

West: Alaska, Arizona, California, Colorado, Hawaii, Idaho, Montana, Nevada, New Mexico, Oregon, Utah, Washington, Wyoming

Figure 7.18 Yard waste composting programs in various regions of the United States, 2018 [36]. U.S. Environmental Protection Agency / Public Domain.

to prohibit disposal of yard waste in landfills and incinerators. In such states, it is illegal to **send yard waste to a landfill or to an incinerator.** Therefore, if any disposal option for such waste is to be provided to residents, it will be a composting program. Figure 7.19 shows, by the shaded areas, the states with yard waste landfill/incineration bans in 1999. The current list is much the same. (In 2013, the state of Michigan in the USA considered legislation that would have lifted the ban on landfilling of yard waste, using the argument that increased methane generation that could be used for energy would result, but the effort failed.)

In the United States, there is very limited composting of other components of the MSW stream. According to the EPA, about **90% of the total MSW composted was yard waste**. The remaining **10% was food and other organic waste**, as will be discussed next. The composting rate for yard trimmings was 63%, while for food waste, it was 4.1%. The EPA's new category of "other food management pathways" accounted for 28.1% of food waste.

In Europe, it is quite common for "source separated organics" to be collected separately from recyclables and other wastes, for diversion to a composting

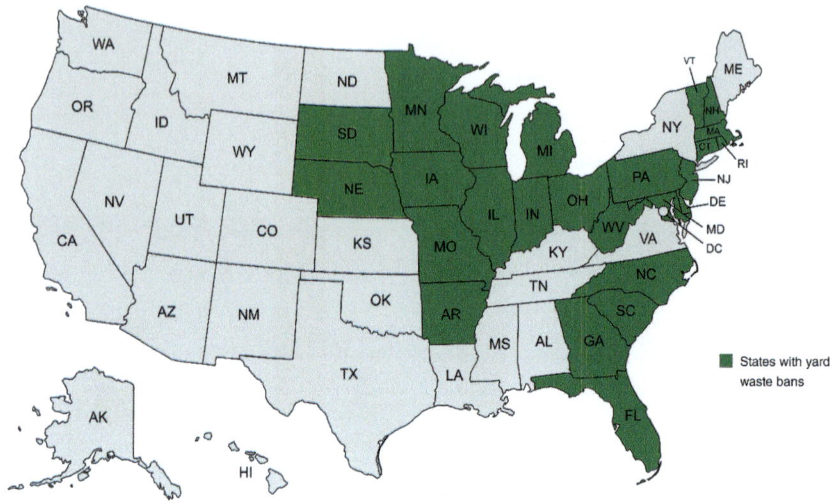

Figure 7.19 States with bans on the landfilling or incineration of yard waste. Based on 2008 data from the 2010 BioCycle "The State of Garbage in America" report. Map created using MapChart (mapchart.net, Maps created with MapChart are licensed under a Creative Commons Attribution-ShareAlike 4.0 International License (CC BY-SA 4.0).

program. Such systems have made some inroads in the United States. ***BioCycle* reported that there were 87 mixed organics recycling programs in the United States in 2014, along with 347 facilities for food waste** [36]. The 2016 EPA report stated that "**food composting collection programs served over 2.8 million households in 2014.**" [37] This seems quite impressive until you learn that this accounted for only 2.4% of U.S. households. However, there has been significant growth in composting of food waste and related products since that time. The most recent BioCycle report, released in 2023, found 410 municipally-supported programs for residential food waste collection, serving 710 communities in 25 states, for a total of 14.9 million households. Some of these programs include certified compostable packaging in the materials that are accepted. The programs differ in type, with some being curbside, some drop-off, and some both. Of the curbside programs, some are year-round and others are seasonal. Much more detail is provided in the report.

By *BioCycle's* January 2019 report, the number of full-scale food waste composting facilities in the United States in 2018 had grown to 185 (N. Goldstein, "Food Waste Composting Infrastructure in the United States," *BioCycle*, Jan. 2019, pp. 23–25) [38]. **The number of U.S. households with access to curbside or drop-off residential food waste collection programs had grown to 3.5% and 4.6%, respectively**. (Note: these numbers cannot be added together, since some households have access to both types of programs.)

The EPA reports that in 2013, there were 12 mixed waste composting facilities in operation, the same number as in 2009, handling about 1,400 tons per day, up from 1,100 tons per day in 2009. Earlier EPA reports said that in 2007 there were 16 such facilities, handling about 1,500 tons per day. The highest processing capacity per person was in the West, followed by the Midwest and Northeast, with the South lagging behind. The 2014 *BioCycle* report mentioned earlier (State of Composting in the United States, B. Platt and N. Goldstein, *BioCycle*, Jul. 2014, pp. 19–27) listed 11 such programs – 2 each in California and Massachusetts and 1 each in Minnesota, Montana, New York, Ohio, South Dakota, Tennessee, and Wisconsin. For more updated information on composting, the readers are referred to *BioCycle* (https://www.biocycle.net/category/composting/) and the United States. Composting Council (https://www.compostingcouncil.org/).

Figures 7.20 and 7.21 compare recycling and composting amounts as well as their growth over time. Note that growth in both composting and recycling has been slow since 2000, in terms of either amounts or rates.

It should be noted that when the EPA issues later reports, it is not uncommon for them to revise the numbers somewhat for earlier years. They also stop reporting values for some of the intermediate years. In a compilation such as the graphs above, in some cases values were revised for certain years in the most recent report, but no values at all were reported for the years surrounding those revised values. That may make it appear that there was an increase or decrease in that particular year, when such may not truly be the case. The changes are small, so they do not affect the overall picture presented.

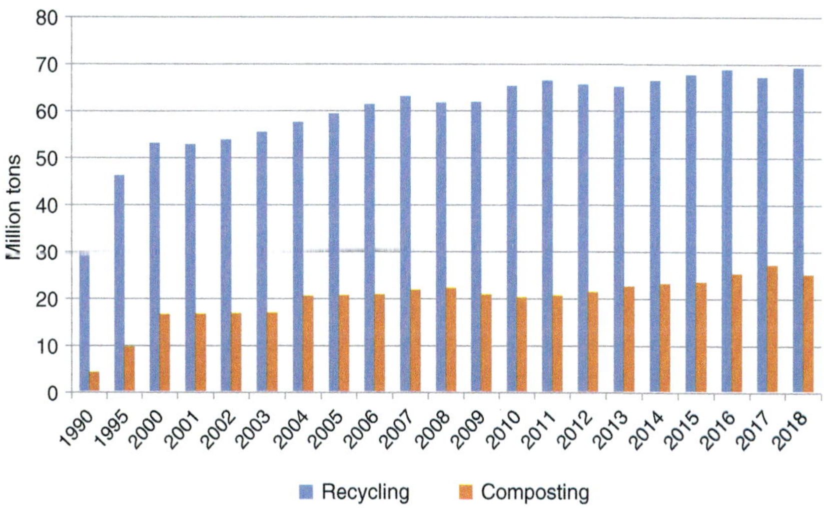

Figure 7.20 Recycling and composting amounts (Data from the EPA report) [6].

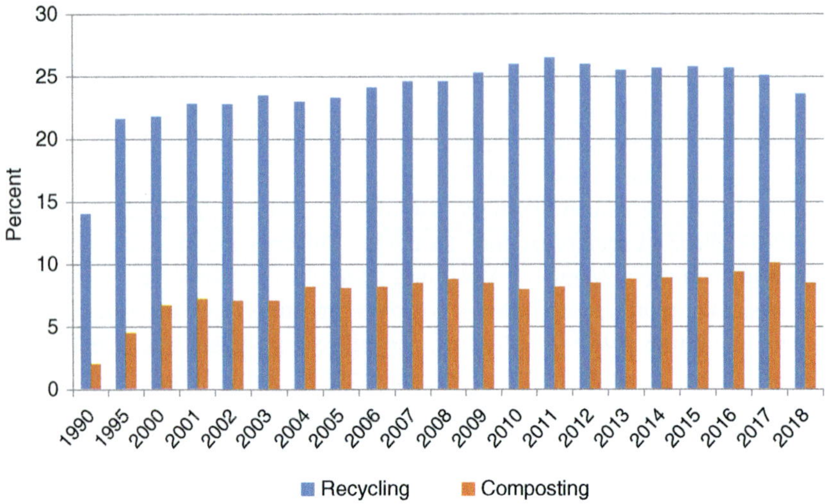

Figure 7.21 Recycling and composting percentages.

7.3.2 Composting Process

Composting is a managed biodegradation process. While **yard waste composting** is typically carried out in outdoor plots, organized (as opposed to backyard) **MSW composting is almost exclusively performed indoors**. Often at least the first part of the process is carried out in containers. The main reason for this is due to the odor that is often generated during this process. Even yard waste composting can create objectionable odors. **The likelihood of odor problems increases greatly if food and other wastes are included.** Since it is very difficult to control odors in outdoor facilities, **this is a major reason for doing the composting indoors**. Of course, problems associated with blowing of materials in the wind, attraction of vermin and birds, and related issues are all increased when food waste is included and when the composting is performed in outdoor facilities.

While, as mentioned, some MSW composting facilities start with **mixed MSW** and separate out a **compostable organics fraction** as well as **recyclables** [39], most facilities rely on pre-separation, or source separation, of the **compostable organics from the recyclables and the rest of the trash.**

One of the most well-known municipal composting programs for mixed organic wastes is the program in San Francisco. In 2009, San Francisco became the first U.S. community to require that **residents separate compostable materials from recyclables and trash.** The program was begun on a pilot scale but was then expanded to encompass the whole city. Residents place their compostable wastes in a separate bin, which is collected at curbside along with garbage and recyclables. An early report on their system and how it was working can be found in "San Francisco Update: 63 Percent Diversion and Rising," *BioCycle*, Feb. 2005, p. 22–29. Each household has three 32-gallon color-coded carts for their waste: black for garbage destined for landfill, blue for mixed fiber and container recyclables, and green for food and yard waste and other compostables. Current information on San Francisco's focus on Zero Waste can be found at Ref. [40].

Collection of source-separated organics for composting has been growing rapidly. In 2007, 42 counties or municipalities with collection systems for source-separated organics were identified [41], up from 30 in 2006. Of these, 17 were located in California and 17 in King County, Washington. The July 2014 *BioCycle* report [42] stated that more than 180 communities had collection programs for residential food wastes, in addition to the growing number of supermarkets, businesses, schools, and other facilities that separate these materials for composting. As mentioned earlier, the January 2019 report from *BioCycle* identified 185 composting facilities serving communities. A number of these facilities reported that they accept compostable paper and plastic products (about 61 and 49, respectively, based on our reading of the graph presented in the full version of the report [43].

The goal of composting is to accelerate the biodegradation of the materials, and, over a reasonably short period of time, to create a material that can be used as a soil amendment. A picture of some finished compost is shown in Figure 7.22.

Composting is most often conducted in an aerobic fashion (in the presence of oxygen). The major products then are **CO_2, H_2O, and the organic matter** that we call **compost**. The obtained humic substances are crucial for soil health, and contain humic acids (organic acid molecules), humin (carbon-based macromolecular substances), essential plant nutrients, including both macro (e.g., N, P, K) and micro (e.g., Fe, Zn, Cu).[2] The compost's chemistry is influenced by the initial waste composition and the composting conditions.

During the composting process, the materials are heated to a temperature that is high enough to kill pathogens, weed seeds, etc. It is important for the composting process that sufficient amounts of water are provided.

In some cases, "composting" is run in an anaerobic mode. Purists will say that the term composting implies that it is aerobic, and that the anaerobic version

2 macronutrients (e.g., N, P, K) and micronutrients (e.g., Fe, Zn, Cu)

Figure 7.22 A photograph of finished compost. Bernard Dejean / Wikimedia Commons / CC BY-SA 4.0.

should be termed anaerobic digestion [44, 45]. We will use the more generic meaning of the term in the remainder of this discussion.

Anaerobic biodegradation (anaerobic composting, anaerobic digestion) requires that the operation be done in a container (in-vessel composting) so that oxygen can be limited. In this case, methane is produced – just like happens in a landfill when it becomes anaerobic. This methane gas, like landfill gas, can be used as fuel. It is relatively uncommon for MSW to be composted anaerobically. It is somewhat more common for animal wastes (such as from dairy cattle) to be handled in this manner. In-vessel composting tends to maximize control over odor-causing substances in the composting materials – a particular asset for animal wastes. Unfortunately, such systems tend to be expensive, so they are not widely used.

Only **biodegradable materials can be composted**. Biodegradable packaging materials include paper, wood, and a few plastics. Glass, metal, and most plastics are noncompostable, and totally, or nearly totally, nonbiodegradable. Referring to a material as compostable is a more stringent classification than just saying it is biodegradable. According to the ASTM and other standards, materials cannot be labeled as compostable unless they are able to fit into a standard composting system [46]. They must degrade within the typical composting times and conditions, and must not leave harmful residues.

The issue of residues is an important one. Compost from mixed MSW or from sewage sludge may contain levels of **heavy metals and other toxic compounds** that prohibit its use on any land that may be used for growing food crops. Compost from source-separated organics, whether yard waste or food waste, is much less likely to be contaminated. In fact, a significant number of California vineyards are reportedly using compost from the San Francisco source-separated organics composting programs (this waste originates from food-related businesses as well as from households.)

7.4 Reuse and Waste Reduction

7.4.1 Reuse

Reuse of packages at the retail level has been uncommon – at least until the introduction of Loop which will be discussed later. At the distribution level, reuse is commonplace in some industries. Much of this difference is due to the ease of returning and preparing the packages for a subsequent use.

For consumer packages, the example that usually comes to peoples' minds is refillable glass beverage bottles. These all but died out in the United States and represent a small minority of beverage packaging. The logistics of collecting the beverage bottles, the cost of using heavier glass to allow them to survive multiple use cycles, cost of transportation, loss due to breakage, requirements for maintaining and operating cleaning/sanitation facilities, etc., all played a role in their decline. However, more recently there has been some resurgence [47]. In 2019, Coca-Cola announced its investment toward returnable glass bottles in Germany [48].

For a while, refillable polycarbonate milk bottles replaced refillable glass in some locations, but these bottles also have mostly disappeared. Refillable 5-gallon plastic water bottles, used in dispensers, are still available, made from either polycarbonate or, increasingly, from PET. In Europe, refillable PET bottles are more widely used than in the United States. A few companies have experimented with refillable bottles for shampoo and other "health and beauty aid" products. Usually these have been companies that see being viewed as "green" as a very important part of their marketing strategy. It was reported in 2018 that hotels in Canada had been shifting toward the use of reusable and refillable shampoo containers as replacements for small single-use containers [49]. For the most part, however, consumer packages are not designed to be returned for reuse. Rather, to the extent that the end-of-life is considered in design, they are tailored to be source-reduced and/or recyclable.

A very interesting new development is the Loop system, https://exploreloop.com/. Loop originated as an offshoot of TerraCycle, a company designed to recycle hard-to-recycle items, including packaging. The idea is that products are packaged in durable, reusable packaging, usually metal or glass. When empty, they are returned and refilled. Consumers pay a deposit on each package, and return packages in a special Loop bag. Loop manages the sorting and cleaning, and then returns the packages to the manufacturer for refill and resale. The venture is quite new, having launched in 2019, and at the end of 2022, they had 150 retail locations across the globe. Initially, they operated as an independent e-commerce platform, but they appear to be shifting their focus toward retail partnerships with store chains such as Walmart in the United States, Metro in France, as well as Aeon in Japan [50], and is currently rolling out to specific locations around the world. Usually the items are shipped directly to consumers, and empties are collected at no charge from the consumers' homes. A google search on Loop reusable packaging can get you to the most recent information on how the venture is going [51, 52]. Loop is working with convertors to design glass, metal and even thick plastic containers suitable for their re-fill model.

As mentioned earlier, reuse is much more common for distribution packaging in some market segments. The automotive industry engaged in a major push to replace one-way corrugated boxes with reusable packaging for parts going to assembly operations. Reuse of pallets is commonplace in many economic sectors. In fact, the EPA calculated that over 13 million tons of wood pallets were refurbished and reused in 2007. However, their estimate in 2009 was 9 million tons. They did not state whether this resulted from recalculation or whether there was actually such a substantial decline. The number reported in 2011 was 10 million tons. Reusable plastic crates for produce have replaced corrugated in some markets. The 2015 EPA report also discusses refurbished wood pallets and the reuse of corrugated boxes but does not provide quantities. The 2018 report (issued in 2020) did not discuss reuse.

A rapidly growing area of reuse is merchandise bags. Many stores now offer inexpensive reusable bags for their customers. Some stores once provided a monetary credit for their use, but this practice seems to be disappearing. In several countries around the world, reusable bags are further supported by taxes or even bans on disposable plastic bags.

Consideration of the environmental benefits and costs of reusable packages must include the impacts associated with the collection and return system and with preparation (cleaning, etc.) of the materials for reuse as well as considering the initial design of the packages, as they may need to be more robust (requiring use of more or different materials) in order to permit reuse.

This is a whole topic in itself, which due to space constraints we are going to have to leave at this point.

7.4.2 Source Reduction

Source reduction has two distinctly different definitions: reducing the amount of material used and reducing the toxicity of the materials used (and their byproducts, etc.). Usually, for packaging applications, we are thinking of the first of these meanings. In that regard, there is a very long history of source reduction in packaging. Almost always, however, the driving force has been cost reduction, rather than any environmental consideration. (I'm not saying this is bad – just that it's a fact.)

The School of Packaging at Michigan State University has long claimed that one of the reasons companies hire packaging professionals is the cost savings that these employees provide through their ability to reduce packaging materials and consequently packaging costs, by proper design and management of packaging.

Once again, unfortunately, we aren't able to do justice to this topic, because of limited space. Some tidbits, for your information:

1994: Crisco vegetable oil came out in newly designed PET bottles that weighed only 39 g for the 48-oz bottles and 30 g for the 32 oz bottles, about 30% less than the previous designs for the PET bottles.

1997: Hershey Foods reduced packaging for their "miniatures" by changing from a paper wrapper over foil to a single wrap – achieving nearly a 50% reduction in the amount of packaging used. A report by the Grocery Manufacturers of America claimed the amount of grocery packaging discarded per person declined by 25% between 1980 and 1993.

2000: Texas Instruments reported at the Take It Back Pacific Rim Conference that the company reduced its transit packaging budget by about 20% through a combination of source reduction, recycling, and use of reusable-returnable packaging systems [53, 54].

2000: The Environment and Plastics Industry Council (EPIC) reported that total waste generated from milk packaging in Ontario was reduced by nearly 20% in 1995 compared to 1968, despite an increase of 22% in annual sales of milk, due to increased use of plastics. Margarine tubs were reported to be source reduced by nearly 30%, and dairy pails by 10%. One restaurant chain made its plastic drinking straws 20% thinner.

2001: The American Plastics Council reported on the "Use Less Stuff" portion of its website that for an 18 oz. PET peanut butter jar, only 9% of the total product weight was packaging, while for an equivalent glass jar, 36% was packaging.

2002: Labatt Breweries of Canada began using refillable amber glass beer bottles with no labels in Toronto, using a type of printing with UV-cured polymer inks that are removed during washing. Bottles were reported to be refilled as many as 15–20 times.

2002: Eco Emballages of France announced it was working with bottled water companies to reduce the weight of a 1.5 L bottle by 33 g, 27% of current weight, by 2005 – largely by switching from PVC to PET. This is on top of a 45.5 to 36.4 g weight reduction between 1994 and 2000.

2002: Kodak Australasia announced a 15% reduction in the weight of its film cartons.

2007: Heinz in the UK reduced the thickness of its can ends by 10%.

2007: American Chemistry Council reported that current 2 L plastic beverage bottles weighed 30% less than they did in the 1970s.

2008: Nestle announced plans to reduce the weight of its half-liter water bottles by an additional 15% by 2010. Its bottles were already among the lightest on the market, having been reduced in weight by 15% in 2007 in an on-going effort to make them as light as possible.

2008: Kraft salad dressings came out in light-weighted PET bottles that achieved a 19% weight reduction compared to the PET bottles that they had previously used. [55]

2008: Impress reported that its new peelable end for retorted food cans provides a reduction in weight of as much as 50% compared to standard easy-open ends.

2009: Ecolean was one of the winners of the DuPont Packaging Awards for its innovative aseptic packaging which was reported to achieve a 50% reduction in the typical weight of an aseptic carton package.

2009: Pepsi announced that it would reduce the amount of PET in Aquafina water bottles by 50% [56].

2009: Arla in the UK announced a reduction in weight of its two-pint HDPE milk bottles, which now weigh only 25 g, compared to the industry average of 27 g. Its 6-pint bottles were reduced in weight by 4 g, and its caps were reduced in weight from 2.3 to 2.0 g.

2009: Bericap announced a new closure line that could provide up to 30% weight reduction compared to alternative closures. [57, 58]

2020: O-I launched a new version of its Bonpas wine bottles for Boisset Group that were lighter and used 20% less glass than those it replaced. [49]

Certainly, this list could go on.

7.5 Recycling

7.5.1 Recycling – General

For recycling to be successful, several components must be in place. In particular, there must be:

1) A collection system for gathering the recyclable materials.
2) A processing system for changing the recyclables into a usable form.
3) Markets for the recyclable materials.

If any of these components are missing, recycling will not occur. There are a number of examples of collection of recycled materials for which there was no market, so the materials sat around in a warehouse somewhere, and were eventually sent to disposal. On the other hand, there are also many examples of companies that have been willing to use recycled materials in producing their products, but have not been able to do so because of a lack of supply (either no supply at all, or no supply of the quality they require).

While it is difficult to discuss end markets for recyclables as a group, and somewhat difficult to talk generically about how recycled materials are processed, the collection of recyclables of various types often has a lot in common. In fact, most recycling collection programs handle more than one type of recyclable material.

Therefore, we will discuss the collection end of recycling first, and then discuss processing and markets for specific materials.

For the most part, we will limit our discussion to packaging materials. We will discuss recycling of other materials in some cases, however, because in some instances packaging is a major market for non-packaging recycled materials (newspapers used in making paperboard cereal cartons, for example).

It is important to note that the recycling rates shown in Figure 7.8 include composting, so would be more properly titled recovery rates, rather than recycling rates. However, it is becoming increasingly common to use the term recycling to include both "traditional" recycling and composting, so we will follow that pattern except where we indicate differently.

Recycling rates differ considerably from state to state. Because of some differences in methodology, as we have discussed earlier in this chapter, *BioCycle's* recycling rates (which like the EPA's rates include composting) differ somewhat from those reported by the EPA. *BioCycle* reported a 2008 MSW recycling rate of 24% [59], compared to the EPA's reported value of 33% for that year [60]. Rates reported for individual states that can be calculated from the *BioCycle* data vary greatly. For example, Minnesota had a recycling rate (without composting) of 43.6%, while Alaska had a rate of only 4.8%. If we look at the tonnage recycled, California was by far the leader, at 24.7 million tons in 2008, with an additional 7.6 million tons going to composting. In a distant second place for recycling tonnage was Pennsylvania at 4.7 million tons.

Recovery rates also differ considerably by material (as do recovered amounts), as shown in Figures 7.23 and 7.24.

We can also note that packaging accounts for a significant portion of all recycling of MSW, whether we consider the total amount or whether we look at the recycling rate, as shown in Figures 7.25 and 7.26.

For materials to be recycled, they must first be collected and processed. We will provide a brief overview of these undertakings below.

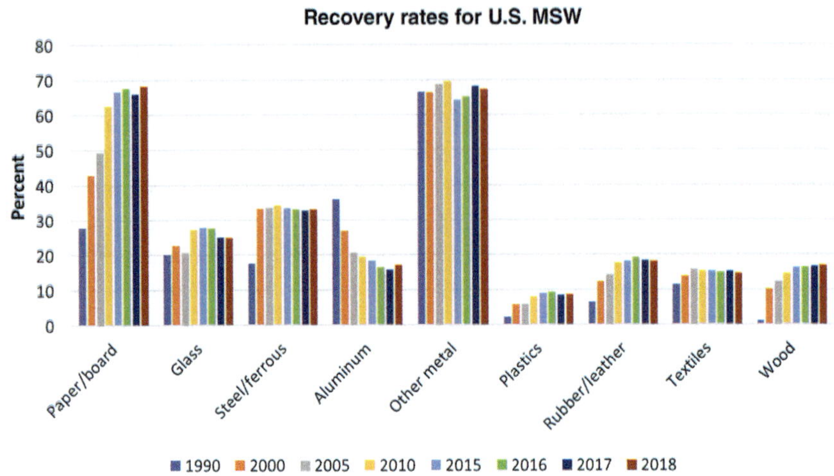

Figure 7.23 Recovery rates for U.S. MSW by material.

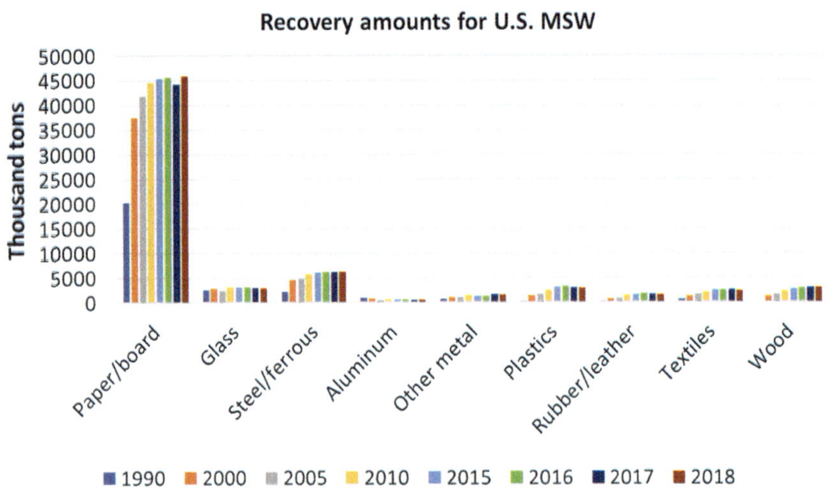

Figure 7.24 Recovery amounts for U.S. MSW by material.

7.5.2 Collection of Recyclables

One approach to the collection of recyclables is to collect all the MSW in a single stream, and then process it in order to separate the recyclable materials. The idea is that this can maximize the recovery of recyclables, since you do not have to rely on individuals diverting recyclables from the waste stream. It also does not require

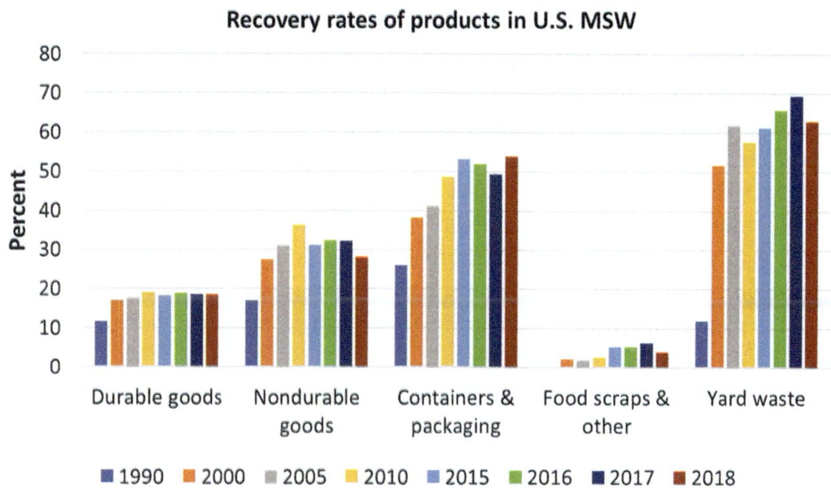

Figure 7.25 Recovery rates for U.S. MSW by product.

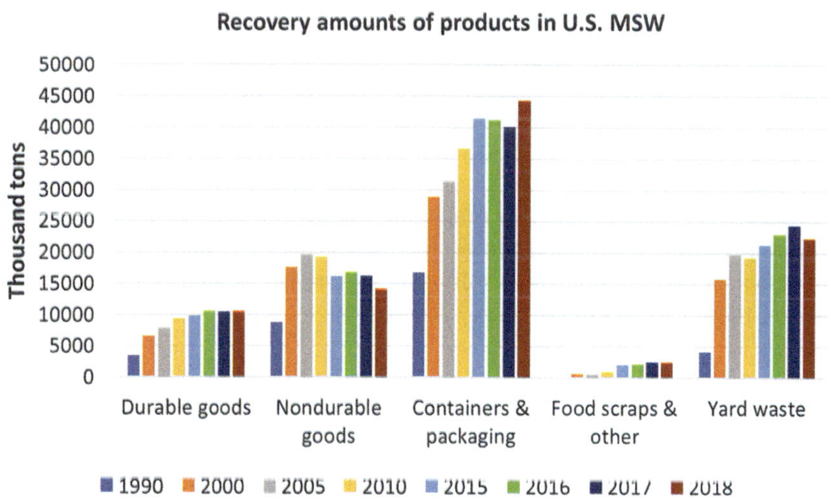

Figure 7.26 Recovery amounts for U.S. MSW by product.

any separate collection operation. For the most part, this approach works for ferrous metals, as they can easily be separated from garbage using electromagnets but does not work well for other materials. The combination of factors such as the low market value for these materials as well as high contamination rates has rendered this approach economically infeasible for most materials. For a few years in the early 2000s, there was increasing interest in this approach. The EPA reports that in 2013 there were 52 mixed waste processing facilities (also known as "dirty MRFs") in the United States, [61] with almost all of these in the West. However, since China (and later other countries) stopped being willing to accept contaminated streams of recyclables, there has been more emphasis on production of clean recycled material streams, and therefore decreased interest in development of dirty MRFs

The alternative to separating recyclables from garbage after the material is collected is to perform that separation prior to collection. This is known as *source separation*, and involves, for the most part, people never putting recyclables in with the rest of the garbage, rather than their trash subsequently being sorted at MRFs to separate out the recyclables [62].

The vast majority of recycling programs in the United States (and in most countries around the world) rely to a considerable degree on source separation so that is what we will concentrate on. Also, in this section, we will concentrate on the collection of recyclables from individuals, rather than from businesses. This is not meant to imply that collection from businesses is unimportant – actually, it is the major factor for some types of materials, as we will see when we look at the individual recyclable packaging materials.

When we think of recycling as a type of human behavior that we would like to encourage, we can recognize that there are two main factors that determine whether someone will engage in recycling. One factor is motivation: the higher their degree of motivation, other things being equal, the more recycling they will do. The other factor is convenience: with all other things being equal, the more convenient recycling is (the less people have to deviate from their accustomed ways to participate), the more they will recycle. Of course, we don't usually isolate these factors from each other, but different approaches do differ in the relative balance they provide between them.

7.6 Motivation

A variety of motivating factors have been investigated. One type of motivation is to mandate recycling by law. Then, if people want to consider themselves, or be considered, law-abiding citizens, they must participate in recycling to the extent that the law requires.

Other motivators are financial. One Illinois community ran a system in which each week a household was randomly chosen, and their garbage was inspected. If there were no recyclables in the family's trash, they won a cash prize. If any recyclables were found in their garbage, they would not win, and the cash would be added to the next week's prize – kind of like the lottery. However, after more than a year without a winner, the community decided this wasn't working very well!

Another type of financial motivator is widely used in Europe and has expanded to Asia, and even to Canada. Under this system, companies must pay a fee on all their packaging to pay for its eventual recycling (or disposal). The fee usually depends on both the type and the amount of the material used [63, 64]. Since this motivator is addressed toward companies rather than individuals, one can argue that it doesn't really fit into this section, but it seemed appropriate to mention it here anyway.

A financial motivator that does apply directly to individuals is deposit systems. As most of you know, in this system, the consumer pays a deposit on the covered containers, and then gets that money back when the containers are turned in for recycling. Therefore, the individual has a direct monetary interest in turning the containers in for recycling. There are currently 11 United States and territories that have Bottle Bill policies. Figure 7.27 shows the deposit states [55].

At bottlebill.org, you can go to the links at https://www.bottlebill.org/index.php/current-and-proposed-laws/usa/additional-links to find information about the law in any particular state. In 2010, Delaware became the first state to rescind statewide deposit legislation, enacting instead a temporary 4 cent recycling fee intended to expand recycling in the state. A report in 2011 states that collections

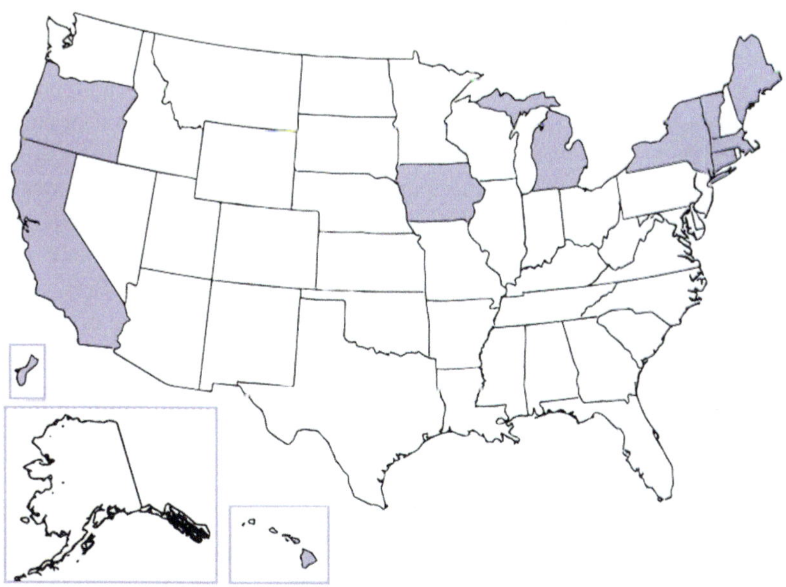

Figure 7.27 Map showing U.S. bottle deposit states. Image used with permission from the Container Recycling Institute (Bottle Bills in the USA, Container Recycling Institute, https://www.bottlebill.org/legislation/usa.htm).

of that fee were running more than $220,000 per month less than expected (http://www.bottlebill.org/news/locations/delaware.htm). The fee went into effect on 1 December 2010, and automatically expired on 1 December 2014. For more information on this topic, please refer to Ref. [65]. Historically, the beverage industry, the suppliers of their containers (with the exception of aluminum can suppliers) and retailers have adamantly opposed deposits. A pro-deposit site is http://www.bottlebill.org/, as mentioned earlier. More recently, some organizations have softened their position on deposits [66]. Increased efforts to improve sustainability, along with increased antiplastic packaging sentiment, seem to leave some companies feeling that bottle bills may be part of the way to ward off legislation that would be even harder for them to deal with – or at least they are sometimes talking like they might be in favor.

Whether one agrees or disagrees with bottle bills, it is hard to argue with the fact that they achieve greater rates of return for covered containers than any other system. Even though redemption rates have been falling somewhat in recent years, in Michigan (with its 10 cent deposit) the redemption rates remain well over 90% – an unheard of rate in most other systems. A report on Oregon's bottle deposit system from 2019 indicates that the increase in deposit amount resulted in a dramatic increase in redemption rates, reaching 90% in 2018 compared to only 64% in 2016 before the deposit was increased [67].

In addition, deposit systems accomplish the primary reason most of them were passed in the first place: reduction of littering of beverage containers. Deposit systems work so well because they provide a powerful motivator – enough to get people to engage in the mild inconvenience of saving up their deposit containers and taking them somewhere that they can be redeemed.

Another way to provide money as a motivator is to institute a buyback system, where consumers are paid for the recyclables they bring in. This approach is common practice in many less developed countries. In the United States, however, it is now limited almost entirely to the buyback of aluminum beverage cans.

Most communities have decided it is undesirable or impractical, or both, to impose deposits on all recyclables, and in many cases the value of recyclables does not justify a buyback system. Therefore, in the absence of the powerful incentive that a deposit can provide, convenience of recycling plays a crucial role.

7.6.1 Convenience

Recycling systems can be set up in ways that provide a great deal of convenience or very little. One common categorization of recycling systems targeted at households is curbside or drop-off.

In a drop-off program, the consumer must deliver the recyclables to some location. This location may be available around the clock, seven days a week or may be available only one day per month for a few hours. It may be centrally located (or provide numerous locations) or it may be in some out-of-the-way location that is hard to get to. Obviously, these factors affect convenience. Michigan State University (MSU) offers a drop-off recycling site that is open to the public, 24 hours a day, 7 days a week. The City of East Lansing, where MSU is located, also operates a drop-off recycling site. In both cases, these sites accept materials that are not included in the city's curbside collection program.

Curbside programs are inherently more convenient than drop-off programs that are otherwise equivalent. In a curbside program, the consumer merely leaves their recyclables at the curb, along with their other garbage. Someone then comes and takes the recyclables away. Usually collection is done on the same days as regular garbage pickup, which further increases convenience.

Programs targeted at the collection of recyclables from businesses obey these same principles but are more hybrid. In the School of Packaging at MSU, for example, we have essentially a drop-off system, but there are a number of recycling container locations around the building. These provide essentially the same level of convenience as a curbside collection system – or arguably even more, since the user doesn't even need to take the recyclables out of the building.

The amount of preparation of the recyclables that is required also has a large impact on convenience, and therefore on participation rates. Virtually all programs require the recyclables to be reasonably clean. Some require various degrees of separation of different recyclable materials from each other. Many increase convenience by permitting most or all recyclables to be put together – this is referred to as commingled collection. Meanwhile, some require that labels be removed or containers be flattened. In general, such demands in some way facilitate or simplify the recycling process, but they also decrease convenience. Programs are always running a fine line between requirements that are needed for efficient management of the program, and the unintended effects of these requirements in reducing participation.

One of the most effective techniques for maximizing participation in curbside recycling is providing a container for recyclables to be placed in. This does several things. It increases convenience by providing a place for the recyclables to be stored, and an easy way of getting them to the curb. It also can increase motivation by bringing peer pressure to bear. If the bright blue (the most common color) recycling bin is not out in front of your house on collection day, all your neighbors will know that you are not recycling.

Speaking of those bins, there is an increasing trend to provide large wheeled carts to residents, in place of the original small open bins. This provides a larger space to collect more recyclables and easier transport to the curb, while also providing a cover to protect the contents against weather.

7.6.2 Education/Publicity

The most well-designed program, with a good balance of motivation and convenience, will not work if no one knows about it. It is essential that there be education of the community about recycling opportunities or requirements, and how to participate. One time is not sufficient, as people move into and out of communities all the time, so an ongoing strategy is required. It has been found that one of the most effective ways to provide this education is in the local school systems, especially at the elementary school level. Elementary school children tend to be very rule-oriented, and to believe what their teachers say. The hope is, also, that if recycling habits are established early, they will remain for a lifetime.

The containers mentioned above also provide an education function. Their presence in a home serves as a reminder of the existence of the recycling program. When a program is initiated, delivery of the container to every targeted home helps ensure that everyone in the community is aware of the program. It is often easy to overlook a mailed notice, or something in the local paper, but it is more difficult to ignore the big blue bin sitting in front of your front door.

7.6.3 Participation Rates Versus Diversion Rates

A couple of terms need to be defined. Recycling programs often quote participation rates or diversion rates as indicators of their success. These two terms are related, but they have distinct differences.

The diversion rate (or recovery rate) is a measure of the amount of a material "diverted" from the waste stream. A 75% diversion rate for milk bottles, for example, would mean that 75% of the milk bottles used were sent to recycling, and only 25% were sent to disposal.

The participation rate is the proportion of households that "participate" in recycling. Participation is usually defined as diverting some materials from garbage to recycling at some time during a designated period, often one month. The argument is made that counting participation for a single week, for example, would underestimate the true participation, as most people will miss putting out their recyclables at least once in a while. Furthermore, many programs encourage residents to put out their recycling bin only if it is full or nearly so, to increase the collection efficiency. Generally, one month is considered a reasonable period of time to allow participation rates to be measured.

Participating households rarely divert 100% of their available recyclables as most people do put some recyclables into the garbage. Therefore, a participation rate of 80% will not translate into a diversion rate of 80%. The diversion rate will be lower – maybe much lower.

7.6.4 Separation/Sorting

For recyclable materials to be useful, they must be separated by type of material. Obviously, there is not much use for a mixture of paper, glass, aluminum, plastic, and so forth. In drop-off recycling programs, this is handled relatively simply by having different containers for the different materials. Of course, failure of people to follow instructions and put only the designated materials into the container can be a significant problem [68]. Usually, for understandable reasons, this is more of a concern at unstaffed sites than those where someone is designated to oversee the process. However, having the site be open only when there is staff present decreases the convenience of this option by a considerable degree. The local East Lansing drop-off recycling sites, for example, are always available – 24 hours a day, year-round.

When materials are collected, as we have mentioned, a common approach is to allow some or all materials to be mixed together in the recycling bin, rather than requiring individuals to sort all the different materials into categories and leave a number of containers out at curbside. Providing commingled collection has been found to significantly increase both participation and diversion rates. Obviously, these materials must be separated at some point. There are two general strategies that are applied.

7.7 MRFs

The most common strategy for sorting recyclables is to collect the commingled materials and deliver them to a material recovery facility, known as a MRF (pronounced "Merf"). In the MRF, the materials will be sorted at least into general categories: mixed paper, newspaper, corrugated, aluminum, steel, clear glass, amber glass, green glass, plastics, etc. The plastics fraction may or may not be subdivided into resin and possibly color categories. Sometimes at the MRF the separation is only "mixed plastics" and separation by resin is done later, at a dedicated plastics sorting facility.

The degree of technology used at the MRF varies considerably. In some facilities, hand-sorting of materials passing on a conveyor belt past picking stations is the main method. Other MRFs rely mostly on automated sorting technologies. In many cases, a combination of both is practiced.

A video of one MRF can be found at Ref. [69] you will see a combination of technology with considerable hand labor. Some of the newer developments in technology are discussed at Ref. [70].

7.8 Comparative Advantages and Disadvantages

Recently, there has been some rethinking of the value of commingled collection (single stream) versus requiring customers to do some separation. Increasing issues with recovery of glass containers from such systems, along with glass contamination of paper streams, have been especially problematic. A couple of articles discussing this issue can be found at Refs. [71] and [72]. It may be that we will see some return to dual-stream collection to avoid some of the issues associated with single-stream systems – at this point it's too early to tell. Another thing that has happened in some communities is a decision to eliminate recycling of particularly problematic materials altogether – usually meaning glass [73, 74].

Case Study: St. Francis Xavier University's Flux Lab and Landfill Emissions

Landfills have been widely used as waste facilities for many years. Although there have been efforts to discourage reliance on landfills, they still tend to proliferate. Although the most evident environmental damage caused by landfills may appear to be the physical waste that accumulates in the landfills as well as the chemicals and microplastics that can leach from these materials, they are also a source of greenhouse gases. Even after landfills are closed, they can still be a source of methane gas for decades after their last use.

In Canada, there are around 3,000 landfills, and Environment and Climate Change Canada (Canada's federal environment department) is seeking to stem the flow of methane from its landfills to help combat climate change. As a greenhouse gas, methane is at least 25 times more effective at keeping heat in the atmosphere than carbon dioxide, and thus it is a significant contributor to climate change. Part of the proposed policy will be to apply new federal regulations to Canada's landfills that receive 10,000 tonnes or more of municipal solid waste each year. Although Canada wanted to craft policy to reduce these emissions, it needed more accurate information about how much methane was actually escaping from landfills in order to guide its policy [75]. To address this issue, Environment and Climate Change Canada reached out to the Flux Lab headed by Dr. Dave Risk at St. Francis Xavier University to help measure these methane emissions that are escaping from landfills. [76]

So far, Dr. Risk and his team have visited 145 landfills of different sizes across Canada. They have a mobile laboratory, with sensors attached to a car's roof as they drive near and around landfills. In one case, they found that a landfill near Halifax, Nova Scotia that had closed in 1997 was still emitting methane gas. These emissions were also occurring despite the fact that the site had been used between 2007 and 2021 by a private company as a source for methane.

In some cases, the methane gas issue is addressed by drilling wells into the garbage contained in the landfill and collecting the gas, which is then burned in a flare to generate electricity.

Although landfills can emit methane for decades or even a century after they close, Dr. Risk and his team did find some promising findings, that help demonstrate some of the benefits of policy decisions made in more recent decades. In Canada's Atlantic provinces, methane emissions have been decreasing since 1990. Provincial policies became stricter in the 1990s, and Nova Scotia's landfills stopped accepting organic waste in 1995.

The information provided by these measurements help demonstrate the legacy left behind by landfills as well as their environmental impacts and can provide valuable information for crafting policy. As stated by Dr. Risk, Canada's oil and gas sectors are already regulated with regard to their methane emissions, and more things can be done in terms of our landfills, such as capturing and monitoring methane escaping from landfills.

7.9 Concluding Remarks

This chapter provided a broad overview of the solid waste issues and how they are disposed/handled in different ways. The following two chapters are dedicated to the recycling processes for individual types of packaging materials. Specifically, Chapter 8 will focus on the recycling of glass, metal, and paper, while Chapter 9 will explore the recycling of plastics, including emerging trends in plastic recycling.

Problem Set for Chapter 7

1 What different categories does the EPA use to classify municipal solid waste (MSW)?

  A Durable goods, nondurable goods, containers and packaging, and other wastes.

  B Durable goods, nondurable goods, containers and packaging, and electronics.

  C Municipal solid waste, hazardous waste, and non-hazardous industrial waste.

  D None of the above.

2 In the United States, what percent of the MSW was recycled, composted, incinerated for energy recovery, and sent to landfills/other disposals in 2015?
  A Recycled 20%, composted 8.9%, incinerated 12.8%, and landfill/other disposals 62.5%.
  B Recycled 10%, composted 8.9%, incinerated 22.8%, and landfill/other disposals 63%.
  C Recycled 40%, composted 5%, incinerated 5%, and landfill/other disposals 52.5%.
  D Recycled 25.8%, composted 8.9%, incinerated 12.8%, and landfill/other disposals 52.5%.

3 Why does the United States utilize landfills more extensively than the European Union for waste management?
  A The United States has more land for landfilling.
  B EU Packaging directives discourage landfilling.
  C EU has better infrastructure for recycling and recovery than the United States.
  D All of the above.

4 Which of the following answers denotes the Solid Waste Management Hierarchy **from the most preferred (left) to the least preferred (right)?**
  A Source reduction and reuse, recycling and composting, energy recovery, treatment, and disposal.
  B Source reduction and reuse, energy recovery, recycling and composting, treatment, and disposal.
  C Energy recovery, source reduction and reuse, recycling and composting, treatment, and disposal.
  D None of the above.

5 The EPA divides solid waste solid wastes into:
  A Three categories.
  B Two categories.
  C Five categories.
  D Four categories.

6 Why is waste incineration common in the European Union while landfilling is more common in the United States?
  A More landfill space available in the United States.
  B EU regulations facilitate incineration as it classifies incineration as recovery.
  C Landfilling is cheaper than incineration.
  D All of the above.

References

1 U.S. Environmental Protection Agency (2023) Wastes - Non-Hazardous Waste - Municipal Solid Waste. https://archive.epa.gov/epawaste/nonhaz/municipal/web/html/ (accessed on October 10, 2023).
2 United States Environmental Protection Agency (2024). Advancing Sustainable Materials Management: Facts and Figures Report. https://www.epa.gov/facts-and-figures-about-materials-waste-and-recycling/advancing-sustainable-materials-management (accessed 26 January 2024).
3 BioCycle (2010). The State of Garbage in America. https://www.biocycle.net/the-state-of-garbage-in-america-4/ (accessed 9 October 2023).
4 United States Environmental Protection Agency (2024). Nondurable Goods: Product-Specific Data. https://www.epa.gov/facts-and-figures-about-materials-waste-and-recycling/nondurable-goods-product-specific-data (accessed 26 January 2024).
5 United States Environmental Protection Agency (2024). Containers and Packaging: Product-Specific Data. https://www.epa.gov/facts-and-figures-about-materials-waste-and-recycling/containers-and-packaging-product-specific (accessed 26 January 2024).
6 United States Environmental Protection Agency (2020). Advancing Sustainable Materials Management: 2018 Tables and Figures. https://www.epa.gov/sites/default/files/2021-01/documents/2018_tables_and_figures_dec_2020_fnl_508.pdf (accessed 18 November 2024).
7 United States Environmental Protection Agency (2024). Containers and Packaging: Product-Specific Data. https://www.epa.gov/facts-and-figures-about-materials-waste-and-recycling/containers-and-packaging-product-specific (accessed 26 January 2024).
8 UN Environment Programme (2024). Integrated Solid Waste Management. https://www.ctc-n.org/technologies/integrated-solid-waste-management (accessed 26 January 2024).
9 What Is Integrated Solid Waste Management? https://www.epa.gov/nscep (accessed 26 January 2024).
10 United States Environmental Protection Agency (2024). Sustainable Materials Management: Non-Hazardous Materials and Waste Management Hierarchy. https://www.epa.gov/smm/sustainable-materials-management-non-hazardous-materials-and-waste-management-hierarchy (accessed 18 November 2024).
11 United States Environmental Protection Agency (2020). Advancing Sustainable Materials Management: 2018 Fact Sheet. https://www.epa.gov/sites/default/files/2021-01/documents/2018_ff_fact_sheet_dec_2020_fnl_508.pdf (accessed 18 November 2024).
12 United States Environmental Protection Agency (2020). Advancing Sustainable Materials Management: 2018 Fact Sheet. https://www.epa.gov/sites/default/

files/2021-01/documents/2018_ff_fact_sheet_dec_2020_fnl_508.pdf (accessed 18 November 2024).
13 United States Environmental Protection Agency (2010). Municipal Solid Waste in The United States: 2009 Facts and Figures. December 2010, Figure ES-5, page 14. https://archive.epa.gov/epawaste/nonhaz/municipal/web/pdf/msw2009rpt.pdf (accessed 26 January 2024).
14 Otto, S.L. (2013). OPINION: Why Not Burn Waste? *Ensia* (6 June). https://ensia.com/voices/why-not-burn-waste/ (accessed 26 January 2024).
15 Factsheet: Extended Producer Responsibility (EPR) For Used Packaging. https://www.europen-packaging.eu/wp-content/uploads/2021/04/EUROPEN-factsheet-on-EPR-for-used-packaging.pdf (accessed 10 October 2023).
16 BioCycle (2010). The State of Garbage in America. https://www.biocycle.net/images/art/1010/bc101016_s.pdf (accessed 17 August 2024).
17 Rischar, H. (2023). EREF study shows MSW tip fees rose 'sharply' in 2022. *Waste Today* (8 June). https://www.wastetodaymagazine.com/news/eref-study-shows-msw-landfill-tip-fees-rose-sharply-in-2022/#:~:text=Average%20weighted%20and%20unweighted%20tip,managed%20at%20each%20reporting%20landfill (accessed 24 January 2024).
18 Statista (2024). Average Cost to Landfill Municipal Solid Waste in The United States from 2020 to 2022, by Region (in U.S. Dollars Per Ton). https://www.statista.com/statistics/692063/cost-to-landfill-municipal-solid-waste-by-us-region/ (accessed 26 January 2024).
19 Edwards, W. and Blough, T. (2024). Waste. *AP Environmental Science*. Millbrook High School, Raleigh, NC. https://mhsapes.weebly.com/day-2-reading1.html (accessed 24 January 2024).
20 Mondal, T., Choudhury, M., Kundu, D. et al. (2023 Jun). Landfill: an eclectic review on structure, reactions and remediation approach. *Waste Management* 164: 127–142.
21 Spikula, D.R. (1997). Subsidence performance of landfills. *Geotextiles and Geomembranes* 15: 395–402. https://doi.org/10.1016/S0266-1144(97)10017-6.
22 Murphy, W.L. and Gilbert, P.A.1985) Settlement and Cover Subsidence of Hazardous Waste Landfills (Final Report), United States Environmental protection Agency. https://nepis.epa.gov/Exe/ZyNET.exe/20007JJF.txt?ZyActionD=ZyDocument&Client=EPA&Index=1981%20Thru%201985&Docs=&Query=&Time=&EndTime=&SearchMethod=1&TocRestrict=n&Toc=&TocEntry=&QField=&QFieldYear=&QFieldMonth=&QFieldDay=&UseQField=&IntQFieldOp=0&ExtQFie.
23 Jain, M., Kumar, A., and Kumar, A. (2023). Landfill mining: a review on material recovery and its utilization challenges. *Process Safety and Environmental Protection* 169: 948–958. https://doi.org/10.1016/j.psep.2022.11.049.
24 Assi, A., Bilo, F., Zanoletti, A. et al. (2020). Zero-waste approach in municipal solid waste incineration: reuse of bottom ash to stabilize fly ash. *J. Clean. Prod.* 245: 118779. https://doi.org/10.1016/j.jclepro.2019.118779.

25 Quina, M.J., Bontempi, E., Bogush, A. et al. (2018). Technologies for the management of MSW incineration ashes from gas cleaning: New perspectives on recovery of secondary raw materials and circular economy. *Sci. Total Environ.* 635: 526–542. https://doi.org/10.1016/j.scitotenv.2018.04.150.

26 Garcia-Lodeiro, I., Carcelen-Taboada, V., Fernández-Jiménez, A., and Palomo, A. (2016). Manufacture of hybrid cements with fly ash and bottom ash from a municipal solid waste incinerator. *Constr. Build. Mater.* 105: 218–226. https://doi.org/10.1016/j.conbuildmat.2015.12.079.

27 Evenergi LLC (2023). EV Basics – EV Batteries. https://www.evenergi.com/ev-basics-ev-batteries/ (accessed 14 October 2023).

28 World Health Organization (2013). WHO Calls for the Phase Out of Mercury Fever Thermometers and Blood Pressure Measuring Devices by 2020. https://www.who.int/news/item/11-10-2013-who-calls-for-the-phase-out-of-mercury-fever-thermometers-and-blood-pressure-measuring-devices-by-2020 (accessed 14 October 2023).

29 United States Environmental Protection Agency (2023). Phasing Out Mercury Thermometers Used in Industrial and Laboratory Settings. https://www.epa.gov/mercury/phasing-out-mercury-thermometers-used-industrial-and-laboratory-settings (accessed 14 October 2023).

30 PVC Tech Corp (2015). CONEG Toxics in Packaging – PVC Tech Corp Compliance Summary. https://pvctech.com/wp-content/uploads/2015/01/pvctech-CONEG-web-summary-02.04.15.pdf (accessed 24 January 2024).

31 Toxics in Packaging Clearinghouse (2023). Toxics in Packaging Clearinghouse Fact Sheet. https://toxicsinpackaging.org/wp-content/uploads/2023/04/TPCH Fact-Sheet-Revised-April-2023.pdf (accessed 14 October 2023).

32 Shen, Y. (2023). Model toxics in packaging legislation: an overview, *Compliance Gate* (23 March). https://www.compliancegate.com/model-toxics-in-packaging-legislation/ (accessed 25 January 2024).

33 United States Environmental Protection Agency (2024). Air Emissions from MSW Combustion Facilities. https://archive.epa.gov/epawaste/nonhaz/municipal/web/html/airem.html (accessed 17 August 2024).

34 Reworld (2021). Waste-to-Energy: What Really Comes Out of the Stack. https://www.reworldwaste.com/news-and-resources/blog/waste-to-energy-what-really-comes-out-of-the-stack (accessed 17 August 2024).

35 Verband Kommunaler Unternehmen e.V. (2023). Waste-to-Energy: An indispensable pillar of waste management and a valuable contribution to the circular economy. https://www.cewep.eu/wp-content/uploads/2021/12/211130_Joint_Position_Paper_Waste_to_Energy.pdf (accessed 18 October 2023).

36 Goldstein, N. (2017). The State Of Organics Recycling In The U.S. *BioCycle* (4 October). https://www.biocycle.net/state-organics-recycling-u-s/ (accessed 18 October 2023).

37 United States Environmental Protection Agency (2016). Advancing Sustainable Materials Management: 2014 Fact Sheet. November 2016, Page 15. https://www.epa.gov/sites/default/files/2016-11/documents/2014_smmfactsheet_508.pdf (accessed 19 October 2023).

38 Goldstein, N. (2019). Food waste composting infrastructure in the U.S.*BioCycle* (4 January). https://www.biocycle.net/food-waste-composting-infrastructure-u-s/ (accessed 24 January 2024).

39 Sullivan, D. (2010). Mixed waste composting facilities review. *BioCycle* 51 (11): 16–20. https://www.biocycle.net/mixed-waste-composting-facilities-review/.

40 United States Environmental Protection Agency (2024). Zero Waste Case Study: San Francisco. https://www.epa.gov/transforming-waste-tool/zero-waste-case-study-san-francisco (accessed 18 November 2024)

41 Yepsen, R. and Goldstein, N. (2007). Source Separated Residential Composting in the U.S. *BioCycle* 48 (12): 27–33. https://www.biocycle.net/source-separated-residential-composting-in-the-u-s/.

42 Platt, B. and Goldstein, N. (2014). State of composting in the U.S. *BioCycle* (July), pp. 19–27. https://www.biocycle.net/state-of-composting-in-the-u-s/.

43 Goldstein, N. (2023). Quantifying existing food waste composting infrastructure in the U.S. *BioCycle*. http://www.biocycle.net/pdf/2019/FoodWasteCompostInfra.pdf) (accessed 19 October 2023).

44 Davidson Environmental (2023). Composting vs. anaerobic digestion. https://www.davidsonenvironmental.ca/composting-vs-anaerobic-digestion/ (accessed 19 October 2023).

45 Lin, L., Xu, F., Ge, X., and Li, Y. (2018). Improving the sustainability of organic waste management practices in the food-energy-water nexus: A comparative review of anaerobic digestion and composting. *Renewable and Sustainable Energy Reviews* 89: 151–167. https://doi.org/10.1016/j.rser.2018.03.025.

46 Narayan, R. and Pettigrew, C.A. (1999). ASTM standards help define and grow a new biodegradable plastics industry. *ASTM Standardization News* (December), pp. 36–42. https://www.ftc.gov/sites/default/files/documents/public_comments/guides-use-environmental-marketing-claims-project-no.p954501-00181%C2%A0/00181-56737.pdf.

47 Drink Milk in Glass Bottles (2023). The Resurgence of Milk In Glass Bottles. http://www.drinkmilkinglassbottles.com/the-resurgence-of-milk-in-glass-bottles/ (accessed 19 October 2023).

48 Packaging Insights (2023). Coca-Cola injects €40M in returnable glass bottles in Germany. https://www.packaginginsights.com/news/coca-cola-injects-%E2%82%AC40m-in-returnable-glass-bottles-in-germany.html (accessed 19 October 2023).

49 Knox, C. (2018). Why those teeny bottles of hotel shampoo might become obsolete. *Canadian Broadcasting Corporation (CBC)* (3 December). https://www.

cbc.ca/news/canada/nova-scotia/cbc-ca-wavesofchange-1.4927787 (accessed 19 October 2023).

50 Wiener-Bronner, B. (2019) How to solve the world's plastic problem: Bring back the milk man. *CNN Business* (24 Jnauary) https://www.cnn.com/interactive/2019/01/business/loop-reusable-packaging-mission-ahead/index.html (accessed 18 November 2024).

51 Pyzyk, K. (2023). Loop touts retail store expansion as standalone e-commerce program sunsets. *Waste Dive* (10 January). https://www.wastedive.com/news/loop-reuse-terracycle-fred-meyer-giant-walmart/639649/ (accessed 19 October 2023).

52 Sanchez, R. (2023). What happened to terracycle's loop and where are they today? *Dieline* (31 July). https://thedieline.com/blog/2023/7/31/what-happened-to-loop (accessed 19 October 2023).

53 Packaging World (1997). Plastics packaging for milk reduces waste. https://www.packworld.com/design/flexible-packaging/news/13329601/plastics-packaging-for-milk-reduces-waste (accessed 20 October 2023).

54 Pyzyk, K. (2023). Loop touts retail store expansion as standalone e-commerce program sunsets. *Waste Dive* (10 January). https://www.wastedive.com/news/loop-reuse-terracycle-fred-meyer-giant-walmart/639649/ (accessed 19 October 2023).

55 Mohan, A. M. (2008). Krafting a better bottle. *Packaging World* (12 May). https://www.packworld.com/design/flexible-packaging/article/13343659/krafting-a-better-bottle (accessed 20 October 2023).

56 Plastics Net (2009). PepsiCo's Aquafina Launches the Eco-Fina Bottle, the Lightest Weight Bottle in the Market. https://www.plasticsnet.com/doc/pepsicos-aquafina-launches-the-ecofina-0001 (accessed 20 October 2023).

57 Stones, M. (2009). Bericap launches new light-weight beverage closure. *BeverageDaily* (28 September). https://www.beveragedaily.com/Article/2009/09/29/Bericap-launches-new-light-weight-beverage-closure (accessed 20 October 2023).

58 O-I (2023). O-I Designs Lighter Weight Bonpas Wine Bottle for Boisset. https://glass-catalog.com/na-en/success-stories/o-i-designs-lighter-weight-bonpas-wine-bottle-for-boisset (accessed 20 October 2023).

59 BioCycle (2010). The State of Garbage in America. https://www.biocycle.net/the-state-of-garbage-in-america-4/ (accessed 20 October 2023).

60 United States Environmental Protection Agency (2009). Municipal Solid Waste Generation, Recycling, and Disposal in the United States: Facts and Figures for 2008. https://archive.epa.gov/epawaste/nonhaz/municipal/web/pdf/msw2008rpt.pdf (accessed 20 October 2023).

61 United States Environmental Protection Agency (2015). Advancing Sustainable Materials Management: Facts and Figures 2013, pp. 135–136. https://www.epa.

gov/sites/default/files/2015-09/documents/2013_advncng_smm_rpt.pdf (accessed 20 October 2023).
62 Waste Management Technology (2014). What Is a Dirty MRF? And Why They Are Being Phased Out Throughout the EU. https://waste-technology.co.uk/what-is-a-dirty-mrf-and-why-they-are-being-phased-out-throughout-the-eu.html (accessed 20 October 2023).
63 EUROPEN (2023). Factsheet: Extended Producer Responsibility (EPR) for Used Packaging. https://www.europen-packaging.eu/wp-content/uploads/2021/04/EUROPEN-factsheet-on-EPR-for-used-packaging.pdf (accessed 20 October 2023).
64 WM Intellectual Property Holdings (2020). WM Report on Recycling. L.L.C. https://sustainability.wm.com/downloads/WM_Report_on_Recycling.pdf (accessed 20 October 2023).
65 Auras, R.A. and Selke, S.E.M. (2022). *Life Cycle of Sustainable Packaging: from Design to End-of-Life*. Wiley.
66 Smalley, M. (2019). In recycling today. *Recycling Today* (12 March). https://www.recyclingtoday.com/news/six-states-propose-bottle-bills-2019/.
67 Profita, C. (2019) Oregob Bottle Deposite System Hits 90 Percent Redemption Rate. NPR (4 February). https://www.npr.org/sections/thesalt/2019/02/04/688656261/oregonbottle-deposit-system-hits-90-percent-redemption-rate (accessed 18 November 2024).
68 Profita, C. (2019) Oregob Bottle Deposite System Hits 90 Percent Redemption Rate. NPR (4 February). https://www.npr.org/sections/thesalt/2019/02/04/688656261/oregonbottle-deposit-system-hits-90-percent-redemption-rate (accessed 18 November 2024).
69 Van Dyk Recycling Solutions (2024). Single Stream Recycling – Tour a Material Recovery Facility (MRF), YouTube. https://www.youtube.com/watch?v=M5nmNKVNCBw (accessed 25 January 2024).
70 Smalley, M. (2019). How MRF operators are designing facilities to combat today's recycling challenges. *Waste Today* (18 June). https://www.wastetodaymagazine.com/news/material-recovery-facility-design-best-practices/ (accessed 25 January 2024).
71 McDonald, J. (2024). Is Single-Stream Recycling Worth It? *Dumpsters.com*. https://www.dumpsters.com/blog/single-stream-recycling (accessed 25 January 2024).
72 Koerth, M. (2019). The Era of Easy Recycling May Be Coming to an End. *FiveThirtyEight* (10 January). https://fivethirtyeight.com/features/the-era-of-easy-recycling-may-be-coming-to-an-end/ (accessed 25 January 2024).
73 Recycling Today (2018). Municipal recycling programs no longer accepting glass, plastics. https://www.recyclingtoday.com/news/recycling-programs-phase-out-glass-2018/ (accessed 25 January 2024).

74 Koerth, M. (2019). The era of easy recycling may be coming to an end. *FiveThirtyEight* (January 10). https://fivethirtyeight.com/features/the-era-of-easy-recycling-may-be-coming-to-an-end/ (accessed 25 January 2024).

75 Environment and Climate Change Canada (2023). Reducing Canada's landfill methane emissions: proposed regulatory framework. https://www.canada.ca/en/environment-climate-change/services/canadian-environmental-protection-act-registry/publications/reducing-landfill-methane-emissions.html (accessed 21 March 2024).

76 Luck, S. (2024).How harnessing methane from landfills could be "easy win" in climate-change fight. *CBC* (21 March). https://www.cbc.ca/news/canada/nova-scotia/garbage-landfill-climate-change-methane-monitoring-1.7143565 (accessed 21 March 2024).

8

Recycling of Metal and Glass

8.1 Overview

Recycling is the process of collecting and processing new products. After reduction and reuse, recycling is the next priority of waste materials. Recycling offers a multitude of benefits for the environment, economy, and communities [1]. It enables a true circular economy, keeping the resources in the loop. For example, it helps to conserve our resources such as trees, minerals, and fossil fuels, thus reducing the ecological toll of new production. Recycling has a critical role in combating climate change. For example, recycling and composting municipal solid waste (MSW) saved over 193 million metric tons of CO_2 equivalent in 2018 [1]. Energywise, recycling is usually highly efficient – for example, a recycled high-density polyethylene (HDPE) bottle is ~50% more energy efficient than its nonrecycled counterpart.

Economically, recycling is a revenue generator. According to the Environmental Protection Agency (EPA) in their 2020 Recycling Economic Information Study, recycling activities contributed to $37.8 billion in wages and $5.5 billion in tax revenues in the United States. This translates to 1.17 jobs per 1,000 tons of materials recycled and significant wage and tax revenue per ton [2].

Recycling can also play a pivotal role in promoting environmental justice. Waste facilities such as landfills often disproportionately affect underserved communities, posing health, and property value risks. The recycling rates of various materials in Europe are shown in Figure 8.1.

In this chapter, we will cover three topics:

- Metal Recycling
- Glass Recycling
- Paper Recycling

Plastics recycling will be covered in the next chapter.

Degradation, Stabilization, and Recycling of Packaging Materials, First Edition.
Muhammad Rabnawaz, Susan E.M. Selke, and Ian Wyman.

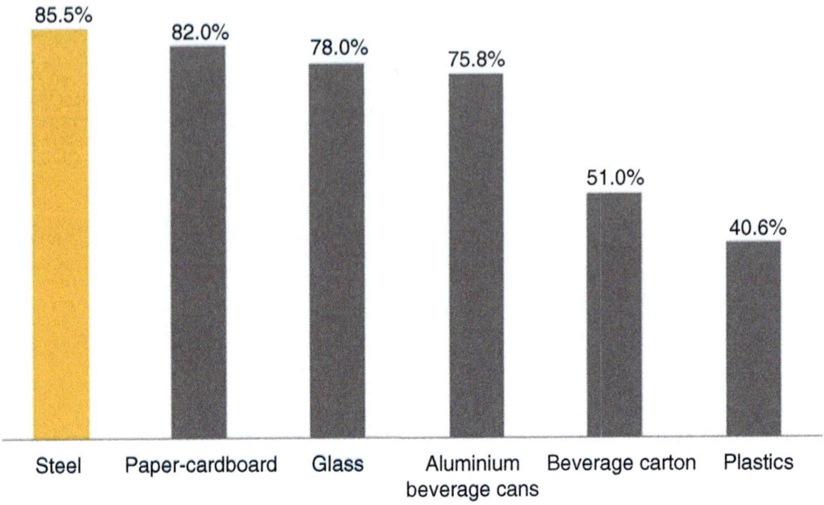

Figure 8.1 Recycling rates of various packaging materials in Europe [3]. *Source:* Image used with permission from Steel for Packaging Europe.

8.2 Metal Recycling

Metal recycling offers several benefits:

- Reducing greenhouse gas emissions,
- Preservation of the environment by reducing the mining of earth for metal raw materials,
- Saving energy,
- Reducing the need for landfills,
- Providing economic benefits, and
- Creating local jobs

In this section, we will discuss the recycling of the two main packaging metals, namely steel and aluminum.

8.2.1 Steel Recycling

Current information on steel recycling in the United States can be obtained from the Steel Recycling Institute (SRI), at Ref. [4].

In theory, steel is 100% recyclable, which means it can be recycled into the same material of the same quality again and again (Figure 8.2).

According to SRI figures, the overall recycling rate for steel was 69% in 2019 based on a 3-year rolling average, and the steel recycling rate varied between 65% and 80% between 2012 and 2019 [6]. The rate for steel can recycling was 70%, slightly below the high of 71% in 2011 and 2012. However, it remained below the rates for automotive steel at 85% and appliances at 82%. The overall steel recycling rate for 2014 was reported to be 86%; rates for individual categories were not reported [7].

Figure 8.2 An example of a steel recycling logo [5]. *Source:* Image obtained from Wikipedia (at https://en.wikipedia.org/wiki/Recycling_codes#/) and distributed under a CC BY 2.5 license (https://creativecommons.org/licenses/by/2.5/).

In some years, the steel recycling rate can exceed 100% – the reason for such a high rate is that there are backlogs of material that can enter the recycling stream. Since the recycling rate is based on the proportion of materials that enter the solid waste stream, it is possible, by diving into these backlogs, to recycle more steel in one year than entered the waste management stream in that year. Of course, that cannot be sustained indefinitely, as eventually the backlogs will disappear. The recycling rates for steel cans, steel in all sectors, and the overall recycling rate, are calculated as follows [5]:

$$\text{Recycling rate for Steel Cans} = \left[\begin{array}{l}(\text{scrap consumed})\,/\\ (\textbf{steel produced})\end{array}\right](\text{old calculation formula})$$

$$\text{Recycling rate for Steel in all sectors} = \left[\begin{array}{l}(\text{scrap consumed})\,/\\ (\textbf{scrap generated})\end{array}\right](\text{old calculation formula}]$$

$$\text{total recycling rate} = \frac{(\text{new} + \text{old scrap consumed} \pm \text{net trades})}{(\text{new} + \text{old scrap generated})}$$

Note: The term **"new scrap"** refers to scrap generated by manufacturers of steel-containing products and removed from the system prior to its end use. New scrap may also be called "prompt scrap" or "manufacturing scrap" (this is also called postindustrial waste). Meanwhile, the term **"old scrap"** refers to scrap that entered service as a consumer product and has been captured for recycling at the end of life (EOL) (this is also called postconsumer waste).

According to the **American Iron and Steel Institute (AISI)**, the U.S. overall steel recycling rate in 2019 was ~72.5% for a 10 year rolling average and 69% for that year with a 3-year rolling average [8].

For steel cans, the recycling rate was impressive and reached 70.9% in 2018 (Figure 8.3).

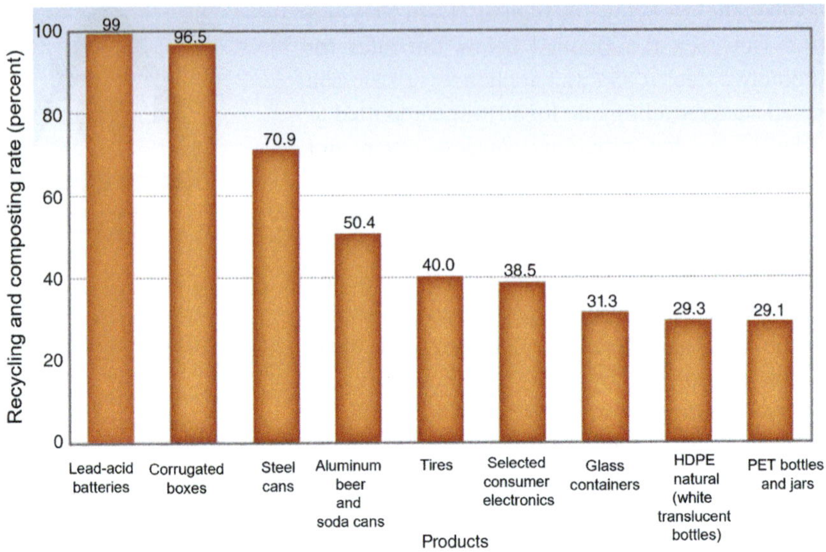

Figure 8.3 Recycling rates of various materials in 2018 [9]. *Source:* [9] / U.S. Environmental Protection Agency / Public Domain.

In Europe, the recycling rates for steel packaging and aluminum cans in 2020 were 85.5% and 73%, respectively. Metal Packaging Europe is targeting an 80% recycling rate of by 2025 for metal packaging [10]. It has recently been reported that the EU's target of reaching a 70% recycling rate by 2025 has been reached.

It should be noted that steel for can-making is often produced using basic oxygen furnaces (BOF), one of the primary methods for steel production. In a BOF, molten iron from a blast furnace is combined with scrap steel and fluxes (such as lime), and then high-purity oxygen is blown into the mixture. The oxygen reacts with carbon in the molten iron to produce carbon dioxide, thereby reducing the carbon content and refining the material into steel. Low-carbon content steel is typically used for can-making, as this type of steel provides excellent formability and is easy to work with.

The CO_2 emissions associated with steel production decline with higher steel recycling rates. The amount of CO_2 prevented corresponds to one and a half times the weight of each steel item that is recycled [11].

Steel is recycled in a predominantly **open-loop fashion**, with collected steel being directed to the nearest appropriate facility. Steel for can-making is produced in BOF, which use about 20–35% recycled content.

8.3 Open-Loop and Closed-Loop Recycling

In discussions of recycling, whether of steel or of other materials, you will often find reference to the terms "open-loop" and "closed-loop."

Closed-loop recycling refers to recycling something into an identical or similar article. Making new aluminum beverage cans out of old aluminum beverage cans is one example.

Open-loop recycling refers to recycling something into a different type of product, often lower grade. Recycling steel food cans into a beam for a house is an example.

Closed-loop recycling is often considered superior to open-loop recycling, but it is not always true. Taking soft drink bottles as an example, the best recycling system is one that continually produces new soft drink bottles from the old ones. However, a number of factors must be taken into account when determining whether closed-loop or open-loop recycling is the best option in a given situation. For example, open-loop recycling may be the preferable option despite a lowering of the grade of the resultant material if the postconsumer containers must be shipped long distances to facilities enabling closed-loop recycling if facilities suitable for open-loop recycling or downcycling are nearby.

8.4 Steel Recycling Process

In steel recycling, magnetic separation is used to separate steel from nonferrous materials. Scrap merchants sort collected steel into various categories so that the appropriate "recipe" can be put together for the intended end-use of the recycled material. The material is shredded, melted, and used to make new products. Waste is generated during this process, especially dust and volatiles. In some cases, valuable metals can be recovered from this waste.

Further details regarding steel and metal recycling processes can be found at Refs. [12, 13]

Steel can be recycled over and over again without losing any of its properties. An interesting fact is that recycling 1 ton of steel helps to save 1.8 barrels of oil, 10.9 million British thermal units (BTUs) of energy, 642 kWh of energy, and 2.3 m^3 of landfill space.

8.4.1 Steel Can Recycling Process

There are several steps involved in the recycling of steel cans, as shown below:

- Step 1: Collection.
- Step 2: Sorting by magnetic separation.

Figure 8.4 An example of an aluminum recycling label. *Source:* https://upload.wikimedia.org/wikipedia/commons/b/bb/41_ALU_Recycling_Code.svg

- Step 3: Preprocessing, such as shredding and baling of the shredded pieces into compact blocks for simpler transport.
- Step 4: Cleaning to remove coatings and labels via chemical or thermal means.
- Step 5: Melting via electric arc furnace (EAF) or basic oxygen furnace (BOF).
- Step 6: Refining involves steel modification and purification.
- Step 7: Solidification into desired shapes such as slabs.

8.5 Aluminum Recycling (Figure 8.4)

8.5.1 Aluminum Beverage Cans

Information about aluminum can recycling rates is available from the Aluminum Association (http://www.aluminum.org/). Such information is also available from the Container Recycling Institute (CRI, http://www.container-recycling.org) as well as from the EPA. The different rates reported have been a point of controversy.

In 2005, the Aluminum Association reported, for 2004, the first increase in the beverage can recycling rate since 1997 – 51.2% compared to 50.0% in 2003. In 2006, they reported that the rate for 2005 had also risen to 52.0%.

The CRI countered that the rate was only 45.1% in 2004, compared to 44.3% in 2003, and that most of the increase was attributable to the increase in redemption value in California. The difference in these reported values illustrates one of the complexities in calculating recycling rates. The Aluminum Association based its recycling rate on the amount of used beverage cans (UBCs) melted in U.S. facilities, plus exports of can scrap, divided by the amount of cans shipped. The CRI countered that it is misleading to include imported scrap cans. This is essentially "crediting" the aluminum industry with recovering a higher proportion of cans than it actually does, so the CRI excludes these imported UBCs from the rate calculation. This discrepancy was not a big concern in the 1980s, when recycling rates were climbing and few UBCs were imported. In 1990, the recycling rate calculated by the Aluminum Association was 2.7% higher than that calculated by the CRI. However, in 2004, the Aluminum Association's calculated rate was 6.1% higher than that calculated by the CRI, as can be seen in Figure 8.5. This difference has continued. For 2010, the most recent year reported by the CRI, the Aluminum Association reported a beverage can recycling rate of 58.1%, while the CRI reported a rate of 49.6%, a difference of 8.5%.

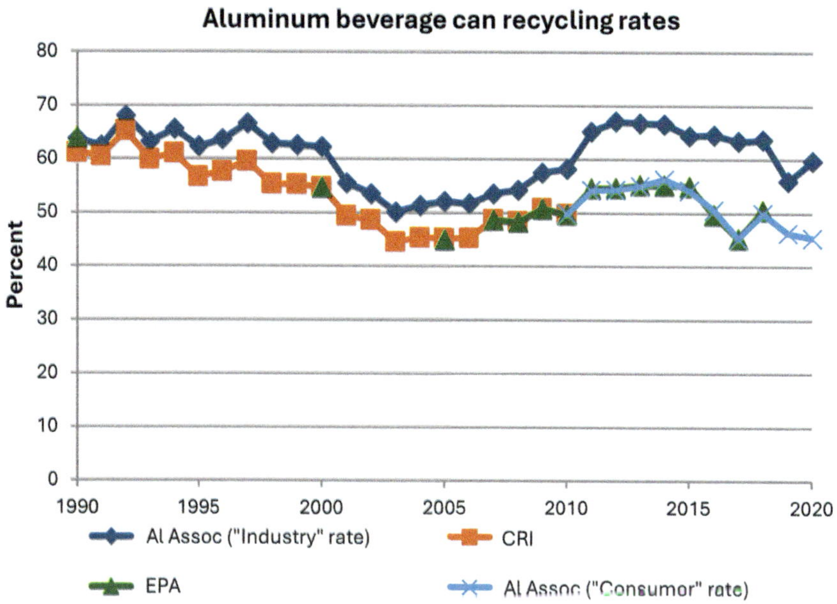

Figure 8.5 Recycling rates for aluminum beverage cans. *Source:* Sources of this data include the U.S. EPA, The Aluminum Association [14], and The Container Recycling Institute.

The EPA determines rates in the same manner as the CRI – not including imported scrap cans – and generally reports rates identical to those of the CRI. Perhaps in response to this criticism, the Aluminum Association in its 2015 report for the first time provided two rates – the "Industry Recycling Rate" as defined above, and the "Consumer Recycling Rate" which appeared to be essentially the same as the EPA and CRI rates [15]. These rates appear to closely match the EPA/CRI rates, as shown in Figure 8.5 [14].

Aluminum recycling has a long history because of the impressive energy savings it provides. A widely quoted statistic is that using recycled aluminum saves 95% of the energy that would otherwise be required to produce aluminum from ore. This statistic is often misused to say that recycling beverage cans saves 95% of the energy that would otherwise be used – but according to the CRI, the savings is about 75%. Where does the difference come from? The savings comes from making the aluminum metal; the energy required for the can-making process is the same. **So, if you look just at the energy used to make the aluminum sheet, starting with recycled aluminum saves about 95% of the energy**, but if you look at the whole can-making process, from raw materials to finished containers, the energy savings is about 75%. In either case, it is substantial, since the production of aluminum requires a lot of energy. **This**

creates a definite economic incentive for can-makers to use recycled aluminum as much as is possible.

8.5.2 Other Aluminum Packaging

Aluminum is used in a few types of food cans, which are likely recycled at about the same rate as steel cans; little direct information is available. Aluminum foil can also be recycled, but its recycling rate is much lower. Thick foil containers can be recycled easily and are accepted in most community recycling programs, mixed in with aluminum cans. However, aluminum foil that is contained within multi-layer structures is generally not accepted by these programs, and has only a very low recycling rate. Aluminum contained in metallized paper or metallized film is not generally regarded as recyclable.

Amounts and rates for recycling of non-beverage-can aluminum packaging materials are provided in the EPA report, and are shown in Figures 8.6 and 8.7, along with the rates for beverage cans. Unfortunately, the EPA stopped reporting the recycling rates for "other (aluminum) cans" or for "foil and closures" beginning in 2010.

8.5.3 Aluminum Packaging Recycling Amounts

8.5.3.1 Aluminum Packaging Recycling Rates

The EPA stopped reporting rates or amounts for recycling of aluminum foil and closures after 2008, and stopped including those amounts in the total aluminum packaging recycling.

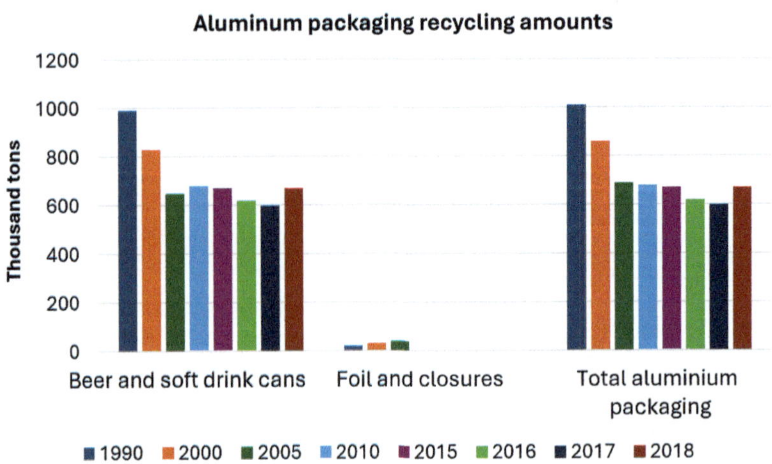

Figure 8.6 Recycling amounts for various types of aluminum for various years between 1990 and 2018.

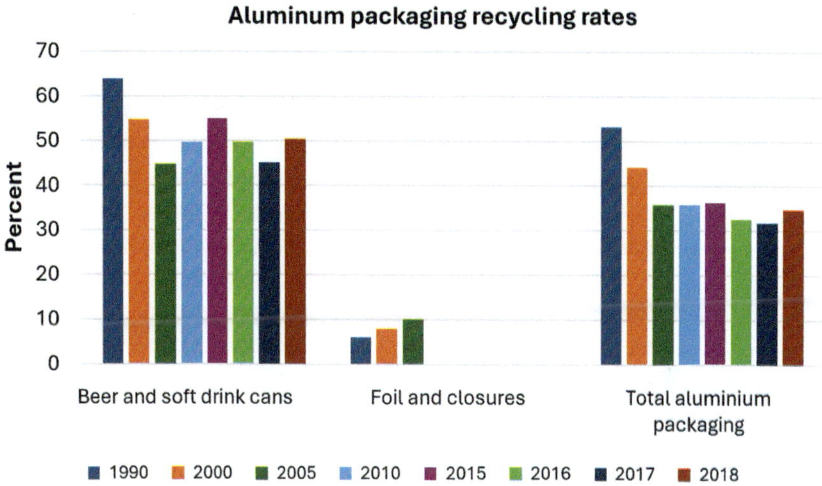

Figure 8.7 Recycling rates for various types of aluminum packaging for various years between 1990 and 2018.

For comparison, the UK aluminum drink can recycling rate in 2018 was reported to be 75% and the overall aluminum packaging recycling rate was 52% [16]. For the European Union plus Switzerland, Norway, and Iceland, the beverage can recycling rate in 2021 was reported to be 76.1% [17]. Rates differ by country, as shown in Figure 8.8.

Aluminum beverage can recycling is a predominantly closed-loop process, as the primary market for UBCs is new beverage cans. In the past, the recovered aluminum went only into the body stock, as the end stock is a slightly different alloy. It is reported that some recycled aluminum does now go into the end stock material, though the proportion is significantly lower than in the body stock.

In aluminum recycling processes, the material is shredded, primarily to prevent any trapped liquid from entering the furnace. It is then melted. During or prior to the melting process, the lacquers are decomposed by heat and are removed. Proper control over the aluminum dust is necessary to prevent problems such as dust explosions. It is noteworthy that 75% of all aluminum ever produced is still in use [19].

8.5.4 Eddy Current Separation

As mentioned earlier, magnetic separation is commonly used for steel and iron recycling. For aluminum, it is eddy currents that separate aluminum from the rest of the packaging materials [20, 21]. When aluminum is exposed to a changing magnetic field, it generates its own magnetic field that opposes the change in the original magnetic field. The resulting opposing magnetic field results in a repulsive

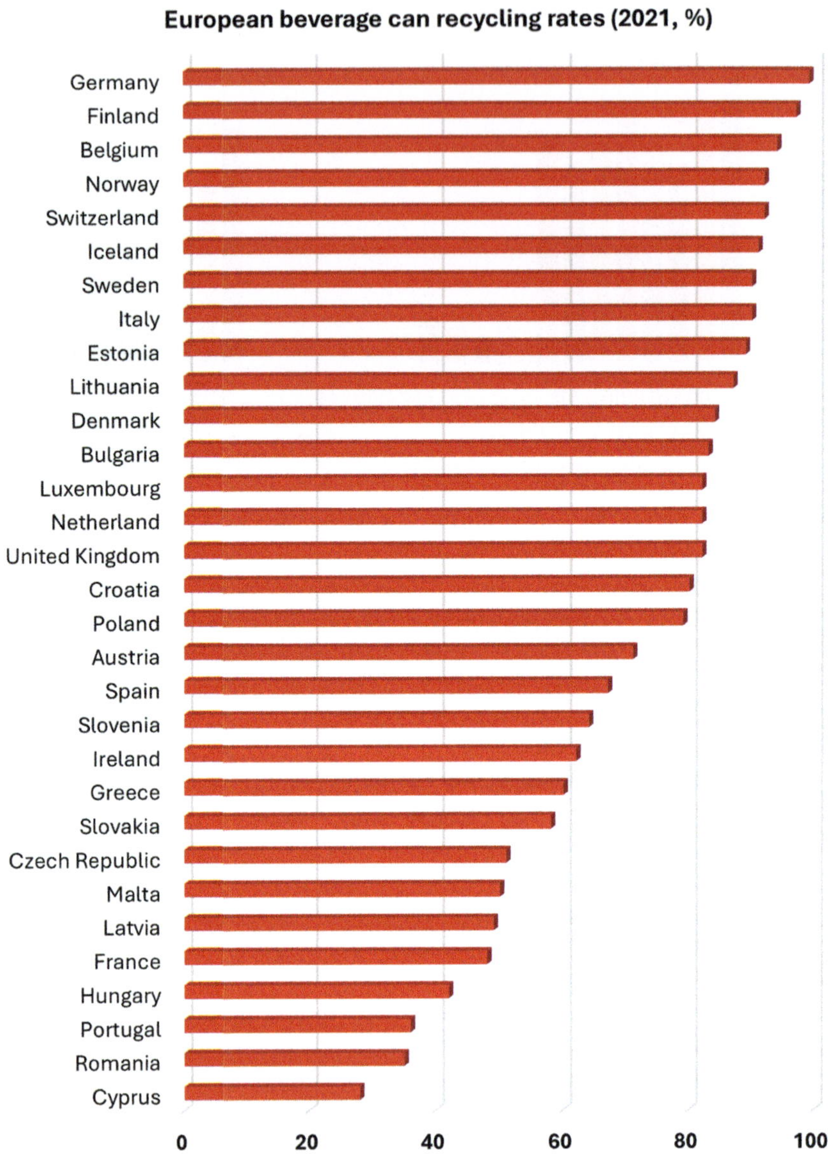

Figure 8.8 European aluminum beverage can recycling rates in 2021 [18].

force, effectively causing the aluminum to be repelled or pushed away from the source of the magnetic field. This principle is used at material recovery facilities (MRFs) to separate nonferrous metals such as aluminum from other waste materials.

8.6 Glass Recycling

8.6.1 Glass Recycling in the United States

Glass recycling is almost exclusively limited to the recycling of packaging glass. Very little nonpackaging glass is recovered. Figure 8.9 shows the types of glass packaging that are recycled in the United States. Note that beer and soft drink bottles are the largest category. No significant amount of glass is recycled from nonpackaging sources.

8.6.2 Glass Packaging Recycling Amounts

Of course, the total amount recycled depends on the amount available for recycling as well as on how effective we are in recovering it. Figure 8.10 shows the recycling rates for various types of glass packaging.

Please note that the peaks for wine and liquor bottles, for food and other bottles and jars (and the decrease in beer and soft drink bottles) that show up in the 2000 data are likely related more to changes in data sources used by the EPA than to actual change in rates. The data for the intervening years between 2000 and 2005 differ dramatically in different years in the EPA series of reports. For example, the 2003 report gives a "food and other" recycling rate of 24.1% for 2002, while the 2006 report gives a rate of only 14.8% for that same category and year. A similar discrepancy occurs in tonnage, with one report of 780,000 tons and another of 330,000 tons. The EPA did not go back and revise the 2000 data like it did for 2001–2004 [22].

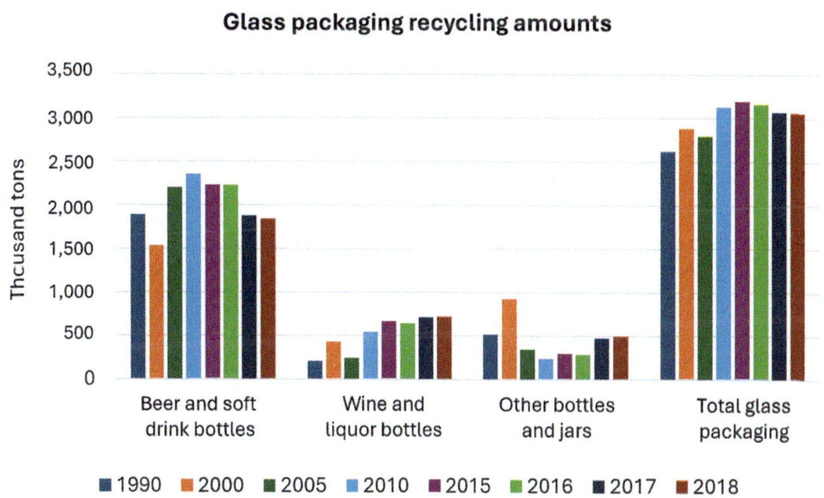

Figure 8.9 Amounts of glass packaging recycled in various years.

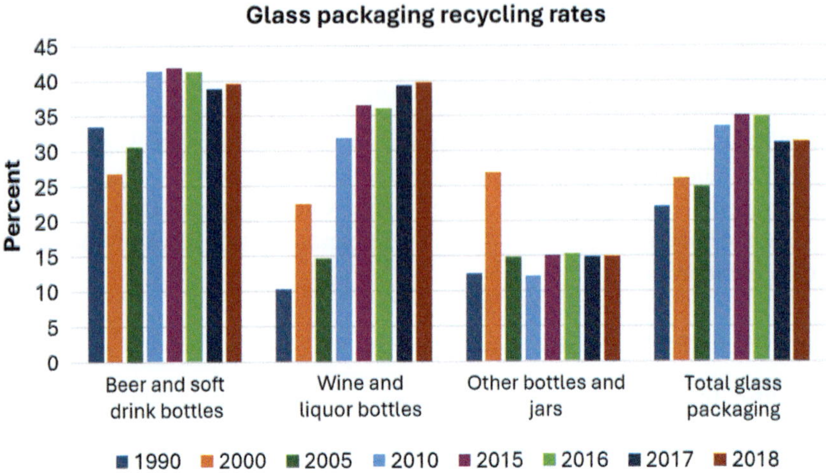

Figure 8.10 Glass packaging recycling rates in various years.

8.6.3 Glass Recycling Elsewhere

Glass recycling rates differ dramatically around the world. The glass recycling rate for Europe for 2021, as reported by FEVE, the European Container Glass Federation (https://feve.org/), was **80.1%** [22].

Rates per country for 2018 are shown in Figure 8.11. Note that these are based on containers collected for recycling, not the actual amount that is recycled

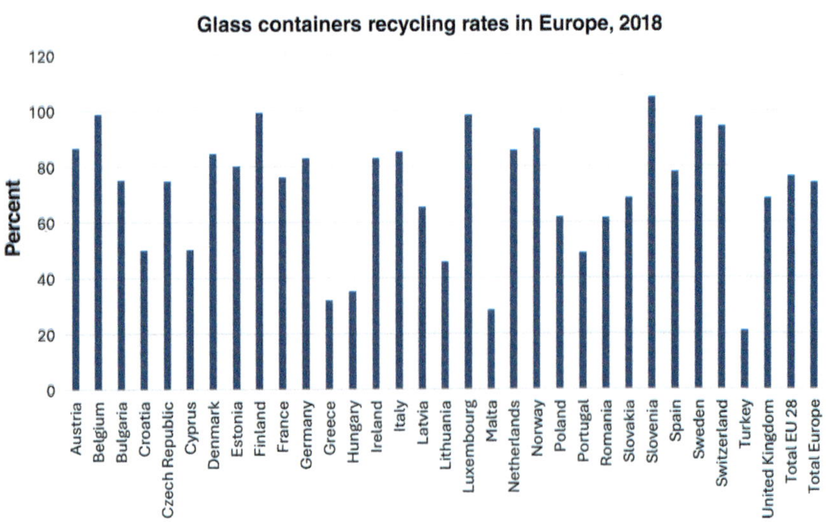

Figure 8.11 Collection rates of glass containers to be used for recycling in Europe.

Table 8.1 Formula used to calculate EU recycling rates for glass containers [23].

Calculation formula:			
EU tonnage of cullet collected (tons) × 100/Consumption = EU recycling rate (%)			
Year	**Consumption**	**Collection for recycling**	**Percent**
2018	18.2	13.5	74.2

(The formula used to calculate the glass container recycling rate and the corresponding value for the EU in 2018 is shown in Table 8.1).

Secondary markets for recycled glass as a construction material have been discussed in a couple of recent review articles, which can be found at Refs 24, 25. There are also uses in other secondary markets.

8.6.4 Glass Recycling

The Glass Packaging Institute's information on glass recycling can be found at http://www.gpi.org/recycling and links from that page [26].

One process for sorting glass cullet is described at http://www.mssoptical.com/material-sortingequipment/glass-sorting-equipment/glass-colorsort/. Note that the sophisticated color-sorting technology described is not available in many, even most, processing facilities. In many MRFs, a major waste stream is contaminated with mixed-color broken glass, for which there are likely no local markets. While such materials can be used as a replacement for some of the aggregate in construction applications, these markets may not be available in a particular location.

8.6.5 Steps Involved in Glass Recycling

There are a number of steps involved in the glass recycling process, as is summarized below:

- **Step 1:** Consumers bring glass to drop-off retail stores/bottle deposit banks, etc. In other cases, curbside collection is conducted by the municipality or by recyclers.
- **Step 2:** Once the desirable amount of glass is available, then it is sorted manually on conveyer belts. (Note: In the case of single stream collection, glass, because of its higher density than other pkg materials, falls through the screens into bins below the conveyor belts.)
- **Step 3:** Once the glass is sorted, then it is crushed into pieces called cullet.
- **Step 4:** A magnet is used to remove ferrous metal from the cullet.
- **Step 5:** Laser technology is used to remove stone and ceramic material from the cullet. (Typically, this is performed with a focused laser beam, which is a monochromatic light source. The laser beam hits a rock or mineral and is absorbed or reflected.)

- **Step 6:** A vacuum is used to remove paper.
- **Step 7:** Nonferrous metals are removed by an air-jet.
- **Step 8:** A final visual inspection is performed.
- **Step 9:** The recycled glass is then used to produce new glass.

Raw virgin materials for glass such as limestone, require heating at 1,500 °C in order to melt. In contrast, recycled material (cullet) will melt when they are heated at a lower temperature of 1,000 °C. This is because the melting of raw virgin materials such as silica sand, soda ash, and limestone requires higher temperatures, often around 1,500 °C, to break their chemical bonds and form molten glass. On the other hand, cullet has already undergone the process of melting and formation once, so its structure is more amenable to re-melting, requiring temperatures of around 1,000 °C. This results in energy savings. In addition, the cullet has already been processed, and it does not produce the CO_2 that would otherwise be generated during the conversion of limestone ($CaCO_3$) into calcium oxide (CaO) (which would be accompanied by one CO_2 molecule for each converted limestone molecule), as shown below.

$$CaCO_3 \rightarrow CaO + CO_2$$

The process for recycling glass back into new containers involves reducing the size of the glass fragments, removing contaminants, and then introducing the **cullet** into the glass furnace along with virgin raw materials. There is uncertainty about the amount of cullet that can be used in manufacturing new glass containers. The glass industry is much more willing to use postconsumer cullet in recent years than used to be the case over 20 years ago. However, cullet is a relatively low value material, and the virgin materials that it replaces are also inexpensive. Since glass is heavy and expensive to transport, opportunities for the use of postconsumer glass to produce new glass are limited. Consequently, it is not uncommon for communities to not collect green glass, as it is particularly low in value. Some communities have eliminated glass recycling altogether.

It is not uncommon for recyclers to view glass in recycling streams as problematic, for several reasons.

First, the market for recycled glass is poor to nonexistent in many parts of the U.S. Glass continues to be a difficult commodity for local solid waste agencies to recycle in a cost-effective manner.

Second, glass breakage and cross contamination at MRFs leads to increased processing costs and increased contamination of other recyclable materials, thus lowering the value of these commodities. While many communities are continuing to instruct their residents to place glass bottles and jars in single-stream carts, some communities have opted to discontinue glass collection in their recycling programs.

It has been reported that the cost of sending collected glass to cullet processors across the United States is in the range of \$10–\$40 per ton. As glass makes up almost 5% of the MSW stream and based on the consideration that state and local agencies have established ambitious zero-waste targets, many agencies are still hoping to develop viable models for glass recycling [27].

Since it is a dense commodity, glass continues to make up a significant portion of landfill diversion rates. However, the limited markets in most regions make it a "loss leader" for many recycling programs [28]. Furthermore, legislative efforts to promote extended producer responsibility can have significant glass recycling impacts. Time will tell if the glass recycling market is half full or half empty.

As glass containers lost some of their market share to aluminum cans, polyethylene terephthalate (PET), and other plastic materials over the last two decades, the glass container industry has consolidated and reduced its capacity. Three companies (Owens-Brockway, Gallo, and Saint-Gobain) supplied about 90% of glass container demand (9.36 million tons, or 60.6 pounds per person per year) in the United States in 2010, estimated at approximately 25 million glass containers, of which nearly 75% was beer bottles, and the remaining was mostly food containers.

We encourage you to read the February 2019 analysis of glass recycling in the United States, published in *Chemical & Engineering News*, at https://cen.acs.org/materials/inorganic-chemistry/glass-recycling-US-broken/97/i6. Glass from multistream collection systems offers higher quality of glass than single-stream collection systems. ~ 90% of glass from multistream systems is successfully recycled compared to only 40% from a single stream. Other perspectives on glass recycling can be found at https://www.recyclingtoday.com/article/busting-myths-about-glass-recycling/ as well as many other sources. Because of the high density and low value of glass, and competition with cheap raw material ingredients, the viability of glass recycling can depend quite a lot on where the glass is located.

8.6.6 What is the Future of Glass?

Glass recycling rates can be increased by introducing multistream collection in the United States. Multistream collection is commonly practiced in Europe, but this approach is not currently common in the United States. With the implementation of extended producer responsibility (EPR) laws (**which are discussed in Chapter 10**), multistream collection is likely to be adopted. Another challenge encountered with glass recycling in the United States is that there are not many MRFs in the United States. (~350) as compared to over 50 times more in the EU [29]. Note that MRFs in the EU are much smaller than those in the U.S. The benefit of having MRFs near to the origin of waste is cost reduction, especially for heavy (such as glass) or very bulky recyclable materials (such as foams). It should be noted that although glass is a single type of material, glass needs to be sorted by color, which adds further cost to the recycling of glass. Mixed colored glasses are often used for fiberglass or as fillers in concrete [30].

Case Study: Glass Recycling

Although glass is highly recyclable and can be recycled for endless cycles, only about one third of glass is recycled in the United States. This stands in stark contrast with some European countries such as Germany and Switzerland, where more than 90% of the glass waste is recycled [31]. When glass is prepared using recycled glass, it is produced using cullet, which is essentially crushed glass from jars and bottles.

There are many benefits that can be achieved by using cullet for glass production. Firstly, less raw material is required to produce glass using cullet than is the case when raw materials such as sand, limestone, or sodium carbonate are employed for the same purpose. Using cullet can also conserve energy, as furnaces can be operated at lower temperatures. Despite cullet being a recycled material, its use can even help improve the quality of the final glass. Cullet does not produce carbon dioxide when it is melted, thereby helping to minimize undesired bubble formation. Cullet can also facilitate the mixing of materials during glass formation, thus preventing unmelted starting materials from forming crystalline deposits, while also inhibiting the formation of streaks or other optical defects. In addition, as cullet does not release carbon dioxide during the glass production process, unlike limestone, which contains carbonates that release carbon dioxide when they are melted. Cullet can help reduce greenhouse gas emissions when it is used as an ingredient. Therefore, the recycling of glass through the use of cullet should be highly desirable, at least from a technical perspective.

One challenge that has impeded recycling efforts in the U.S. is that it can be difficult to obtain clean cullet that is suitable for glass production without a lot of costly processing treatments. The main reason for this issue is that single-stream curbside collection is used by most municipalities in the U.S. for the collection of recyclables. Consequently, the glass is mixed with other recyclable materials such as plastic, metal cans, newsprint, and cardboard. In addition, individuals often mistakenly mix non-recyclable items with recyclable materials, or fail to separate non-recyclable components of an item from its recyclable components. This means that costly sorting is required at MRFs before any recycling can begin.

In contrast, multi-stream waste collection is more widely used in Europe. Drawbacks of multi-stream processing are that it can be more costly and requires more participation from consumers, who separate the recyclable items according to their category (such as glass, metals, plastic), rather than the sorting being performed at the MRF. On the other hand, a key advantage of multistream waste collection is that the recyclables obtained tend to be cleaner, with less sorting needed at the waste facilities. In fact, glass recovered via the multistream approach can often be sent directly to cullet processors without requiring sorting. Consequently, the glass recycling rates from multi-stream collection systems are 90%, which are much higher than the corresponding rate of 40% for single-stream systems. The lack of access to high-quality cullet is one of the challenges that has impeded glass recycling efforts in the U.S.

The weight of glass and the cost of transporting it are another issue. Glass is a relatively heavy material, and it can be expensive to transport it for longer distances. In the U.S., MRFs, cullet suppliers, and cullet often tend to be further apart than is typically the case in Europe. These transportation costs can be prohibitive, which makes it less economically viable for glass recycling to take place. Another challenge is that landfilling costs tend to be lower in the U.S. than in Europe, which provides less financial motivation for glass recycling than is the case in Europe.

Despite these challenges, there have been some encouraging developments. At the state level, bottle bills have been passed in some states, where people must pay a deposit when they purchase beverages, but get a refund for the deposit when they return bottles for recycling. Consequently, states with bottle bill policies have significantly higher glass recycling rates than the national average. Meanwhile, some communities have also established bottle-recovery facilities. Solutions have also been developed by businesses. Ripple Glass was founded as a cullet supplier as part of Boulevard Brewing's efforts to encourage recycling of their beer bottles. They placed glass numerous collection bins throughout the Kansas City area, encouraging residents to fill them with glass bottles.

Case Study: SGD Pharma's Lightweight Glass Bottles

Glass has many desirable attributes. It is chemically inert, can offer good barrier properties, and has desirable aesthetics. The desirable aesthetics of glass bottles or jars makes them a popular choice for the packaging of premium, high-end items such as wines, certain medicines, and cosmetics.

While glass has many desirable characteristics, a couple of its shortcomings include its breakability as well as its heavier weight compared to other packaging materials. In late 2023, the glass packaging manufacturer SGD Pharma announced that they would be producing lightweight 200 mL bottles (with a brimful capacity of 215 mL) marketed under the name Nova. These new bottles would weigh only 140 g, while in comparison the 200 mL standard Dorado bottles (also manufactured by SGD Pharma) weigh in at 178 g [32]. A 20% reduction in carbon dioxide emissions is achieved through this weight reduction. While the reduction of materials consumed during production helps to reduce the emissions, there will also be fewer emissions associated with transportation due to the lower weight of these bottles.

Typically, a cosmetics bottle produced through the blow and blow process would have a weight-to-volume ratio (the brimful volume divided by the empty bottle weight) in the range of 1–3, with a lower ratio corresponding to a lighter weight. Impressively, the weight-to-volume ratio of SGD Pharma's Nova bottle is significantly lower, at 0.65 [32]. In addition to a weight reduction, SGD Pharma worked to optimize the glass distribution, thereby enhancing its mechanical properties. Consequently, despite being lighter weight, these bottles are designed to be able to withstand stresses that they may encounter during transportation and distribution.

Case Study: Reusable Bottles

The reuse of bottles is an important aspect of the sustainability equation. One example of the reuse concept being put into practice is the reusable glass jars that are in use by the German coffee company Cofi Loco, which is employing reusable glass bottles for its products, such as coffee beans. These bottles are equipped with a patented lid that allows the flavor and aroma of the coffee to be preserved. The beans are able to mature within the bottle, rather than being stored overnight to allow pressure to be released. With these bottles, the coffee beans are placed in the bottles immediately after they are roasted, and sealed. The specially designed lid allows any excess pressure to be released from the bottle, while preventing air from getting in [33, 34].

In December 2020, these bottles became available at 150 Edeka stores in Germany and are also sold through Cofi Loco's online store. Cofi Loco emphasizes sustainability as a major aspect of its brand, with these reusable glass bottles helping to further reinforce that theme.

8.7 Summary

Glass and metals are recyclable without loss of quality. In the United States, one-third of packaging glass is recycled. In some EU countries, the glass recycling rate approaches 95%. Glass has a lower environmental footprint when it is recycled compared to the production of new glass from virgin raw materials. Glass will continue to be used in packaging, although its use will likely become more focused toward luxury brands or high-value products such as wines, medicines, vaccines, and so forth [35]. The market for some beverage containers will also shift to metal cans because of their better environmental perception than plastics.

Problem Set for Chapter 8

1 Why is recycling aluminum more environmentally friendly than producing new aluminum from its ore?
   - A Because more energy is required to extract aluminum from its ore.
   - B Because $CaCO_3$ is used in the aluminum extraction process that becomes converted into CO_2.
   - C There is no significant difference between obtaining aluminum from ore or by recycling.
   - D None of the above.

2 Discuss why using recycled glass is more carbon dioxide efficient compared to producing it from raw materials.

3 In the United States, in 2015, the most recycled packaging materials **by percentage** among the following was:
   - A Glass.
   - B Wood.
   - C Steel.
   - D Aluminum.

4 Which of the following answers is true about MRFs?
   - A In MRFs, mixed recyclables are separated from each other.
   - B The existing MRFs have no problem with the separation of plastic bags from other recyclables.
   - C Dirty MRFs are used to separate single streams recyclable from each other.
   - D a and c.

5 Which of the following metals is not banned by Model Toxics Laws in Packaging?
   - A Nickel.
   - B Cadmium.
   - C Mercury.
   - D Lead.

6 Which of the following statements is true about closed-loop recycling?
  A Closed-loop recycling refers to recycling something into an identical or similar article.
  B Making new aluminum beverage cans out of old aluminum beverage cans is one example.
  C Making fiber for carpet from recycled PET bottles is an example of closed-loop recycling.
  D a and b.

References

1 United States Environmental Protection Agency. (2024). The U.S. Recycling System. https://www.epa.gov/recycle/recycling-united-states (accessed 25 January 2024).
2 National Recycling Coalition. (2024). Together, We are Recycling!, https://nrcrecycles.org/ (accessed 25 January 2024).
3 Association of European Producers of Steel for Packaging (APPEAL). (2024). Recycling rates of main packaging materials in Europe 2020. https://www.apeal.org/statistics/ (accessed 26 January 2024).
4 American Iron and Steel Institute and Steel Manufacturers Association. (2021). Technical Report: Determination of Steel Recycling Rates in the United States. https://www.steel.org/wp-content/uploads/2021/08/AISI-and-SMA-Steel-Container-Recycling-Rates-Report-Final-07-27-2021.pdf (accessed 20 November 2024).
5 Recycle logo FE, steel. (2006). Author: User Moebius1. Wikipedia https://en.wikipedia.org/wiki/Recycling_codes#/media/File:Recycling-Code-40.svg (accessed 20 November 2024).
6 American Iron and Steel Institute. (2021). Fact About American Steel Sustainability. https://www.steel.org/wp-content/uploads/2021/11/AISI_FactSheet_SteelSustainability-11-3-21.pdf (accessed 20 November 2024).
7 Sustainability. (2024). American Iron and Steel Institute. https://www.steelsustainability.org/recycling (accessed 20 November 2024).
8 American Iron and Steel Institute and Steel Manufacturers Association. (2021). Technical Report: Determination of Steel Recycling Rates in the United States. https://www.steel.org/wp-content/uploads/2021/08/AISI-and-SMA-Steel-Recycling-Rates-Report-Final-07-27-2021.pdf (accessed 20 November 2024).
9 United States Environmental Protection Agency. (2020). Advancing Sustainable Materials Management: 2018 Fact Sheet. https://www.epa.gov/sites/default/files/2021-01/documents/2018_ff_fact_sheet_dec_2020_fnl_508.pdf (accessed 20 November 2024).

10 Metal Packaging Europe. (2024). Metal packaging and the UN SDGs. https://metalpackagingeurope.org/sustainability (accessed 26 January 2024).

11 Steel for Packaging Europe. (2024). Data Room: Steel Packaging Facts. https://www.steelforpackagingeurope.eu/data-room/ (accessed 20 November 2024).

12 Azo Materials. (2012). Editorial Feature: Steel Recycling Principles and Practice. https://www.azom.com/article.aspx?ArticleID=7979 (accessed 20 November 2024).

13 Rinkesh. (2024). Metal Recycling: How to Recycle Metal and its Importance. Conserve Energy Future. https://www.conserve-energy-future.com/recyclingmetal.php (accessed 20 November 2024).

14 The aluminum can advantage: sustainability key performance indicators. (2021). https://www.aluminum.org/sites/default/files/2021-11/2021_KPI_Report_1.pdf

15 The Aluminum Association. (2021) The Aluminum Can Advantage. http://www.aluminum.org/aluminum-can-advantage (accessed 20 November 2024).

16 Alupro. (2019). Aluminium drink cans hit 75% recycling rate, an increase of 38% in just eight years. https://alupro.org.uk/aluminium-drink-cans-hit-75-recycling-rate-an-increase-of-38-in-just-eight-years/ (accessed 26 January 2024).

17 Aluminium beverage can recycling in 2021 at a new record level of 76%. (2024) https://european-aluminium.eu/wp-content/uploads/2024/02/EA-MPE_BevCan-2021-Recycling-Results_Press-Release_23-February-2024final.pdf (accessed 12 May 2024).

18 Moore, D. (2019). Aluminium beverage can recycling in Europe hits record 74.5%. Circular Online. https://www.circularonline.co.uk/news/aluminium-beverage-can-recycling-in-europe-hits-record-74-5/ (accessed 26 January 2024).

19 Norsk Hydro ASA. (2024). The benefits of using aluminium for beverage cans. https://www.hydro.com/globalassets/download-center/publications/210218-hydro-papers-cans-nov2020-eng.pdf (accessed 26 January 2024).

20 GTEK Magnet Co. (2023). Working principle of eddy current separator. https://gtekmagnet.com/how-does-an-eddy-current-separator-work/ (accessed 17 October 2023).

21 Jujun, R., Yiming, Q., and Zhenming, X. (2014). Environment-friendly technology for recovering nonferrous metals from e-waste: eddy current separation. *Resour. Conserv. Recycl.* 87: 109–116. https://doi.org/10.1016/j.resconrec.2014.03.017

22 FEVE - The European Container Glass Federation. (2023). EU's glass value chain confirms glass collection rate steady progress at 80.1%. https://feve.org/eu-glass-value-chain-80-collection-rate/ (accessed 19 August 2024).

23 FEVE – The European Container Glass Federation. (2018). Latest glass packaging recycling rate steady at 76%. https://feve.org/glass_recycling_stats_2018/ (accessed 20 November 2024).

24 Bisht, K. and Ramana, P.V. (2018). Sustainable production of concrete containing discarded beverage glass as fine aggregate. *Construct. Build. Mater.* 177: 116–124. https://doi.org/10.1016/j.conbuildma.

25 Gowtham, R., Manikanda Prabhu, S., Gowtham, M., and Ramasubramani, R. (2021). A review on utilization of waste glass in construction field. *IOP Conf. Ser.: Mater. Sci. Eng.* 1130: 012010. https://doi.org/10.1088/1757-899X/1130/1/012010.

26 Glass Packaging Institute. (2024). Why recycle glass?https://www.gpi.org/why-recycle-glass (accessed 26 January 2024).

27 Rogoff, M.J. and; Gardner, B. (2016). High costs and limited markets put a crack in glass recycling. MSW Management. https://www.scsengineers.com/wp-content/uploads/2016/05/High_Costs_Limited_Markets_Put_Crack_in_glass_Recycling.pdf (accessed 27 January 2024).

28 Glass recycling, Rumpke waste & recycling. (2024). https://www.rumpke.com/for-your-home/recycling/glass-recycling (accessed 27 January 2024).

29 Alves, B. (2024). Number of recycling facilities in the European Union 2010-2020. Statista. https://www.statista.com/statistics/1316189/number-of-recycling-facilities-european-union/ (accessed 20 November 2024).

30 Taha, B. and Nounu, G. (2008). Properties of concrete contains mixed colour waste recycled glass as sand and cement replacement. *Construct. Build. Mater.* 22: 713–720. https://doi.org/10.1016/j.conbuildmat.2007.01.019.

31 Jacoby, M. (2019). Why glass recycling in the US is broken. Chemical & Engineering News. https://cen.acs.org/materials/inorganic-chemistry/glass-recycling-US-broken/97/i6 (accessed 25 March 2024).

32 SGD. (2024). Pharma unveils NOVA – a new lightweight glass bottle. Glass Online. https://www.glassonline.com/sgd-pharma-unveils-nova-a-new-lightweight-glass-bottle/ (accessed 26 March 2024).

33 Messe Düsseldorf GmbH. (2022). Refill and reuse. https://www.interpack.com/en/Media_News/Tightly_Packed_Magazine/COSMETICS_PACKAGING/News/Refill_and_Reuse (accessed 20 March 2024).

34 Orsolya, G. Coffee from the reusable bottle. Trade Magazin. (2020). https://trademagazin.hu/en/kave-ujrahasznosithato-uvegben/ (accessed 20 November 2024).

35 Rabnawaz, M. *Sustainable Packaging Materials: Winning Solutions*. Carl Hanser Verlag GmbH Co KG.

9

Paper and Paperboard Recycling

Products made of paper or paperboard materials are classified by the Environmental Protection Agency (EPA) as nondurable goods or alternatively as containers and packaging. Some examples of nondurable goods are newspapers, office paper, tissue paper, as well as paper plates or cups. Items like milk cartons, corrugated boxes, as well as bags and sacks are examples of containers and packaging. Packaging is both an important source of recycled paper and paperboard as well as an important market for these materials, whether they are sourced from packaging or from other paper-based materials.

In 2018, the recycling rate for paper and paperboard was 68.2%, with around 46 million tons of these materials being recycled that year [1]. This recycling rate was higher than those for other materials found in municipal solid waste (MSW). Meanwhile, 4.2 million tons of paper and paperboard were incinerated for energy generating, accounting for 6.2% of all paper and paperboard waste. In 2018, landfills received 17.2 million tons of MSW paper and paperboard. This means that 25.5% of paper and paperboard ended up in landfills in 2018 (Figure 9.1).

1960–2018 data on paper and paperboard in MSW by weight (in thousands of U.S. tons).

Degradation, Stabilization, and Recycling of Packaging Materials, First Edition.
Muhammad Rabnawaz, Susan E.M. Selke, and Ian Wyman.

Management pathway	1960	1970	1980	1990	2000	2005	2010	2015	2017	2018
Generated	29,990	44,310	55,160	72,730	87,740	84,840	71,310	68,050	67,010	67,390
Recycled	5,080	6,770	11,740	20,230	37,560	41,960	44,570	45,320	44,170	45,970
Composted	—	—	—	—	—	—	—	—	—	—
Combustion with energy recovery	—	150	860	8,930	9,730	7,800	4,740	4,450	4,490	4,200
Landfilled	24,910	37,390	42,560	43,570	40,450	35,080	22,000	18,280	18,350	17,220

Source: Paper and Paperboard information is from the American Forest & Paper Association.

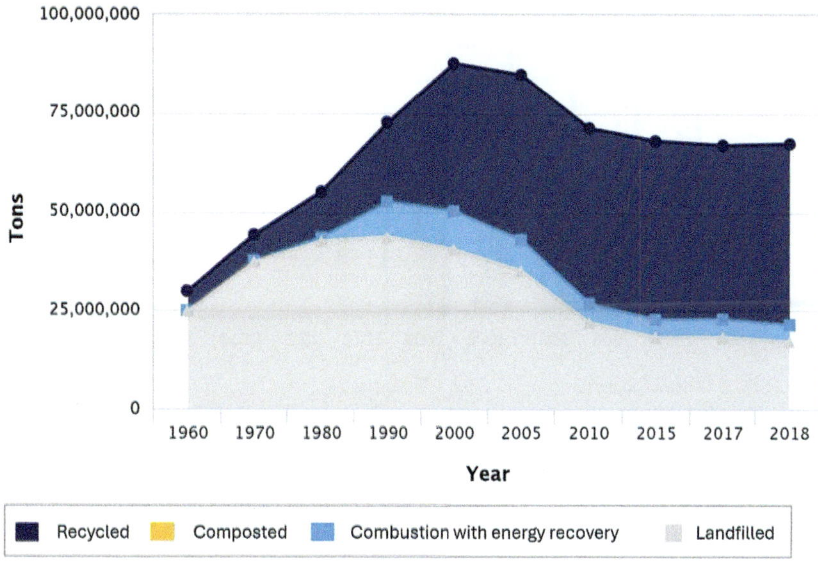

Figure 9.1 Waste management scenarios for paper and paperboard materials [1] / U.S. Environmental Protection Agency / Public Domain.

Figures 9.2 and 9.3 show the recycling amounts and rates for paper and paperboard in nondurable goods. Note that **newspapers and office paper** account for the largest total amounts, as well as having the highest recycling rates in the nondurable category.

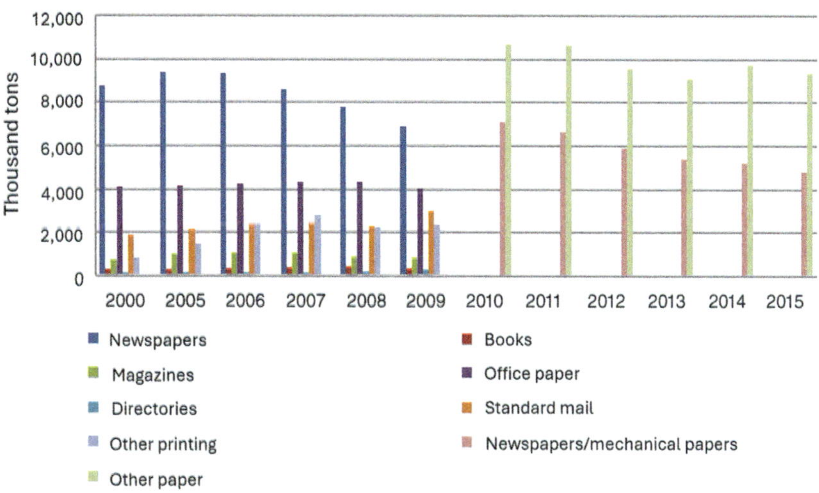

Figure 9.2 Recycling of paper and paperboard from nondurables.

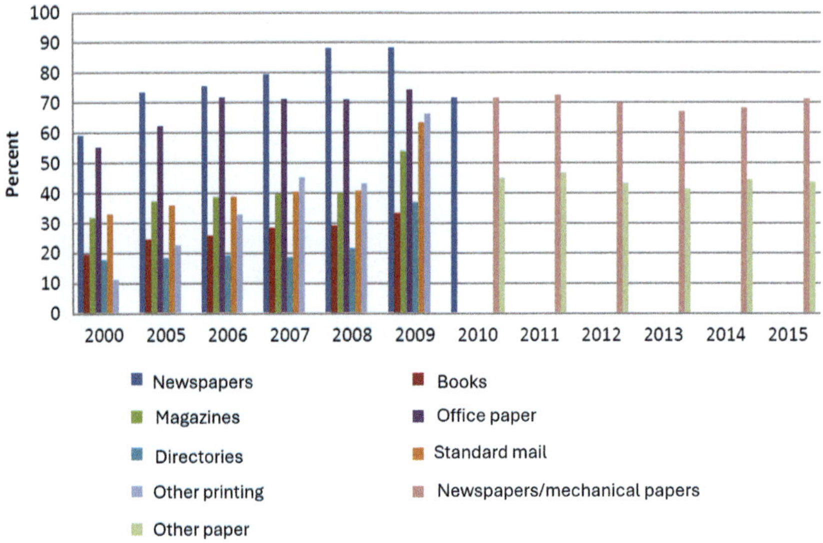

Figure 9.3 Recycling rates for paper and paperboard from nondurables.

Since 2010, the EPA no longer separates paper in nondurables into as many categories as it did in the past; rather, it lumps newspapers and directories into the "Newspapers/Mechanical Papers" category and all the others into the "Other Paper Nondurable Goods" category. Also observe the rather steep decline in total amounts recycled and the modest decline in recycling rates (together, a signal of decreasing production) in these categories through 2013. There was a small increase in amounts of "other paper" that was recycled in 2014, but a decline again in 2015. The rates for newspaper/mechanical paper recycling increased somewhat in both 2014 and 2015. Meanwhile, the recycling rates for other paper increased in 2014 but fell slightly in 2015.

In the packaging fraction of MSW, corrugated boxes account for the vast majority of recycled paper and paperboard and also have the highest recycling rate by a wide margin. In 2015, the recycling rate for corrugated boxes was 92.3%, breaking the previous record of 91% in 2011 (Figures 9.4 and 9.5). The EPA has changed its categories for paper packaging materials, dividing them into corrugated boxes and "other paper packaging."

Recycling information from the American Forest and Paper Association (AF&PA) is available at https://www.afandpa.org/statistics-resources. They provide more recent information than does the EPA. The AF&PA had established a goal to exceed 70% recovery by 2020, and they reported that the overall 2018

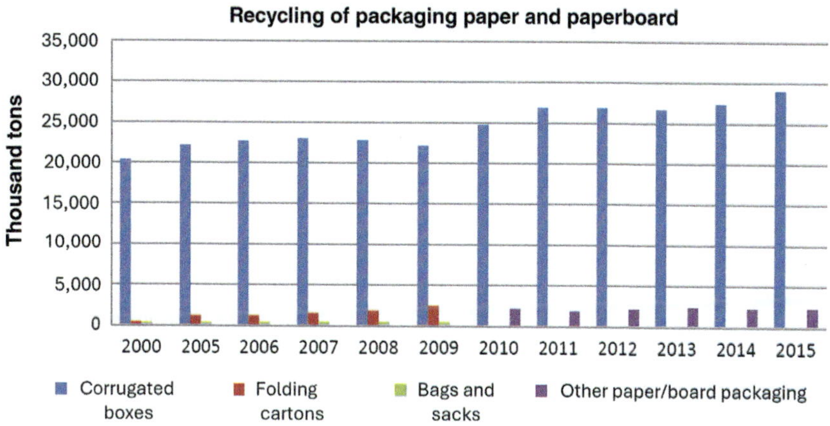

Figure 9.4 Amounts of paper and paperboard packaging recycled in the United States during various years between 2000 and 2015.

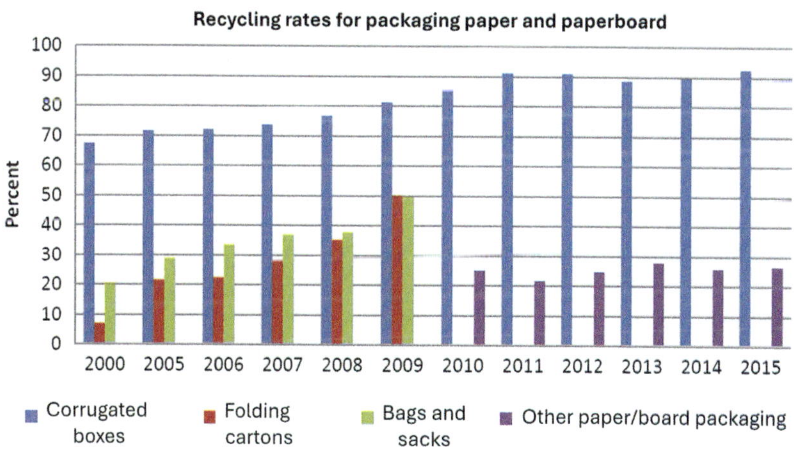

Figure 9.5 Recycling rates for paper and paperboard packaging in the United States for various years between 2000 and 2015.

paper/paperboard recycling rate set a new high at 68.1%. The paper recycling rate had declined slightly in 2020, but had recovered to 67.9% in 2022. The tonnage of paper recycled has not increased to the same extent as the recycling rate, due to **falling usage**. There are nice charts on the page referenced above (specific links to these charts are provided in the footnotes cited below) that show that the tonnage recycled, though it grew substantially in 2018 to 52,676 thousand tons, remains below the record high of 54,325 thousand tons in 2007, and the tonnages recycled in 2020 and 2022 were lower than those recycled in 2018 [2]. Detailed charts from AF&PA are shown with data points for every two years, with charts for both even and odd years available at Refs [3, 4].

The AF&PA reported that the recycling rate for corrugated (OCC – old corrugated containers) was 96.4% in 2018, after falling in 2017. The rate had declined in 2020, but has made a partial recovery to 93.6% in 2022. The tonnage recycled also grew to 35,923 thousand tons, a new high.

For newspapers/mechanical papers, in contrast, the AF&PA reported that total recovery in 2018 fell by 21%, causing a decline in the recycling rate to 66.2% for 2018, compared to 77.3% in 2017. Most of this decline was attributed to a drop in exports. While this rate recovered to over 70% in 2020, it has declined to approximately 55% in 2022 [5].

9.1 Sorting Phase

To start the recycling of paper materials, it is important to separate wastepaper into categories such as newspaper, office paper, and magazine paper. This sorting is performed because each type of paper will undergo different kinds of treatments during the recycling process.

9.2 Processing Phase

During the processing phase, the sorted paper is soaked in water in an industrial pulper; optionally, chemicals are also added. The temperature is typically kept at 55 °C to help break down the paper into fiber quickly, and the pH kept at around 7.0.

9.3 Processing Phase: Pulp Screening and Cleaning

The next step is to screen and clean the pulp. The purpose is to filter out minor contaminants such as plastic residues and adhesives. Further, the pulp may be centrifuged in expansive cone-shaped cylinders to eliminate heavier contaminants such as metal staples via centrifugal forces, while lighter contaminants accumulate at the cone's core for removal. Sometimes, the centrifugal force is adjusted to separate plastic impurities from the fiber, and in such cases now plastic is the lighter impurity.

9.4 Processing Phase: Deinking

After screening and cleaning, the recycled pulp must be de-inked. De-inking is accomplished through a synergy of mechanical processes, such as shredding and chemical interventions. Minute ink particles are flushed out using water, whereas larger ink particles are removed. During this stage, some materials, such as "stickies," are also removed via a flotation method [6, 7]. During flotation, air bubbles and surfactants are introduced into the pulp, causing ink and "stickies" to adhere to the bubbles, which are subsequently skimmed off to isolate the cleansed pulp.

9.5 Processing Phase: Refining, Color Stripping, and Bleaching

During this stage, the pulp is mechanically agitated to trigger fiber swelling and enable fiber separation into individual fiber threads. This is also the stage during which recovered paper can be treated for color enhancement by the removal of dyes, such as by using chemical agents. If a white color is desirable, then bleaching agents such as hydrogen peroxide or chlorine dioxide can be used to improve the whiteness and brightness of the paper.

9.6 Processing Phase: Papermaking

At this stage of the recycling process, the purified and thoroughly dispersed recycled pulp is ready to be converted into recycled paper. The recycled fiber can be used alone or combined with virgin fiber, depending on the end application. Virgin fibers are longer and stronger and thus enhance the strength of the paper. The pulp is homogenized with water and chemicals such as sizing agents to achieve a very dilute solution; essentially, 99.5 wt% is water. This aqueous dilute solution is then injected into the headbox of the paper-making machine. Due to the porous nature of the mesh, the water starts to drain from the pulp, facilitating bonding and turning the recycled fiber into a new recycled paper sheet. The sheet is further heated under press rollers to remove any residual water.

Loss of fibers during recycling is significant, typically around 10%. Moreover, the fibers that remain are, on average, shorter. This is the primary reason that recycled paper is generally not as strong as virgin paper. Blending virgin and recycled fiber can enhance the paper's strength. Using a higher basis weight of paper can also improve the performance.

9.7 Recyclable Versus Repulpable Paper

The distinction between "repulpable" and "recyclable" paper is based on the suitability of the recycled pulp for making a good quality paper.

Repulpable Paper: This is paper that can be broken down into pulp through a repulping process. However, not all repulpable paper is suitable for recycling into new paper products of high quality. For instance, laminated paper can be repulped, but the presence of plastic can prevent it from being recycled into high-quality new paper.

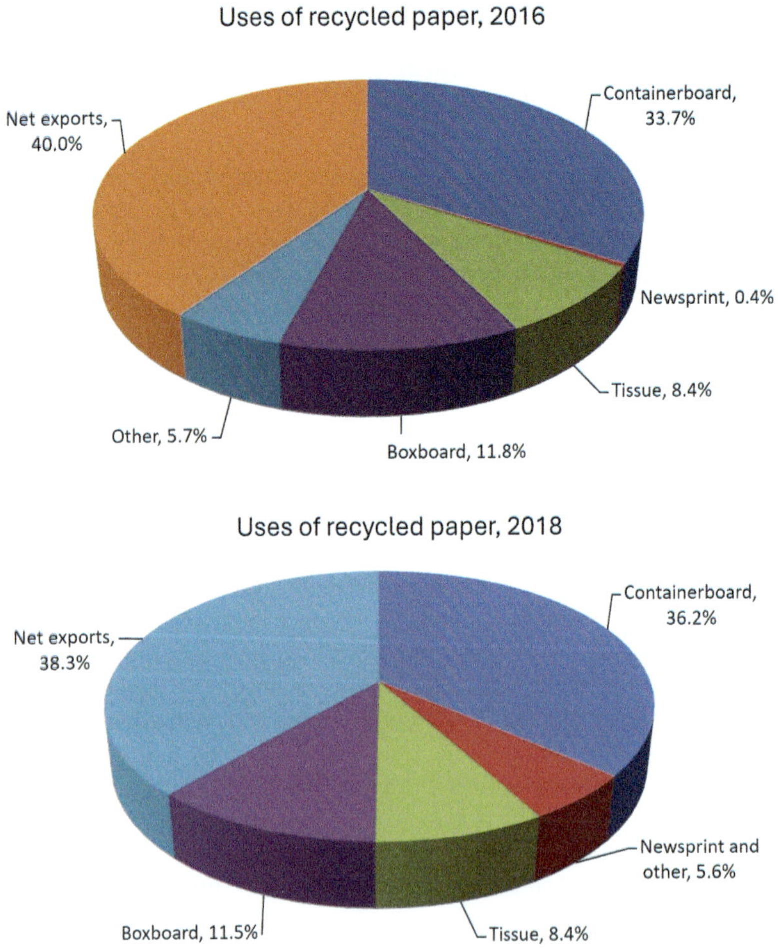

Figure 9.6 Uses of recycled paper in 2016 and 2018. *Source:* AF&PA.

Recyclable Paper: This refers to paper that can be not only repulped but also processed into new paper products that maintain a high quality.

An example of repulpable, but non-recyclable paper, is laminated paper. The lamination material contaminates the paper pulp, making it unsuitable for recycling into new, high-quality paper. On the other hand, wet-strength paper is an example of non-repulpable paper because the added chemicals or processes that give it wet strength also prevent it from being successfully repulped and recycled.

9.8 Uses of Recycled Paper

Paper recycling generally operates in an open-loop fashion, with downgrading of a portion of the recovered fibers into less demanding applications commonplace.

Figure 9.6 shows uses of recycled paper in 2016 and 2018 as reported by the AF&PA. The share going to newsprint has decreased substantially – and you may have noticed your local newspaper getting both smaller and thinner. Less than 0.5% of recycled paper was used for newspapers in 2016. In the 2018 report, the AF&PA combined newsprint with other paper (printing, writing, kraft packaging paper, construction paper, and board, molded pulp, and so forth).

Old corrugated boxes (OCC) are used predominantly in the manufacture of new boxes (the containerboard shown in Figure 9.6). Another major use is in the manufacture of recycled paperboard, such as that which is commonly used to make folding cartons (boxboard, in Figure 9.6). The category "boxboard" includes folding cartons but also gypsum wallboard facings. Exports are also an important market; exports to the Far East are particularly significant, though changes in import policies in China and elsewhere have had some profound effects on this market [8].

Figures 9.7–9.9 show the uses of particular grades of recycled paperboard in 2007 as reported by the Paper Industry Association Council, which has now been

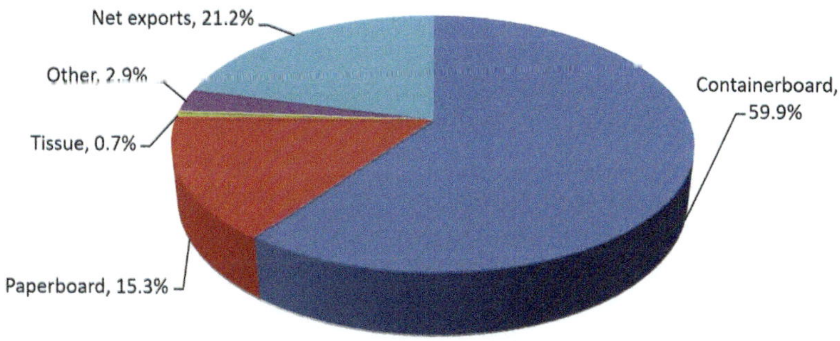

Figure 9.7 Uses of OCC in 2007 (as percentages).

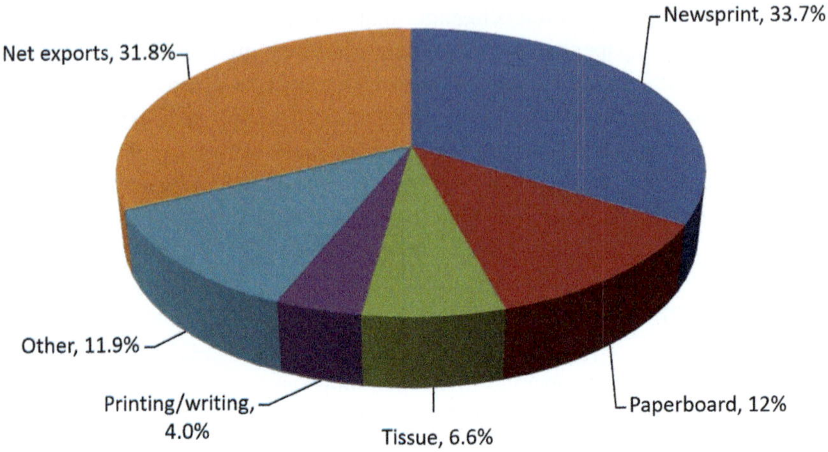

Figure 9.8 Uses of old newspapers in 2007 (as percentages).

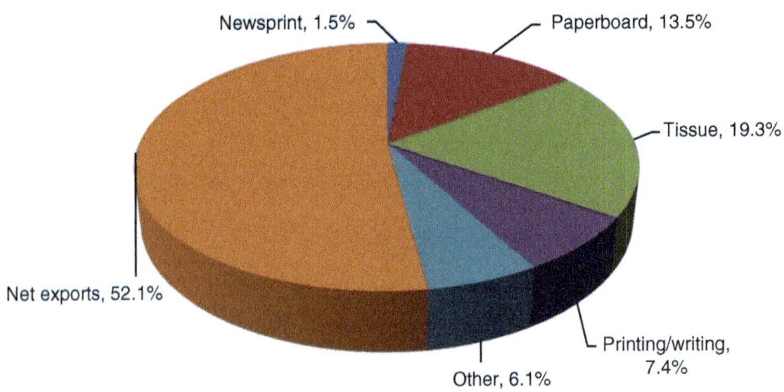

Figure 9.9 Uses of recycled printing and writing paper in 2007.

merged into the AF&PA. We have not been able to find more recent information, but it is very likely that there have been some significant changes, such as less newspaper going into newsprint and likely more corrugated going into new corrugated. Similar information for Europe is provided in Section 9.8.1.

In the past, the main use of recycled newspapers was to make new newsprint. However, that use has declined significantly since the 2007 data shown below. Recycled paperboard is a significant market, and a lot of recycled newsprint is exported. A major export destination is Canada, which is a large manufacturer of both recycled and virgin newsprint, a significant amount of which is then imported by the United States.

Recycled printing and writing paper is also used to manufacture recycled paperboard, but export is the largest single market.

Notice that the second most common use of recycled printing and writing paper after exports is to produce tissue paper, while relatively little of this stream goes back into the manufacture of printing and writing paper. As we can see from the figures above, paper recycling is characterized by a significant mixture of open-loop and closed-loop recycling.

It is important to recognize that, in contrast to the recycling of metals and glass, the recycling and use history of paper fibers have a significant impact on the strength of the paper made from the fibers. Degradation of the fibers, and hence loss of performance attributes, is to a significant degree unavoidable in the recycling of paper.

9.8.1 Paper Recycling in Europe and Other Areas

The Confederation of European Paper Industries (CEPI) (https://www.cepi.org/) reports the use of recycled paper by sector, as shown in Figure 9.10 [9]. Containerboard, carton board, writing paper, as well as wrappings and other packaging are the most common uses for recycled paper, accounting for approximately 70% of the total. Of course, some of this is due to the sheer volume of

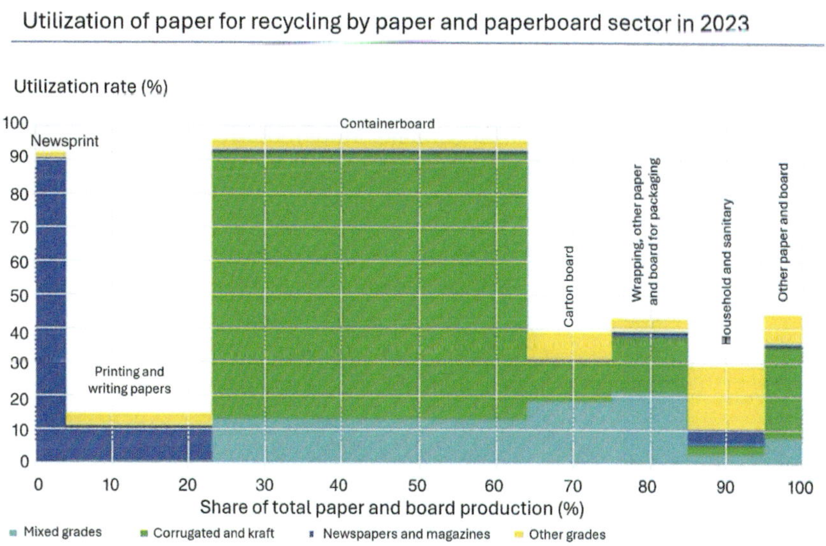

Figure 9.10 Uses and origin of recycled paper in Europe, by sector, for the year 2023. *Source:* Image obtained from and used with permission from the Confederation of European Paper Industries (CEPI).

packaging paper, but as can be seen in Figure 9.10, the utilization rate (the proportion of all production that is from recycled fiber) is also very high, particularly in the "containerboard" category. The "containerboard" (also referred to as "case material") generally refers to the paper or cardboard used to make the exterior of products such as books and packaging. This can include various types of paperboards or cardboard that are more robust than typical paper sheets. Note that this figure also shows the various grades of recycled paper used in each category.

CEPI reported an overall paper recycling rate in Europe of 71.6% for 2018, and signatories of the European Declaration on Paper Recycling had a goal of reaching 74% by 2020 [10]. After increasing to 73.3% in 2020, the overall recycling rate had apparently declined slightly to 71.4% in 2021 [11]. Rates by country were not reported.

The European Recovered Paper Council, in their Monitoring Report 2017, http://www.paperforrecycling.eu/download/900/, reported a 2017 recycling rate for paper in Europe of 72.3%. The rate for paper and paperboard packaging was 82.1%, compared to 79.5% for steel packaging, 74% for glass packaging, and 73% for aluminum packaging. More recently, in 2021, the respective packaging recycling rates were 82.5%, 74.8%, 74.9%, and 39.7% for paper/paperboard, metal, glass, and plastic packaging [12]. In 2023, 19 European countries had recycling rates of greater than 70% (up from 15 countries in 2022), while 4 countries were under 60% in that year (down from 7 countries in 2022) [12]. The growth in paper recycling in Europe is shown in Figure 9.11. The overall recycling rate for all paper and paperboard consumed grew to 73.8% of all paper and paperboard that was consumed in 2020, but this rate decreased slightly in 2021 and 2022 to 72.8% and 71.1%, respectively. This decrease was attributed to the energy crisis as well as the impact of the war in Ukraine. More recently, there was a significant recovery in the paper and paperboard recycling rate to 79.3%. While this

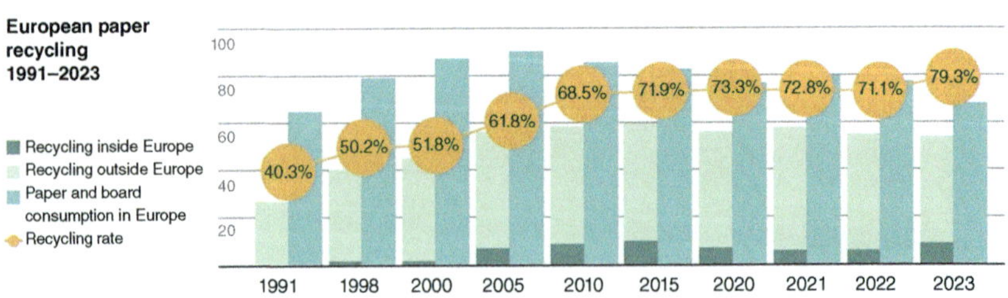

Figure 9.11 European paper recycling, 1991–2023. *Source:* Monitoring Report 2023: European Declaration on Paper Recycling 2021–2030, Confederation of European Paper Industries (CEPI), https://www.paperforrecycling.eu/download/1704/?tmstv=1721407323; Reproduced with permission.

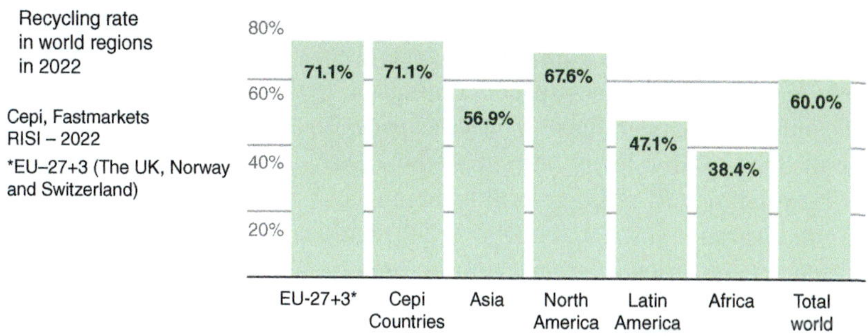

Figure 9.12 2022 paper recycling rates by region. *Source:* Monitoring Report 2023: European Declaration on Paper Recycling 2021–2030, European Paper Recycling Council (EPRC)/Confederation of European Paper Industries (CEPI), https://www.paperforrecycling.eu/download/1704/?tmstv=1721398811. Reproduced with permission.

impressive increase was attributed partially to recycling mill investments in that year, but it was also noted that this increase was also partially attributable to a low level of apparent consumption during that year. The report noted that the annual recycling rate can vary for a given year due to external economic and geopolitical factors. Signatories of the European Declaration on Paper Recycling have committed to reaching a 76% recycling rate by 2030 [13].

Recycling rates by world region for 2022 were reported as shown in Figure 9.12.

9.9 Contamination Issues

Another way in which paper recycling differs from that of metal and glass is the issue of contaminants that may migrate into products that are packaged in recycled materials. For glass and metal, this is not much of a concern. Any organic contaminants in those materials would be destroyed by the very high temperatures involved in melting and reforming the materials. There are potential concerns about contaminants such as heavy metals, but these are typically addressed by controlling the source and composition of the starting materials.

It is more difficult to remove contaminants from paper. Microbiological contamination of recycled paper is typically not a concern, as the time/temperature combinations encountered during the drying process will kill most organisms. However, these temperatures are not high enough to drive off or destroy various types of other contaminants, and these can potentially migrate into the package contents, including foods.

This issue hit the press in 2010 when concerns were raised in Germany about mineral oil in cartons made from recycled paper. The origin of the mineral oil was thought to be inks used in newspaper printing [14, 15]. Mineral oils are certainly not the only possible migrants from recycled paper. In 2006, there were reports of contamination of dry foods with trimethyldiphenylmethanes from recycled paper [16] and in 2004 of fluorescent whitening agents [17]. The U.S. Food and Drug Administration (FDA) has established regulations to address these issues, as have agencies in a variety of other countries. Here are some more links to the contamination of recycled paper [18–21].

Generally, dry, nonoily foods are less susceptible to contaminant migration than liquids or oily foods. That is the reason why the studies above have concentrated on dry foods – direct contact between recycled paper and food products is generally not allowed for liquids or oily foods, so those situations did not arise.

There are reports that food companies have changed some of their practices in the use of recycled paper for food packaging, as a result, in particular, of the mineral oil concerns [22]. Of particular concern is that there is evidence of migration through film liners in some cases. One major tactic in dealing with this problem is to either change the inks used in printing newspaper, or to remove newspaper from the recycled paper mix.

In the US, the presence of mineral oil in recycled paper has not been much of an issue. Most U.S. newspapers use vegetable oil-based inks (often soy) rather than mineral oil [23].

Case Study: Voluntary Standard for Repulping and Recycling Corrugated Fiberboard

Treated corrugated fiberboard containers are widely used in packaging, such as the shipping of produce, meat, and seafood. However, in some cases, the ways in which they are treated to impart them with water resistance can adversely affect their recyclability. These materials are often treated in different ways by a given company, with some of these ways of imparting water resistance having more or less of an impact on the recyclability than other methods. Unfortunately, it is often not readily apparent at the recycling facilities which particular treatment was used to impart barrier properties on a given piece of fiberboard. In recognition of this issue, the Fibre Box Association (FBA) and the AF&PA established a joint committee to assess the repulpability and recyclability of corrugated fiberboard so that a minimum level of repulpability and recyclability could be established for moisture barrier-treated and coated cardboard, with those meeting these criteria being labeled as recyclable and repulpable [24]. The committee thus established a set of Voluntary

Standard tests to evaluate the repulpability and recyclability of fiberboard and cardboard. In essence, the purpose of these tests was to encourage the adoption of treatment methods for corrugated fiberboard that can impart water resistance or other desirable attributed while still ensuring that the material is recyclable and repulpable, with these approaches replacing methodologies that are more detrimental to the recyclability and repulpability. In addition, the tests were intended to provide a repeatable methodology for simulating commonly used repulping and recycling methods so that the impacts of the repulping and recycling treatments on the corrugated fiberboard, and in turn on their impacts on containerboard mill processes and final products can be evaluated. With this test established, manufacturers of cardboard or fiberboard packaging can voluntarily complete these tests and can register their treatment with the FBA if they have demonstrated that their treatment or coating methods pass the Standard test, as certified by an FBA-approved lab.

In summary, the repulpability test involves the repulping of corrugated treated fiberboard in a modified waring blender and a British disintegrator in water at a pH of 7 and at 125 °F. The repulped material is subsequently separated with a screen that has 0.10-inch slots (or smaller), and the fiber recovery is then determined as a percentage with respect to the amount of fiber that was initially introduced into the repulping system.

During the recyclability tests, a minimum of 20% of treated corrugated fiberboard is mixed with untreated corrugated fiberboard using a laboratory-scale pulper at a pH of 7 and a temperature of 125 °F. For comparison, completely untreated corrugated fiberboard is pulped under similar conditions. Each of the pulped samples is then successively passed through a pressure screen with a basket that has 0.062 inch holes, either the same screen again or one with 0.010 inch slots, and a reverse centrifugal separator. Subsequently, 3.0 g handsheets are produced from the final stage accepts. These handsheets are pressed and dried before their performances are evaluated. The criteria used for these performance evaluations include the slide angle, short-span compression strength (STFI), burst strength, and water droplet penetration.

Corrugated fiberboard is classified as repulpable and recyclable based on these tests if it satisfies three main criteria, including fiber yield, has an acceptable operational impact, and offers acceptable product requirements. The fiber yield obtained from the repulpability test must be either 80% based on the total weight or 85% based on the dry fiber introduced to the pulper. The operational impacts are considered to be acceptable if the repulping can be performed without requiring an acid wash to clean the screen used in the repulpability test or dismantling the pressure screens for cleaning before completing the recyclability test. In addition, there should be no visible evidence of deposition on any parts of the disintegrator during the repulping test or in the repulper during the recyclability test. The product requirements are considered to be fulfilled if the handsheets made from the recyclability tests do not exhibit any significant difference from those of handsheets made from the control sample. In addition, the spot count should be 15 counts or less, or no more than 30% higher than the counts of the control, with an average of 0.4 mm^2 or less averaged over three sheets. The decrease in the slide angle of handsheets made from recyclability test samples should be no greater than 15% in comparison to handsheets made from control samples. In addition, the burst strength and STFI should have no more than a 10% decrease in comparison with the control. And last, the water droplet penetration of handsheets made from recyclability test samples should not exceed that of the control handsheets by any more than 200 seconds.

Case Study: NextGen Consortium (Paper Cups)

The NextGen Consortium was founded in 2018 to address the issue of cup waste. This initiative involves a threefold strategy that includes: (1) advancing reusable cups; (2) employing materials science technology to ensure that the materials used in cups are easier to recover; and (3) developing the waste recovery and recycling infrastructure so that cups can be more readily recovered and recycled. A challenge encountered with paper cups is that they contain a polymer layer (often polyethylene) that lines their interior to help prevent drinks from leaking through the paper and to help retain heat. The presence of this coating or liner can impede recycling efforts, and currently, only 11% of communities in the United States accept paper cups in their recycling streams. However, there are 30 paper mills in the United States that are willing to accept bales containing recovered cups, thus demonstrating that there is some demand for this type of waste. The NextGenConsortium along with Moore & Associates and FPI have reported that a further 15 mills in North America are either able to process paper cups or are willing to test their acceptability [25]. Although paper cups account for a small proportion of the overall waste stream, more and more beverage companies and retailers are seeking recycling solutions for single-use paper cups.

In a typical paper recycling process, paper is fed into a pulper and mixed with water as well as other chemicals that help the paper material to be broken down. The paper is agitated so that it can be separated from nonpaper components (such as coatings or liners, adhesives, or other materials that may have been mixed with the paper). A slurry is thus obtained, which is then passed through screens to remove any undesired content. The recovered paper fiber is then cleaned and introduced into a paper machine. This recovered fiber may be either mixed with virgin (unrecycled) paper fiber or processed as completely recycled fiber, depending on the final application. It is noteworthy that paper cups tend to be composed of long cellulose fiber, which tends to be more valuable than short cellulose fibers, as the longer types of fibers are more versatile than the former. This can represent another motivation for finding effective ways to recycle paper cups.

In one example, the packaging company Sonoco began accepting paper cups at one of their mills located in Hartsville, South Carolina. Prior to accepting these cups, Sonoco conducted evaluations in cooperation with the NextGen Consortium. These tests were performed initially at the laboratory-scale with single- and double-side coated cups, and subsequently at a larger scale. During the large-scale tests, almost 20 tons of cupstock and cups were fed into their pulper as well as other forms of mixed paper. The results of these tests indicated that their Hartsville mill would be able to accept paper cups without encountering problems.

Although Sonoco hopes to expand paper cup acceptance to its other plants across the globe, they noted that testing must be performed at each plant before the paper cup acceptance is expanded to their other facilities. Scott Byrne, the Director of Global Sustainability Services at Sonoco, noted that each mill tends to be unique, with different equipment, operating conditions, and with different material compositions that are received. Therefore, it is critically important to test against the operating conditions of a given mill in order to determine what approach provides the best performance for each mill.

9.10 Concluding Remarks

Paper is a highly desirable material from an environmental perspective, and it is becoming widely adopted as an alternative to plastic packaging. One of its key shortcomings is that it tends to have poor water resistance due to its hydrophilic nature as well as its porous nature, which also imparts it with poor grease resistance. While these shortcomings can be addressed through the use of coatings, the coatings can pose issues during the repulping and recycling processes in some cases. Research is progressing to develop sustainable coatings which can be readily separated from the paper fibers. Ideally, it is highly desirable for these coatings to be environmentally friendly and free of potentially harmful substances such as per- and polyfluoroalkyl substances (PFAS). The selection of suitable inks and sizing agents that can accommodate paper repulping and recycling is also an important consideration. In contrast with glass or metals which can be recycled for unlimited cycles, paper fibers tend to degrade during recycling, so that there are some limitations with the recyclability of paper, as approximately 10% of the fibers are lost in a typical recycling process and the remaining fibers tend to be shorter. Although paper products can be prepared using solely recycled fiber, the quality of the resultant paper tends to be lower than that of paper prepared using virgin fibers. In many cases, virgin fibers are often used in combination with recycled fibers for paper production so that the virgin fibers can help to strengthen the paper. Consequently, paper recycling tends to involve a combination of open-loop and closed-loop processes. Despite these challenges, paper is a renewable, highly versatile, lightweight choice for many packaging applications, and it is anticipated to have an expanding role as environmental considerations continue to grow in prominence.

Problem Set for Paper Recycling

1 Which of the following packaging materials is compostable?
 - **A** Glass.
 - **B** Metal.
 - **C** Polyethylene.
 - **D** Uncoated paper (and some coated paper).

2 Which of the following answers is true about the recycling of paper and paperboard?
 - **A** Paper and paperboard are recycled via a closed loop only.
 - **B** Paper and paperboard are recycled via open loop only.
 - **C** Both closed- and open-loop recycling is used for paper and paperboard recycling.
 - **D** None of the above.

3 What is the difference between repulpable and recyclable paper?
 - **A** Repulpable refers to the ability of paper to be broken down to its pulp under standard repulping conditions without ensuring its suitability for producing high-quality recycled paper.
 - **B** Recyclable paper can be broken down to its pulp under standard repulping conditions, with the resulting pulp being suitable for creating high-quality recycled paper.
 - **C** Both a and b are correct.
 - **D** None of the above.

4 How is the deinking of recycled paper performed?
  A Flotation.
  B Surfactant treatment.
  C Both a and b.
  D None of the above.

References

1 United States Environmental Protection Agency. Paper and paperboard: material-specific data. https://www.epa.gov/facts-and-figures-about-materials-waste-and-recycling/paper-and-paperboard-material-specific-data (accessed 17 October 2023).
2 American Forest & Paper Association. Paper recycling graph. https://www.afandpa.org/statistics-resources/paper-recycling-graph (accessed 1 October 2023).
3 American Forest & Paper Association. (2022). Unpacking continuously high paper recycling rates. https://www.afandpa.org/news/2022/unpacking-continuously-high-paper-recycling-rates (accessed 30 January 2024).
4 American Forest & Paper Association. (2023). U.S. Paper and Cardboard Recycling Rates Continue to Hold Strong in 2022. https://www.afandpa.org/news/2023/us-paper-and-cardboard-recycling-rates-continue-hold-strong-2022 (accessed 30 January 2024).
5 American Forest & Paper Association (AF&PA). (2023). Recycling of old newspapers and mechanical papers graph. https://www.afandpa.org/statistics-resources/recycling-old-newspapers-and-mechanical-papers-graph (accessed 30 January 2024).
6 U.S. Department of Energy. (2023). Removal of wax and stickies from OCC by flotation. https://www1.eere.energy.gov/manufacturing/resources/forest/pdfs/removal_waxstickies.pdf (accessed 17 October 2023).
7 Tandon, R., Bhardwaj, N.K., Mathur, R.M., and Kulkarni, A.G. (2001). Contaminants in recycled paper and their removal. *IPPTA* 13: 29–38.
8 Smalley, M. (2018). Markets for mixed paper. Recycling Today. https://www.recyclingtoday.com/article/making-marketable-mixed-paper-bales/ (accessed 18 October 2023).
9 Monitoring report 2023: European declaration on paper recycling 2021-2030. (2024). Brussels. https://www.paperforrecycling.eu/download/1704/?tmstv=1721398811 (accessed 18 July 2024).
10 Confederation of European Paper Industries (CEPI). (2019). Key statistics 2018: European pulp & paper industry. p. 23. https://www.cepi.org/wp-content/uploads/2020/10/Final-Key-Statistics-2018.pdf (accessed 20 November 2024).
11 Confederation of European Paper Industries (CEPI). (2022). Key statistics 2021: European pulp & paper industry. p. 5. https://www.cepi.org/wp-content/uploads/2022/07/Key-Statistics-2021-Final.pdf (accessed 20 November 2024).

12 European Paper Recycling Council (EPRC). (2024). Monitoring report 2023: European declaration on paper recycling 2021-2030. https://www.paperforrecycling.eu/download/1704/?tmstv=1721398811 (accessed 20 November 2024).

13 European Paper Recycling Council (EPRC). (2023). Monitoring report 2022. pp. 1–2. https://www.paperforrecycling.eu/download/1471/?tmstv=1697640151 (accessed 18 October 2023).

14 German Federal Institute for Risk Assessment (BfR). (2023). New EFSA risk assessment: some mineral oil residues in food remain a health concern. https://www.bfr.bund.de/cm/349/new-efsa-risk-assessment-some-mineral-oil-residues-in-food-remain-a-health-concern.pdf (accessed 18 October 2023).

15 EPEA Factsheet on MOSH/MOA. Healthy Printing. (2023). https://www.healthyprinting.eu/mosh-and-moah/ (accessed 18 October 2023).

16 Sturaro, A., Rella, R., Parvoli, G. et al. (2006). Contamination of dry foods with trimethyldiphenylmethanes by migration from recycled paper and board packaging. *Food Addit. Contam.* 23: 431–436. https://doi.org/10.1080/02652030500526052.

17 Milanova, E. and Sithole, B. (2024). Paper, film & foil converter (PFFC). https://www.pffc-online.com/flexpack/2658-the-place-whitening-agents (accessed 30 January 2024).

18 Triantafyllou, V.I., Akrida-Demertzi, K., and Demertzis, P.G. (2007). A study on the migration of organic pollutants from recycled paperboard packaging materials to solid food matrices. *Food Chem.* 101: 1759–1768. https://doi.org/10.1016/j.foodchem.2006.02.023.

19 Asensio, E., Peiro, T., and Nerín, C. (2019). Determination the set-off migration of ink in cardboard-cups used in coffee vending machines. *Food Chem. Toxicol.* 130: 61–67. https://doi.org/10.1016/j.fct.2019.05.022.

20 Fengler, R. and Gruber, L. (2020). Migration and permeation of mineral oil components from paper-based food contact materials into foods – A critical comparison of analytical methods. *Food Packag. Shelf Life* 25: 100537. https://doi.org/10.1016/j.fpsl.2020.100537.

21 Parry, S.J. and Aston, D.S.J. (2004). Migration of inorganic contaminants into dry food from packaging made from recycled paper and board. *Food Addit. Contam.* 21: 506–511. https://doi.org/10.1080/02652030410001677808.

22 Harrington, R. (2011). Food companies changing packaging on mineral oil migration risk. Food Manufacture. https://www.foodmanufacture.co.uk/Article/2011/03/08/Food-companies-changing-packaging-on-mineral-oil-migration-risk (accessed 18 October 2023).

23 Sackett, S. (2016) Is newspaper safe for your garden? Permaculture Research Institute. https://www.permaculturenews.org/2016/03/14/is-newspaper-safe-for-your-garden/ (accessed 20 November 2024).

24 Fiber Box Association (FBA). (2013). Voluntary standard for repulping and recycling corrugated fiberboard treated to improve its performance in the presence of water and water vapor. https://www.fibrebox.org/corrugated-recycles-symbol (accessed 20 November 2024).

25 NextGen Consortium, Closed Loop Partners. (2023). Closing the loop on cups: collective action to advance the recovery of paper cups in the U.S.https://www.closedlooppartners.com/research/closing-the-loop-on-cups-collective-action-to-advance-the-recovery-of-paper-cups-in-the-u-s/ (accessed 21 March 2024).

10

Plastics Recycling

10.1 Introduction

Plastics offers numerous benefits to society. Some of the key benefits include: **(1) Resource Preservation:** One major benefit of plastic recycling is reducing the demand for new raw materials, which helps in preserving our natural resources; **(2) Energy Saving:** Recycling, particularly mechanical recycling, conserves energy. Reports indicate that using recycled plastics saves about 50% more energy than virgin plastics (of course, they vary from plastic to plastic). In 2019, it was reported that energy savings of 79%, 88%, and 8% were achieved by recycling PET, HDPE, and PP, respectively [1]. Chemical recycling, currently, is more energy-intensive than mechanical recycling; **(3) Economic Benefits:** Only a few materials are more cost-effective, making the manufacturing process more affordable and price-competitive when using recycled inputs; **(4) Reduction in Greenhouse Gas Emissions:** On average, every ton of recycled plastic prevents the emission of around 1–2 tons of CO_2 equivalent; and **(5) Environmental benefits:** Recycling significantly reduces the volume of waste destined for landfills and incinerators. This results in decreased soil and air contamination, including the reduction of microplastic, which has enormous environmental and health benefits.

10.2 Plastic Recycling Rates

Despite its benefits, plastic recycling rates have been low in the United States and worldwide. For instance, the plastic recycling rate was ~5% in the United States in 2021 and ~9% globally in 2019 [2]. This can be attributed to inadequate infrastructure for the collection and sorting of plastic waste and a limited market for many

Degradation, Stabilization, and Recycling of Packaging Materials, First Edition.
Muhammad Rabnawaz, Susan E.M. Selke, and Ian Wyman.

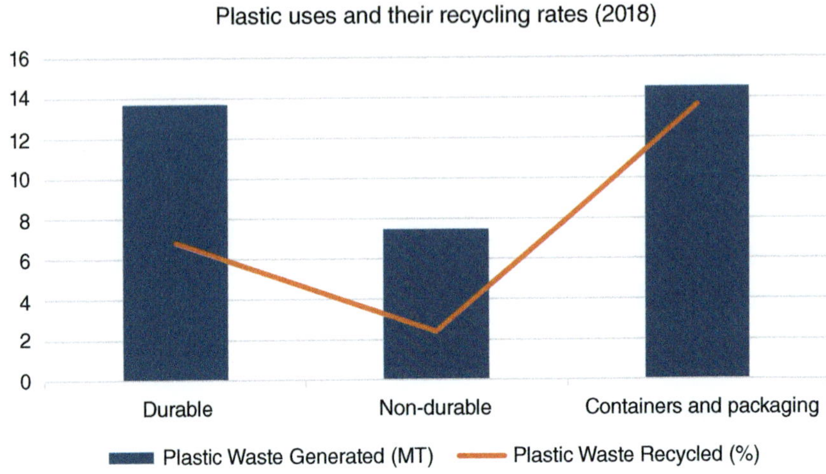

Figure 10.1 Plastic waste in durable, nondurable goods, and packaging containers, along with their respective recycling rates (2018). The left *y*-axis indicates quantity in million tons as well as the percentage recycled.

post-consumer plastics. However, the situation started to shift in favor of plastic recycling in 2018, though the full impact remains to be seen.

As is the case for paper, plastic recycling is predominantly an open-loop process. Recycled plastics typically go into various applications, which are often very different from their original use. Recycling rates for plastic packaging are higher than for most nonpackaging plastics. The exception is polypropylene (PP) battery cases, which have a recycling rate of over 90%. In the United States, plastics in durable goods had an overall recycling rate of 6.8% in 2018, and plastics in nondurables had a rate of 2.4%, while the rate for plastics packaging was 13.6% (Figure 10.1). The uses of various resins in different types of packaging containers is shown in Figure 10.2.

Trends in the amounts of recycled plastic packaging over the last few years are shown in Figure 10.3.

10.3 Recycling of Plastics Packaging

There are significant differences in rates for specific types of packaging plastics, as shown in Figure 10.4. To a large extent, plastic packaging recycling used to be limited to bottles and jars. Recycling of other plastic containers was limited but growing. A significant fraction of stretch wrap was recycled, but there was little recycling of other packaging films. Plastic bag recycling opportunities had been

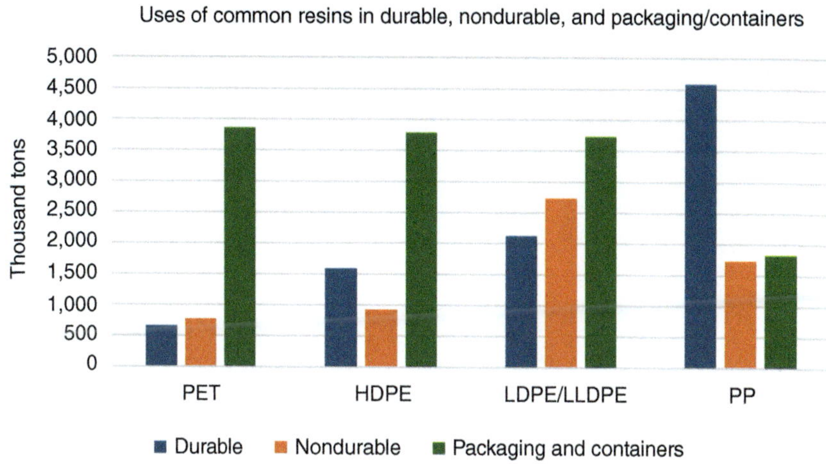

Figure 10.2 Use of plastics by resin in durables (D), nondurables (ND), and packaging/containers (P&C). This data is from the EPA document: Advancing Sustainable Materials Management: 2018 Tables and Figures: United States Environmental Protection Agency, December 2020.

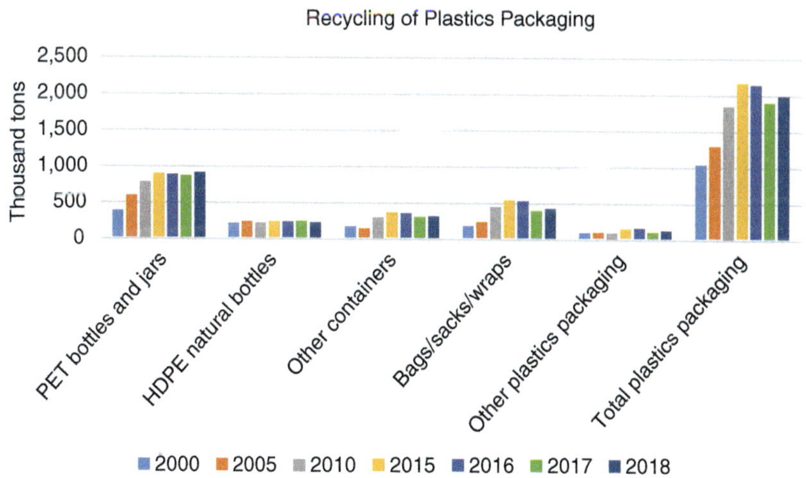

Figure 10.3 Recycling of plastic packaging in the United States.

reasonably plentiful in the early 1990s but had later largely disappeared. This picture has changed quite a bit in the last few years, as we will discuss, but the majority of recycled plastics still come from bottles.

The blue bar shown in Figure 10.5 represents the recycling rates of individual resins with respect to all plastics in the container and packaging category. For

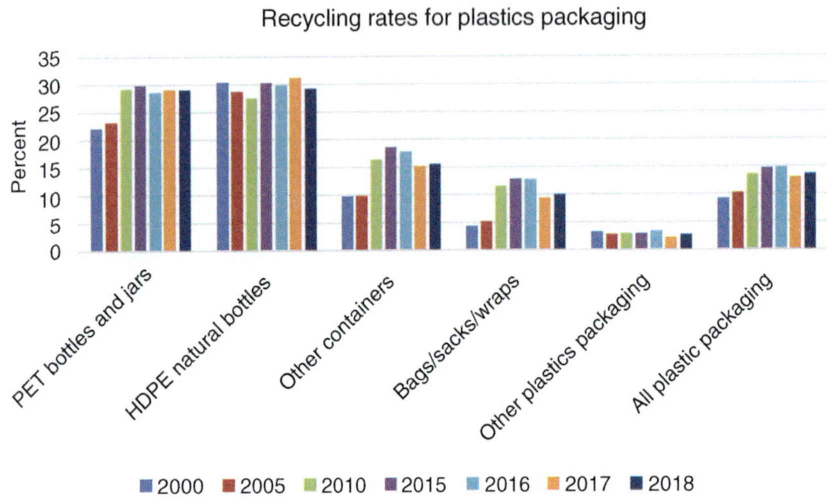

Figure 10.4 Recycling rates for plastic packaging in the United States.

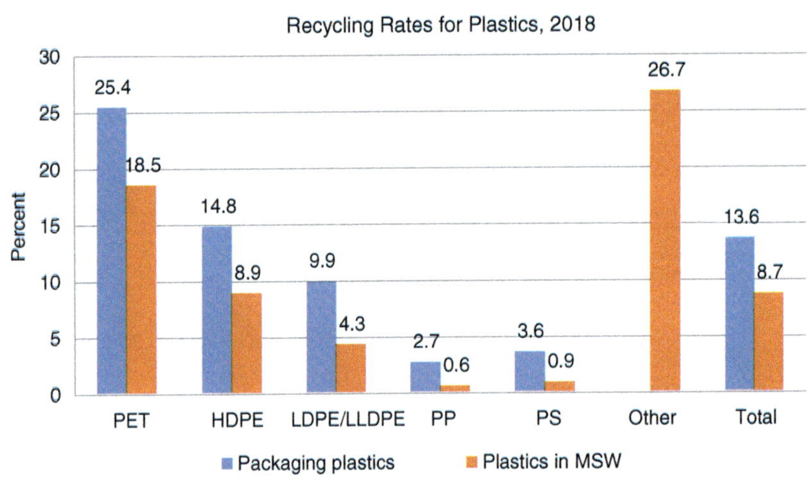

Figure 10.5 U.S. recycling rates for plastics and for packaging plastics in 2018 (EPA).

instance, 3.8 million tons of poly(ethylene terephthalate), PET, were generated from packaging and containers, and 0.98 million tons were recycled. This results in a recycling rate of 25.4% for PET within the packaging and container category.

Meanwhile, the orange bar, in Figure 10.5, represents the recycling rate of individual resins with respect to all plastics in MSW. For example, 5.3 million tons of PET were produced in MSW from all categories, and 0.98 million tons were recycled; this results in a PET recycling rate of 18.5% with respect to plastic waste in MSW [3].

Bottle recycling information, produced by the American Chemistry Council (ACC) and the Association of Plastics Recyclers (APR) [4] contains data for high-density polyethylene (HDPE), PP, and PET bottle recycling for 2018, while more recent data for the year 2021 is available at Ref. [4]. A report on PET recycling by the National Association for PET Container Resources (NAPCOR) provides additional information on PET recycling for 2020. At the time of this writing, reports are available for 2021 and 2022 as well, but only for purchase, while earlier reports are free to download [5].

The APR has a web page that has links to complete historical sets of reports on plastic recycling at Ref. [6]. In addition, the APR provides guidance to help manufacturers design plastics packaging such as bottles in a way to help maximize their recyclability (https://plasticsrecycling.org/apr-design-guide).

For a number of years now, the ACC has been promoting the **"all plastic bottle"** idea for curbside collection programs, rather than the previously common arrangement of only collecting PET and HDPE bottles. Some studies have found that asking for "all plastic bottles" significantly increases the collection of the desired PET and HDPE bottles, while not increasing, and perhaps even decreasing, the amount of contamination with other types of bottles. Some caution is warranted in these conclusions, however. The parallel movement to single stream collection of recyclables (all recyclable materials mixed together) has been cited as the probable cause of the increase in contamination levels of baled plastic bottles that has resulted in decreased yields of usable material (see ACC and NAPCOR reports, for example).

ACC conducted a study on access to recycling in the United States, which was published in May 2011 and updated in 2013. They reported that **94%** of the U.S. population had access to PET and HDPE bottle and cap recycling, either through curbside or drop off programs. The population with access to recycling for rigid plastic containers (other than bottles) had grown to 60% [7].

ACC published a series of annual reports on recycling of other rigid plastics through 2018, with the most recent the "2018 National Postconsumer Non-Bottle Rigid Plastic Recycling Report" [8].

It includes nonpackaging as well as packaging. They reported at least 1.3 billion pounds of such plastics were recovered in 2018, a decrease of 3.5% compared to 2017. U.S. purchases increased but exports declined by 35% and were 63% below the 2014 exports high. ACC reports that the majority of the material collected was polyolefins, especially HDPE and PP, as shown in Figures 10.6 and 10.7.

A similar ACC report was issued on plastic bag and film recycling for 2018 [9, 10]. It reported a total of about 1 billion pounds of postconsumer bags and film were recycled in 2018, up by 54% from 2005 – but down substantially in 2016 and 2017, as shown in Figure 10.8, mostly due to a decline in exports [10].

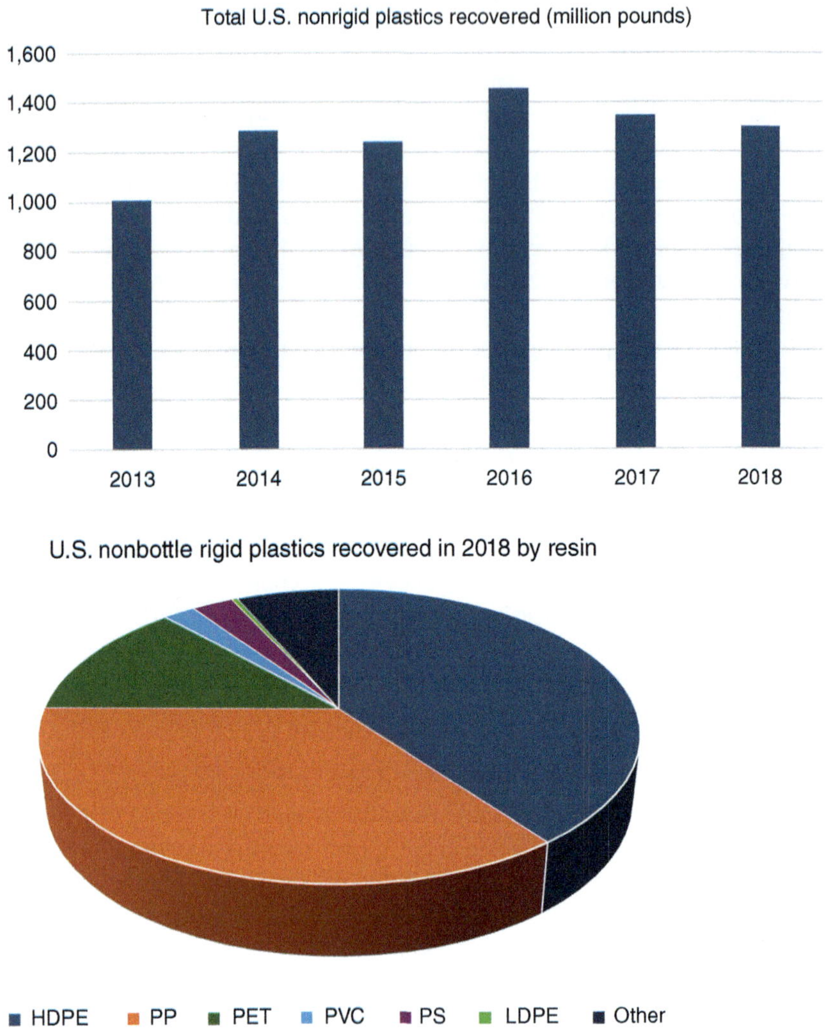

Figure 10.6 Nonbottle rigid plastics recycling, shown as totals for the years 2013–2018, and percentage by resin for 2018 American Chemistry Council.

The breakdown by category is shown in Figure 10.9.

In 2021, ACC, APR, and the Foundation for Plastic Recycling combined the plastic bottle, container, and film recycling reports into a single report titled "2019 U.S. Post-consumer Plastic Recycling Data Report" prepared by Stina Inc. [13]. At

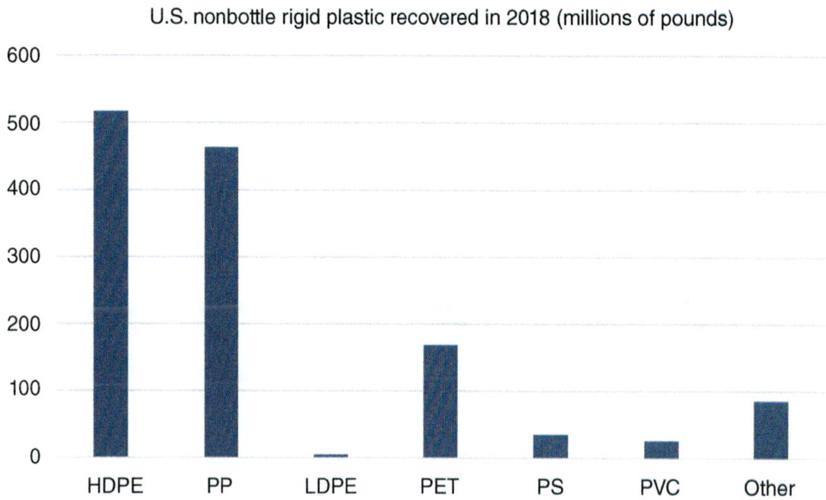

Figure 10.7 U.S. Post-consumer Non-Bottle Rigid Plastic Recovered in 2018 By Resin (2018 National Post-Consumer Non-Bottle Rigid Plastic Recycling Report, Prepared by More Recycling for the American Chemistry Council, August 2020).

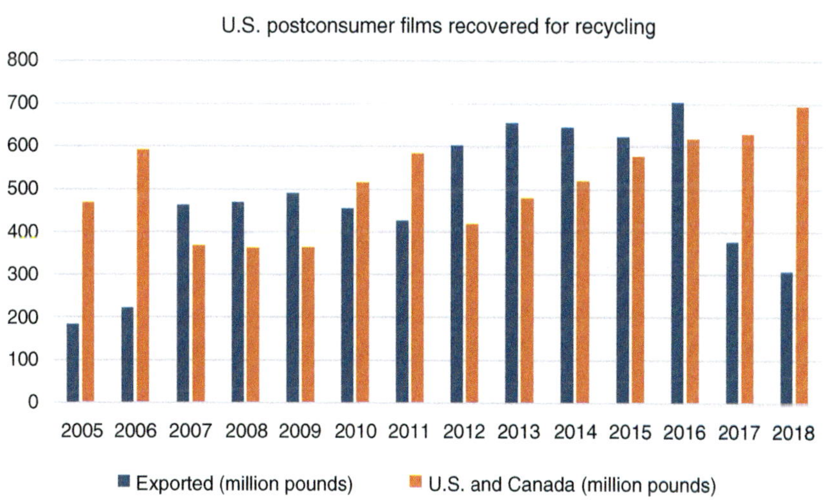

Figure 10.8 Total plastic bag recycling [11].

the time of this writing, 2022 data is available through the U.S. Post-Consumer Plastic Recycling Dashboard [14]; 2022 totals are shown in Figure 10.10.

Figure 10.11 shows EPS recycling amounts as reported by the EPS Industry Alliance (EPS-IA). As can be seen, there was a large decline in 2017 but considerable

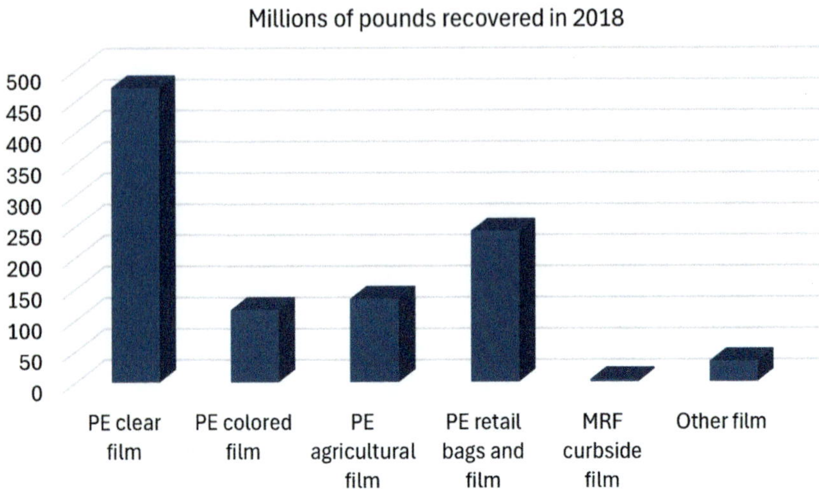

Figure 10.9 Plastic film and bag recovered by category [12].

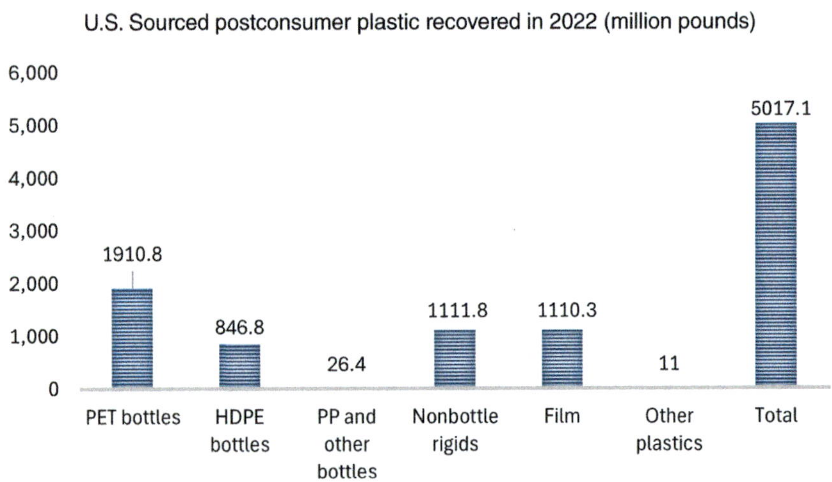

Figure 10.10 Postconsumer plastics recycled in 2022 (in millions of pounds) [14].

recovery in 2018. The data presented is based on responses from 54 EPS manufacturers and independent recyclers, and EPA-IA therefore considers these minimum amounts, as not all recyclers responded.

For most applications, plastics must be separated by resin type. Mixtures of plastic often tend not to have very good mechanical properties. As is the case with glass, color is also an issue with plastics.

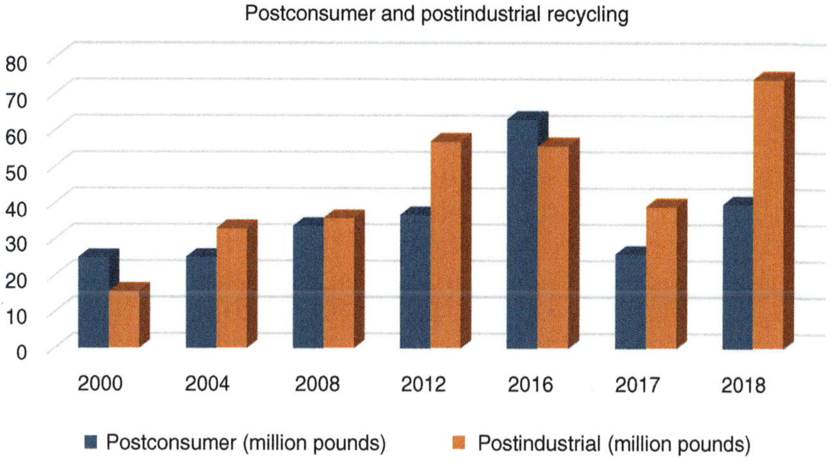

Figure 10.11 EPS recycling, both post-consumer and post-industrial (EPS-IA).

Sophisticated systems have been developed for sorting plastic bottles by resin and by color. This technology works well on whole bottles but not as well on chipped plastics. Since it is quite expensive, such sorting by resin is generally done only in designated plastics sorting facilities. In many places, hand-sorting is used to separate PET and HDPE bottles, which can easily be differentiated by the high transparency of PET and cloudiness or opacity of HDPE. The plastic resin identification codes are also an aid in doing this sorting but are intended more for telling consumers which bottles to put out for recycling and which to not put out – which is increasingly irrelevant with the growth in "all plastics" collection.

Information about the SPI resin identification codes can be found at Ref. [15]. This resin coding system was originally developed by **SPI**, the Society of the Plastics Industry, and enclosed the numbers indicating resin type inside a triangle made up of three chasing arrows as shown in Figure 10.12.

According to the current ASTM standard that specifies the Resin Identification Code (RIC), it consists of an equilateral triangle (a triangle that has all its sides equal in length), a Resin Identification Number, and an Abbreviated Term for the

Figure 10.12 Example of a label for PET from the SPI Coding System. Tomia / wikimedia commons / CC BY 2.5.

polymeric material. More information can be found at the ANSI Blog: Resin Identification Codes (RICs), as Specified by ASTM D7611 https://blog.ansi.org/?p=160542 [16]. As quoted there:

"Resin Identification Codes are not "recycle codes." The Resin Identification Code is, though, an aid to recycling. The use of a Resin Identification Code on a manufactured plastic article does not imply that the article is recycled or that there are systems in place to effectively process the article for reclamation or re-use. The term "recyclable" or other environmental claims shall not be placed in proximity to the Code [17]."

Revisions to the standard were announced in 2013, including a change from the "chasing arrows" triangular logo to a solid triangle, as shown in **Figure 10.13** [19, 20].

10.4 What Is the Impact of Impurities in Plastics Mechanical Recycling?

Typically, blending minor quantities of PP with HDPE has minimal impact on material properties, where "minor" refers to around 5% of the matrix/blend. However, when the PP content exceeds 10%, the adverse effects become substantial. In contrast, even trace amounts of polyvinyl chloride (PVC) contamination can render PET unusable. This is due to PVC generating acids that break down the ester bonds of PET.

10.5 United States Plastic Scrap Exports

It can be informative to look at trends for United States plastic scrap in recent years (Figure 10.14). United States plastics scrap exports shrank by 35% in just 1 year between 2017 and 2018. The full-year export figures can be found at the United States Census Bureau. These statistics highlight the countries that increased imports in 2018 and emphasize the minimal volume that was shipped to China that year. The most significant decrease was in plastic scrap, which went from 3.7 billion pounds in 2017 to 2.4 billion pounds in 2018. It is worth noting that the plastic exports were 4.3 billion pounds in 2016 for the United States. These plastic exports have fallen further in 2019, 2020, and 2021 (to approximately 1.5, 1.4, and 1.2 billion pounds, respectively), with a significant drop occurring in 2022 to 0.95 billion pounds [23]. The most substantial decline came from the reduction of exports to China. For instance, at one point in 2017, China was importing 33.4% of plastics, but this percentage had declined to only 4.5% in 2018. Some changes were especially notable – Malaysia's share of United States plastic

Resin identification number	Resin	Resin identification code – option A	Resin identification code – option B
1	Poly(ethylene terephthalate)	1 PETE	01 PET
2	Highdensity polyethylene	2 HDPE	02 PE-HD
3	Poly(vinyl chloride)	3 V	03 PVC
4	Low density polyethylene	4 LDPE	04 PE-LD
5	Polypropylene	5 PP	05 PP
6	Polystyrene	6 PS	06 PS
7	Other resins	7 OTHER	07 O

Figure 10.13 Resin Identification Codes (RICs), as specified by ASTM D7611 [18]. RICs for various polymer resins are shown above. With this labelling system, the symbols have a solid triangle that surrounds a number corresponding to the type of resin. An abbreviation for the resin is also provided beneath the triangle. Teleman84 / Wikimedia Commons / CC BY 4.0.

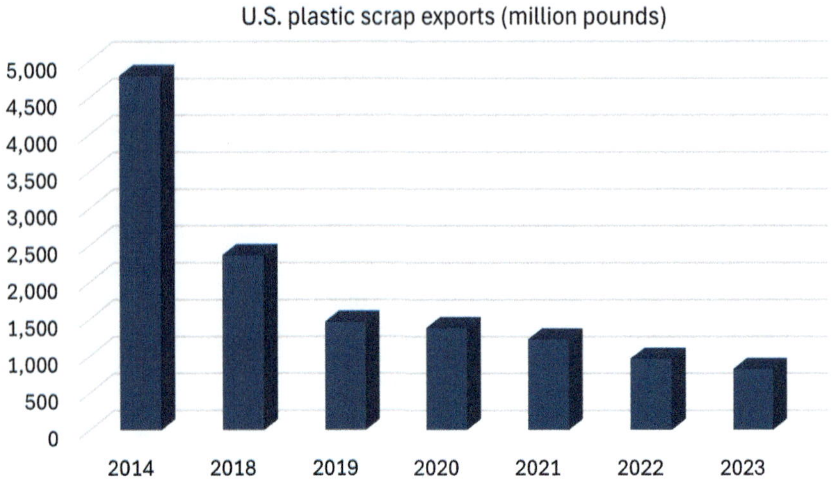

Figure 10.14 Plastic scrap exports during 2018–2023, with the record-high year of 2014 also shown [21, 22].

surged from 7.3% to 18.8%. By weight, the largest importers of United States scrap plastic were Malaysia (443 million lbs.), Canada (294 million lbs.), India (294 million lbs.), Hong Kong (267 million lbs.), Thailand (235 million lbs.), Vietnam (165 million lbs.), Taiwan (114 million lbs.), China (106 million lbs.), Mexico (101 million lbs.), South Korea (78 million lbs.), Indonesia (60 million lbs.), and Turkey (52 million lbs.). In 2022, the leading ten destinations (listed in order from higher to lower tonnage) for United States plastic scrap were Canada, Mexico, India, Malaysia, Indonesia, Vietnam, Germany, Turkey, El Salvador, and Pakistan [21]. For 2023, the corresponding list changed slightly to Canada, Mexico, India, Malaysia, Vietnam, Indonesia, Spain, Germany, Turkey, and Pakistan (the top five destinations are shown in Figure 10.15). The decline in United States plastic scrap exports continued in 2023, with the total United States scraps falling from 962 million tons in 2022 to 918 million tons in 2023 [22].

10.6 Plastic Recycling Elsewhere

10.6.1 European Plastic Recycling Numbers

Europe has made significant progress in its waste management efforts. For example: recycling of plastics is at 42.1%, energy recovery is at 39.3%, and only 18.5% of plastic waste goes to landfills [25]. Between 2006 and 2018, the recycling of plastic waste from packaging had increased by 92%, energy recovery for plastic packaging waste had increased by 84%, and landfilling of plastic packaging waste had decreased by 54%.

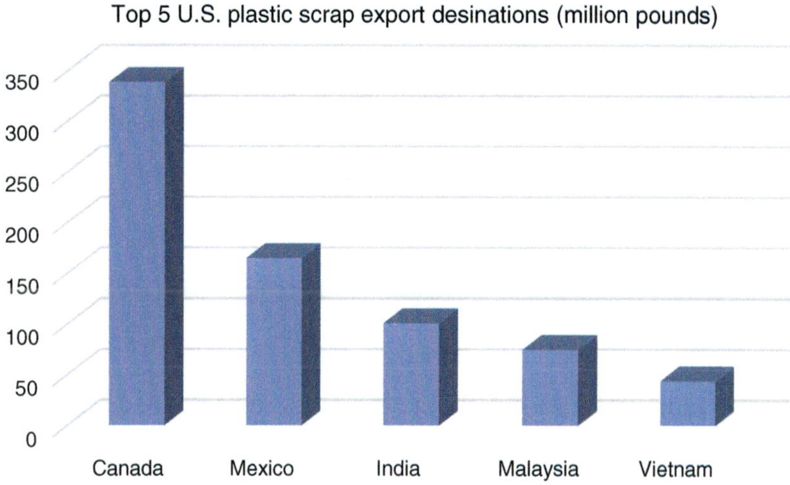

Figure 10.15 Top five U.S. plastic scrap export destinations in 2023 [24].

10.7 Global Plastic Recycling Rates

Globally, only 9% of plastic is recycled, 50% goes into regulated landfills, 19% ends up in incinerators, and 22% was sent to unregulated landfills, burned in the open atmosphere, or leaked into the environment in 2019 according to an OECD report, as shown in Figure 10.16 [26]. In addition, global plastic waste generation had reached 353 million tons in 2019, which was more than twice that generated in 2000. Overall, Europe is a global leader in recycling because of its better infrastructure thanks to its EPR policies.

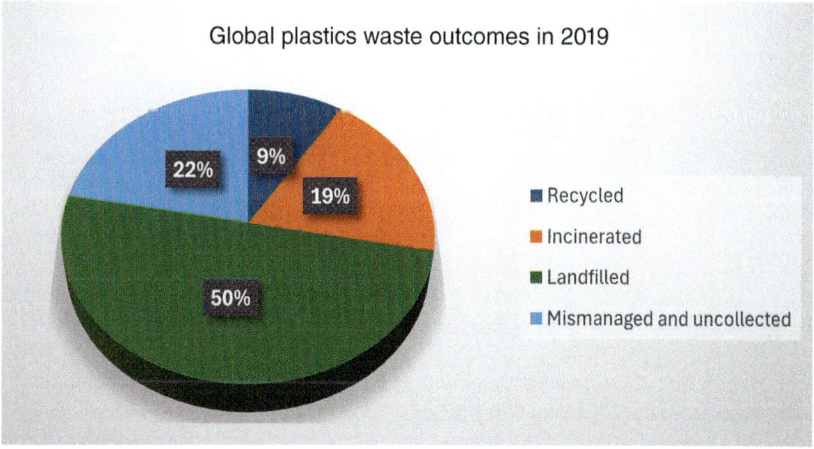

Figure 10.16 Global plastic waste outcomes in 2019 [26].

10.8 CO_2 Footprint of Different Ways of Plastic Disposal

Figure 10.17 shows CO_2-equivalent emissions of different end-of-life (EoL) treatment technologies for various plastic waste streams, indexed to incineration (100 wt%) without energy recovery. Mechanical recycling and dissolution-precipitation are the most CO_2-efficient methods.

10.9 Terminology in Plastic Recycling

10.9.1 Postindustrial Versus Postconsumer Plastics

Postindustrial plastic waste refers to plastic scrap that is generated during templating, from defective products, and trimming inside an industry and is never used by consumers. There are two primary categories of postindustrial plastic waste:

- Those that require treatment before reuse for the same purpose, such as trimmed parts of multilayer plastics.
- Those that are directly reusable for the same purpose, such as scrap from single-layer plastics.

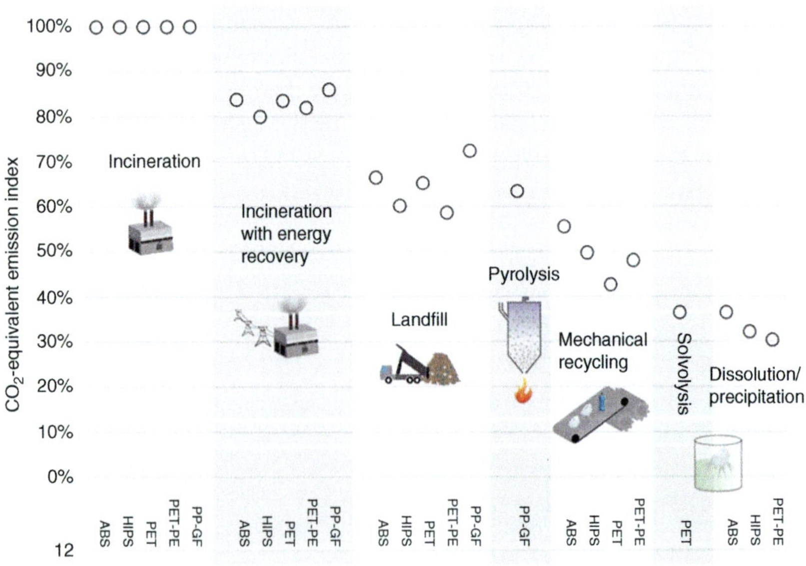

Figure 10.17 CO_2-equivalent emission indices for various EoL scenarios for different plastic waste streams [27] / John Wiley & Sons / CC BY 4.0.

The practice of postindustrial plastic recycling is known as primary recycling.

Postconsumer recycled (PCR) plastics denote plastic materials that are discarded by consumers after their intended use, having finished their consumer life cycle. Examples of post-consumer plastics include used PET bottles. The practice of PCR recycling by mechanical recycling is known as secondary recycling.

10.9.2 Chemical Versus Mechanical Recycling

Mechanical recycling: Mechanical recycling involves reusing waste plastic by cleaning and melting to produce postconsumer recycled (PCR). The PCR is then used as a substitute for virgin plastics and often used in combination with virgin resin. The conversion of used bottles to new bottles, used plastic bottles to shirts, used plastic bottles to lumber/buckets, and so on are examples of mechanical recycling.

Chemical recycling/advanced recycling refers to the recycling process where postconsumer plastic is broken down into small molecules, which can then be used as feedstocks for new plastics, feedstock chemicals, or even fuel [28]. Also, advanced recycling is a term used by industry to sidestep the negative connotations many consumers have associated with the word "chemical" in chemical recycling. A comparison of mechanical and chemical recycling is provided in Figure 10.18.

Note: A newer term, molecular recycling [29, 30], has also been introduced recently and used for technologies that use solvents, heat, enzymes, and even sound waves to purify or break down plastic waste to create polymers, monomers, oligomers, or hydrocarbon products.

Dissolution-Precipitation Approach: As shown in Figure 10.19, this approach relies on the use of solvents to selectively dissolve one polymer, followed by the addition of another solvent to precipitate that polymer. The solvent is then recovered and reused. This idea was initially developed for multilayer plastics that are difficult to recycle. In the case of multilayer plastics, each layer is selectively dissolved and then allowed to precipitate for separation. While the material obtained can behave similarly to virgin material, the process of solvent recovery and separation can be energy-intensive and involves the use of organic solvents [31].

10.9.3 Extraction Approach

Purification by extraction: Here, additives are extracted from plastics via selectively dissolving the additive, while the polymer remains undissolved. The purpose of extracting the additives is to remove any unwanted impurities. In certain cases, during the extraction conditions (high temperature and pressure), the polymer is

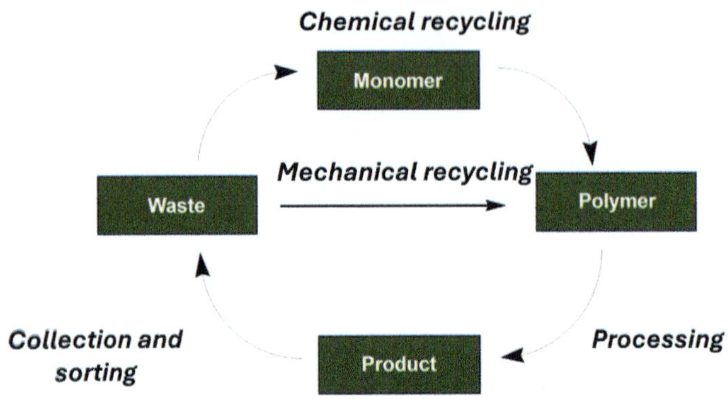

Figure 10.18 Mechanical versus chemical recycling.

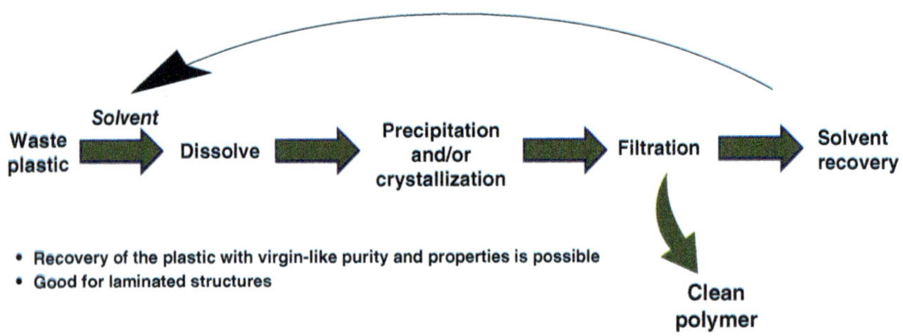

Figure 10.19 Flow scheme depicting the dissolution-precipitation approach.

soluble but upon cooling the polymer precipitates out, leaving additives in the solvent. This approach has been tested for PP. A great example is "PureCycle," where low-boiling hydrocarbon solvents are pressurized at high temperature to remove additives from postconsumer PP.

10.9.4 Pyrolysis

Thermal pyrolysis: Plastics are heated in the absence of oxygen to achieve degradation into low molecular weight molecules [32].

Thermocatalytic pyrolysis: Plastics are heated in the presence of a catalyst and in the absence of oxygen to achieve degradation into low molecular weight molecules [33–35].

Thermochemical pyrolysis: Plastics are heated in the presence of chemicals to achieve degradation into low molecular weight molecules [36].

10.9.5 Chemical Upcycling

Recycling of waste plastics by degradation into value added feedstock products/ chemicals of higher value.

10.9.5.1 Challenges Associated with Plastics Recycling

There are a number of challenges that are encountered with plastics recycling, such as those listed below:

- Large number of types of plastics
- Use of polymer blends and multilayer plastics
- Plastics often have poor mechanical properties when they are blended. For example, with more than 10% HDPE in PP or vice versa, the mechanical properties of the blend begin to degrade unless they are restored with compatibilizers.
- Lack of cost effective sortation processes.
- High collection and transportation costs, particularly for flexibles and foams.
- Compositional fluctuations in recycled plastics
- Contamination, which lowers the value of products produced from recycled plastics
- Food contact compliance issues
- Low cost of virgin plastics and lack of incentives for using recycled plastics; although some incentives are coming such as the UK's recently implemented Plastic Packaging Tax which is applied to plastic packaging which has less than 30% recycled plastic [37].

10.9.5.2 Prices of Recycled plastics as of October 2023

For 2023, a summary of the prices for various types of plastics are shown below. However, the need for decontamination often makes the final cost of recycled plastics more than their virgin counterparts.

Plastic type	Price in cents/lb
PET	9–11
Natural HDPE	26–27
Colored HDPE	15–16
Mixed plastics #1–7	1–2
PP	5–6
PS	2.5–3.5

10.10 Emerging Trends in Recycling

10.10.1 Challenges in Mechanical Recycling of Plastics

Currently, key barriers that are impeding growth in mechanical recycling include:

- The heterogeneous nature of recycled materials that makes the processing difficult.
- Contamination preventing food contact applications.

10.10.1.1 Path Forward to Solve this Problem

- Better infrastructure for waste collection (e.g. multistream collection) funded through EPR or similar schemes.
- Improved sorting methods.
- Introduction of tracking requirements for food-grade recycled materials.

10.10.2 Digital Watermarking

One potential approach to aiding separation of plastics is to incorporate a type of additive, known as a chemical marker, to the plastic to enable rapid sensing and sorting [38]. Chemical markers are advantageous because they work with all types of plastics. However, it is often necessary to remove these chemical markers during recycling. A recent innovation is the use of digital watermarking [39–41]. Brands can apply these digital watermarks and link data to the cloud (Figure 10.20). This data includes information about the packaging's composition and material. Digital watermarking codes can be detected by high-definition cameras[1] and specialized mark detection software, like that from Digimarc [42]. The captured images are processed using specialized software and once the pattern is detected, the software deciphers the code. The digital watermark typically contains a unique identifier that corresponds to a database entry. This identifier is used to access information from a connected database or cloud service. The retrieved data can include the type of plastic, the recycling class, manufacturing information, or instructions for the sorting process. Note: Some QR codes can be created using fluorescent dyes, but they require a fluorescence detector. However, these dyes need to be removed during recycling.

1 High-Resolution Imaging: Cameras equipped with high-resolution sensors take images of the plastic item with sufficient resolution to capture the fine details of the watermark pattern.

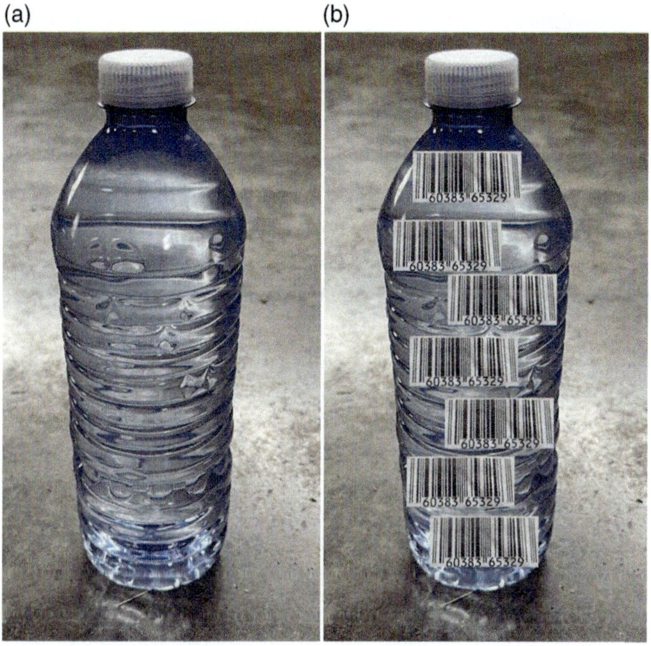

Figure 10.20 Example of a water bottle with digital watermarking. To the human eye, the digital watermarking is invisible (left) but can be detected or "read" by scanners (right).

10.10.3 Near IR sorting

Near infrared spectroscopy (NIR, which operates in the wavelength range of 0.7–2.5 μm) is an effective technique for the automated sorting of plastics. Near IR light is shined on waste plastic products on a conveyor belt equipped with NIR detectors [43]. The detector is part of a spectrometer, which analyzes the light to identify the unique spectral fingerprint of each plastic. This spectral fingerprint is created due to fundamental molecular vibrations. The software of the spectrometer compares the spectral data against a database of already known spectra for various types of plastics. Then a mechanical system is activated to separate the plastic item from the rest of the material on the conveyor. This is often achieved using blasts of air to divert the item into the correct bin or chute. The advantages of this technique are that it offers reliable and rapid identification with a low environmental impact. **Limitations:** One of the drawbacks of NIR sorting is that it does not distinguish color. The classification of black plastics is the main limitation of NIR technology because it absorbs NIR light. Another challenge is encountered with flake separation if the two sides of the plastics are composed of different materials, such as in the case of multilayers. Also, for whole bottles, full body shrink sleeve labels pose a daunting challenge as the NIR reads the label rather than the bottle plastic.

10.10.4 Monomaterials

Multimaterials (either multiplastics or plastic with other materials such as paper and plastic) are difficult to recycle. One such example is metalized films, where the metal does not melt at the plastic processing temperature. Multiplastic packages contain different plastics that are not compatible with each other, and thus the recycled material that is produced from these packages tends to have poor mechanical properties and loses its value. Meanwhile, some plastics such as nylon or PET are problematic for catalytic pyrolysis.

Paths forward in this direction include:

- All-polyethylene (such as DOW Retain) with compatibilizers. Here, after use, the multilayer structures are just mechanically recycled thanks to the use of a tie-layer that acts as a compatibilizer at the end of life.
- All-polyester multilayer plastics (MLPs) have been developed by Dr. Rabnawaz's group at Michigan State University. In this case, polyesters offer chemical and mechanical recycling as EoL solutions. For example, Dr. Rabnawaz and coworkers have developed PET/PGA/PBAT trilayer plastics. These types of plastics do not require any tie-layer due to the ester-based nature of each layer. This has brought the number of individual layers down from seven (for polyolefins) to three.

10.10.5 Moving Towards Fewer Plastics

Currently, a wide range of plastics (and their combinations) are used in the packaging sector. To simplify the recycling process, packaging plastics should be limited to a narrower range of high-volume polymers such as HDPE, PET, PP, and LDPE.

PP is widely used in the EU, while US manufacturers prefer PE for flexible packaging.

10.10.6 Additives for Mechanical Recycling

Pigments: NIR sensors cannot detect black items because they absorb NIR light waves so that no light is reflected back to the sensors. Solution: Alternative black pigments to carbon black do not interfere with optical sorting methods and support the mechanical recycling of packaging. Emerging solutions include oxides such as black iron oxide, Mars black, manganese black, and synthetic organics like iron gallotannate and nigrosine.

10.10.7 Additives to Enhance Properties of Recycled Materials

Recycled plastics face various challenges; for instance, mixed plastics often exhibit poor mechanical properties and low optical clarity, presenting significant hurdles. This has spurred extensive research into compatibilizers, which enhance both the mechanical strength and optical properties of these materials. Similarly, polymers such as polyesters suffer chain cleavage during reprocessing and recycling; additives like chain extenders are utilized to restore the properties of these recycled plastics.

10.10.8 Flexible packaging

MRFs struggle with sorting flexible materials and face operational disruptions, particularly from plastic bags and the handling of lightweight flexible materials. Current trends in flexible packaging include: (1) enhancing the design of flexible packaging to be more recycling-friendly, and (2) developing dedicated mechanical recycling infrastructure.

10.10.9 Path Forward in Mechanical Recycling

To improve the mechanical recycling of plastics the following strategies could help.

1) Increase collection (multistream).
2) Develop effective and efficient identification and sorting processes.
3) Develop cleaning methods to provide a clean stream of packaging waste materials.
4) Design for recycling will help mechanical recycling.
5) Switch to fewer plastics from the existing large number of plastics.

10.11 Trends in Chemical Recycling

Mechanical recycling is an economical and climate friendly approach for plastic recycling. However, that approach has heterogeneity and contamination issues. Chemical recycling offers homogenous and contamination-free recycled materials, utilizing low-quality waste plastics. Key benefits of chemical recycling include the compatibility of this approach for mixed plastics, contaminated plastics, multilayer materials, agriculture films, and so on. In short, chemical recycling works best where mechanical recycling does not work. The U.S. EPA is yet to classify pyrolysis and gasification as chemical recycling processes (they currently consider these approaches to be equivalent to incineration). However, the ACC is pushing hard to make a case for pyrolysis and gasification to qualify as chemical recycling if the pyrolysis product is not used for fuel.

Currently, the challenges in chemical recycling include cost, contamination of feedstocks, and the need for energy-intensive processes.

10.11.1 Depolymerization to Monomers

Plastic waste reversion into monomers is an ideal recycling pathway as it will lead to virgin plastics. However, there are challenges associated with the depolymerization of plastics into monomers, such as:

- PE and PP are difficult to revert to monomers, at least economically and energy efficiently.
- The commodity polymers PET (100%) and PS (80+%) can be reverted into their monomers at high conversion efficiencies of 100% and >80%, respectively. However, PET depolymerization is an energy-intensive process and PS collection poses a challenge.
- PVC cannot be reverted to its monomer (0% monomer yield).

Commercial activities at some scale exist for:

- PET and PBT chemical recycling → need improvement on the economics.
- Chemical recycling of polyamides.
- Chemical recycling of PS packaging → collection of recycled feedstocks is a challenge.
- ExxonMobil is building a reactor in Texas with a capacity to process 36,000 metric tons of hard-to-recycle plastic each year [44].

10.11.2 Conversion of Plastic Waste into Petrochemicals (chemicals)

Plastics conversion to petrochemicals is favorable economically as well as environmentally as compared to plastic conversion into fuel. Consequently, there has been investment from major petrochemical companies with a focus on plastic waste to petrochemicals. **For example,** Saudi Aramco is shifting its focus

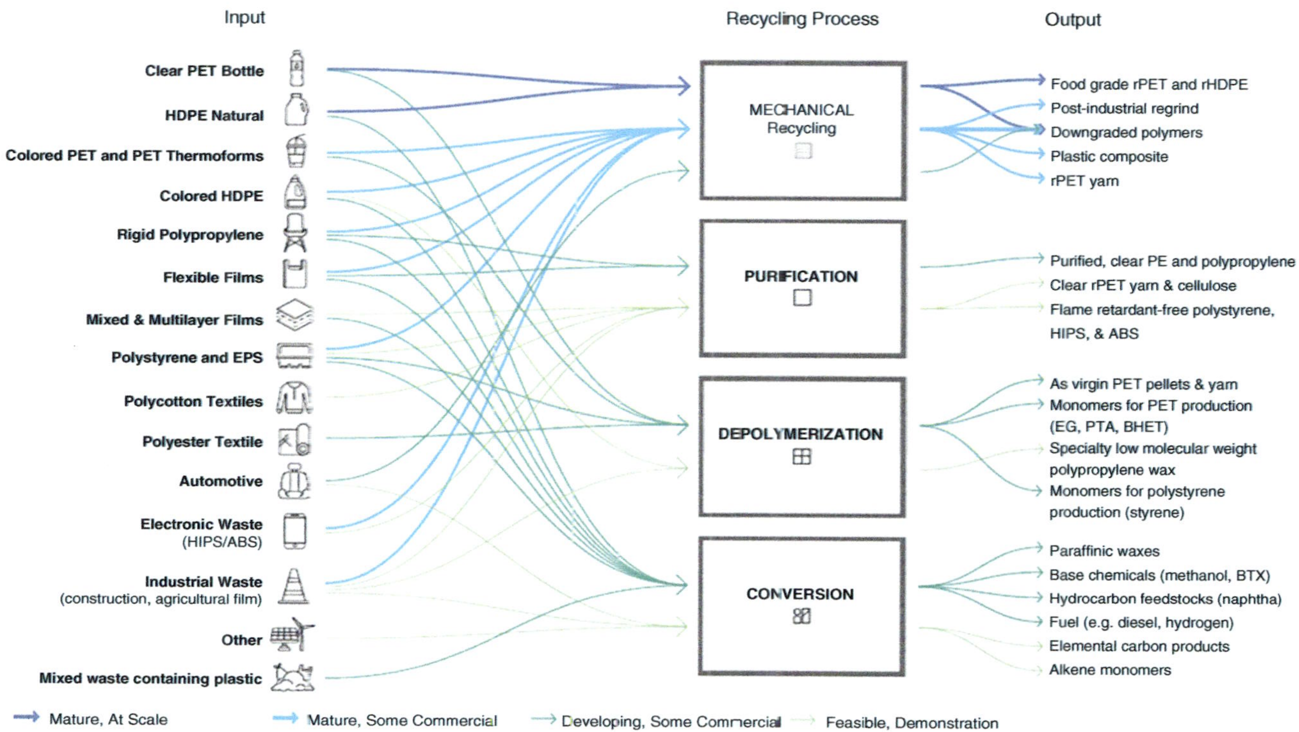

Figure 10.21 Overview of the recycling landscape for various types of plastic. Luu, Paula., 2021 / Closed Loop Partners.

from "plastic-to-fuel" to "plastic-to-chemicals" conversion [45–47]. **Benefits:** A key benefit of this strategy as that it leads to circularity, as materials remain in the loop.

10.11.3 Summary of the Plastic Recycling Landscape

There are various pathways and processes involved in plastic recycling such as through mechanical recycling, purification, depolymerization, and conversion stages, leading to different outputs (Figure 10.21) [48, 49]. For instance, PET and HDPE natural bottles can be mechanically recycled into food-grade recycled PET (rPET) and recycled HDPE (rHPDE) and these technologies are mature. Some recycling processes are mature and commercially available, while others are still in the development or demonstration phases. Only mechanical recycling is employed commercially on a large scale. The remaining approaches are being developed and used to some degree at a commercial scale.

10.11.4 PET Recycling

For many years, the major use for recycled PET was fiber, as shown in Figure 10.22. An overview showing the various uses of recycled PET is provided in Figure 10.23. The amount of PET used for food and beverage containers has grown substantially in the last four decades, fueled initially in part by pressure from consumer and environmental organizations on Coca Cola and Pepsi to use recycled content in their beverage bottles, and in 2020, use of rPET in bottles of all types exceeded fiber uses for the first time. **Availability of collected PET** for recycling continues to be a major issue for the industry, fueled by a decrease in market share for carbonated soft drinks.

As stated earlier, the NAPCOR report also discusses PET thermoform recycling, although those amounts were not included in the report data until 2020. In discussion of this recycling stream, they state "While PET thermoforms are technically recyclable with PET bottles, they can be problematic from a performance and yield perspective" [52] – adding that part of the issue is due to concerns about mingling of PET thermoforms with "look-alikes" of oriented polystyrene (OPS), PVC, polyethylene terephthalate glycol (PETG), or poly(lactic acid) (PLA), along with problems associated with inks, labels, and adhesives that can affect quality. Inclusion of PET thermoforms in curbside bales, for processors that do accept these materials, is typically capped at about 10% of the bale weight. An increasing number of reclaimers, however, are now separating out PET thermoforms from

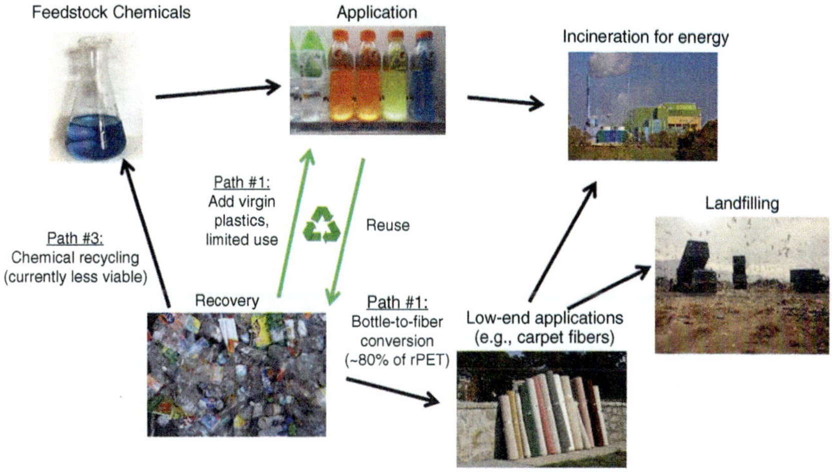

Figure 10.22 PET recycling via bottle-to-bottle conversion (Path #1), bottle-to-fiber conversion (Path #2), and chemical recycling (Path #3) are shown. Path #2 is an example of downcycling. Recycling symbol obtained from Siera Wild / Wikimedia Commons / CC BY SA 4.0. Image of carpets obtained from Ildar Sagdejev / Wikimedia Commons / CC BY SA 4.0. Image of the soft drink bottles obtained from Gee KongLawn / Wikimedia Commons / CC BY SA 4.0. Photograph of the Erlenmeyer flask obtained from Mjolk1 / Wikimedia Commons / CC BY 4.0. Photograph of the recovered plastic bottles obtained from Gee KongLawm / Wikimedia Commons / CC BY SA 4.0. Photograph of the waste-to-energy incineration facility obtained as Wikimedia from Norbert Nagel / Wikimedia Commons / CC BY SA 3.0. Photograph of the landfill obtained from Truckdriver53 / Wikimedia Commons / CC BY SA 4.0.

mixed PET bales. and processing them successfully. The report states that 14 million pounds of recycled PET from thermoforms was used to make new thermoform packaging, and increasingly companies are promoting circularity in thermoform supply chains.

Challenges with the mechanical recycling of PET include:

- **Thermal degradation of PET** – leading to formation of impurities such as acetaldehyde, etc.
- **Catalytic degradation of PET** – loss in performance and decrease in melt viscosity due to chain scission.

10.11.5 Chemical Recycling of PET

Chemical recycling is the preferred method for processing contaminated PET feedstocks (Figure 10.24). For instance, carpet fibers, which cannot be recycled

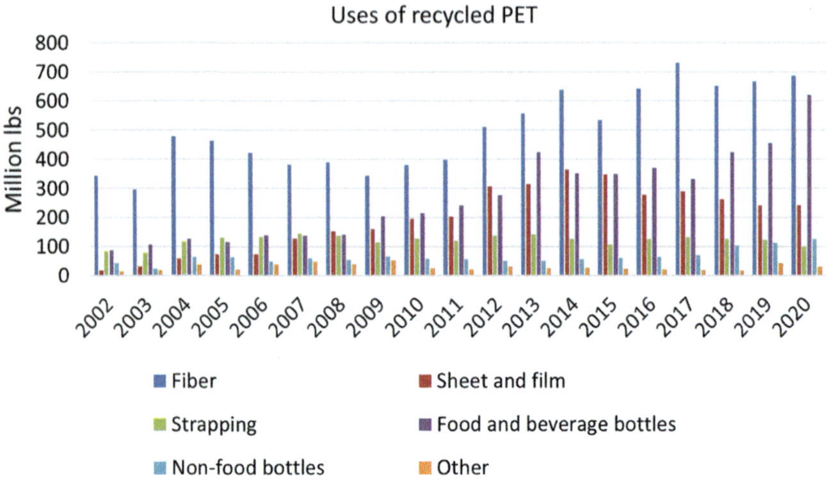

Figure 10.23 Uses of recycled PET. *Source:* NAPCOR, PET container recycling report, 2021 [50, 51].

mechanically, can be effectively recycled through chemical processes to avoid landfilling [53–55]. Chemical recycling maintains the material properties after repolymerization, provided that the resulting monomers are adequately purified.

The most common methods include:

Methanolysis: PET is broken down into dimethyl terephthalate and ethylene glycol using an excess of methanol under heat [56].

Hydrolysis: PET is depolymerized into terephthalic acid and ethylene glycol with an excess of water under heat.

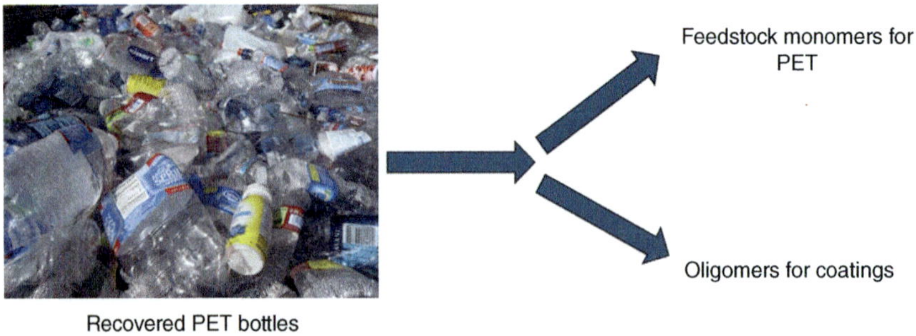

Figure 10.24 Chemical recycling by solvolysis. Inset image of the recovered bottles obtained from Wikimedia Commons Streetwise Cycle / Wikimedia Commons / Public Domain. The author of this image is Streetwise Cycle and the image has been released into public domain.

Glycolysis: PET is converted into bis(2-hydroxyethyl) terephthalate) (BHET) and ethylene glycol using excess ethylene glycol under heat [57].

Dimethyl terephthalate, terephthalic acid, and BHET are suitable for creating virgin plastics. It is important to note that these recycled feedstocks can be mixed in varying proportions with petrochemical derived PET monomers to manufacture new PET that contains a specified percentage of recycled content. **Environmental benefits:** The production of PET via chemical recycling saves 1.5 tons of CO_2 per ton of plastic recycled [58]; Food compliance: high-quality polyester suitable for food contact applications. The US FDA gave approval in April 2020 for Indorama Ventures to use chemically recycled PET in food packaging.

Many depolymerization processes for PET polymers are under development at pilot scale.

There are three main types of pilot-scale PET depolymerization processes:

- Glycolysis processes which produce BHET;
- Methanolysis processes which produce dimethyl terephthalate (DMT); and
- Hydrolysis processes which produce terephthalic acid (TPA).

10.11.6 Key Technologies/initiatives in PET (Polyesters) Chemical Recycling

A number of key technologies have been developed

- BP Infinia Consortium
- Eastman polyester renewal glycolysis and methanolysis processes
- Gr3N Demeto (depolymerization by microwave technology) [59]
- IBM VolCat [60]
- Ioniqa/Unilever/Indorama/Coca-Cola process [61]
- JEPLAN BRING Technology [62]
- Loop Industries [63]
- Poseidon Plastics (UK) [64]
- CuRe Technology BV/Coca-Cola Polyester Rejuvenation Technology [65]
- Carbios enzymatic biorecycling [66]
- MSU Technology [67]

10.11.7 Challenges PET Chemical Recycling

There are certain challenges that impede the chemical recycling of PET. One such hurdle is the need for facile routes to recover high purity monomers from depolymerized PET.

Due to the challenges associated with high purity monomer recovery, chemical recycling is **2–3 times more costly** than the mechanical recycling of PET. Some of the criticisms of chemical recycling refer to its higher energy consumption and lack of maturity compared to mechanical recycling [68]. However, there are efforts being pursued to address this issue. For example, Petcore Europe is working towards promoting the circularity of PET recycling in Europe.

10.11.8 Polyurethanes

Approximately, 15–20% of all produced polymers are utilized as thermosets [69, 70], which are notoriously challenging to melt and recycle. Significant efforts are directed towards the reprocessing of thermosets. A common example is urethane, which is often used in packaging for its foam-in-place applications. Yet, its resistance to melting and reprocessing makes it difficult to recycle (Figure 10.25). Progress is being made in developing methods for multiprocessing thermosets. Dr. Rabnawaz's team in the School of Packaging at Michigan State University has developed a method to convert thermoset polyurethane into a reprocessable material using partial decrosslinking and a catalyst that reverses the bonding (Figure 10.26). This innovation could significantly enhance the recycling of post-consumer polyurethanes, offering substantial benefits for both the community and the environment [71].

10.11.9 Polystyrene

Globally, more than 15 million tons of polystyrene (PS) was produced in 2019 and this polymer is widely used in the packaging and nonpackaging sectors. PS has unique chemistry that enables its depolymerization into its parent monomer to be achieved in high yields (~66%). Research in the School of Packaging at Michigan State University has shown that just by using table salt as a catalyst, the monomer yield can be increased to 83–84% [72].

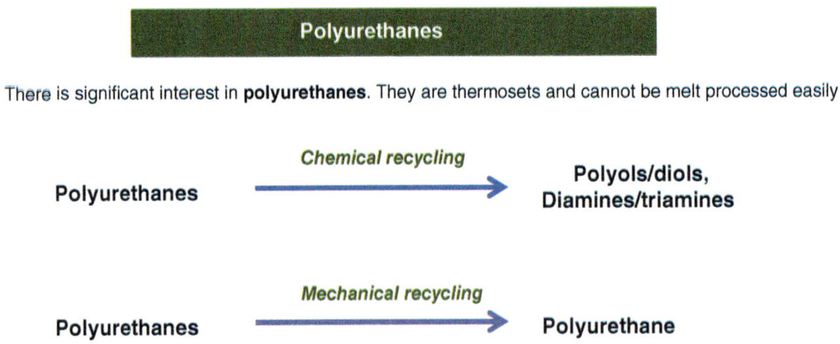

Figure 10.25 Chemical and mechanical recycling routes for polyurethanes.

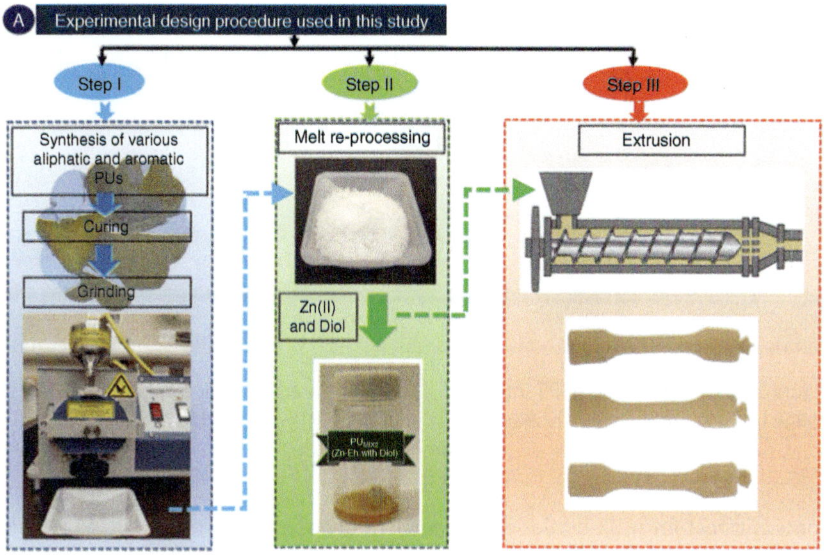

Figure 10.26 Flow scheme for the melt reprocessing of polyurethane [71]. Reproduced from [71] / with permission of royal chemistry Society.

Styrenics Circular Solutions in Europe was set up in 2018 (supported by 11 major styrenics manufacturers and included members along the PS value chain) with the following **Goals in mind:**

- To demonstrate PS chemical recycling feasibility
- Develop new chemical recycling technology for PS
- Create value-added products from PS
- To achieve solvent-based purification of PS
- Improve mechanical separation.

Bottlenecks: A barrier impeding the recycling of PS include the low collection volumes of this plastic and the widespread down-cycling of this polymer (particularly via mechanical recycling)

Recently, PS was converted into chemicals such as benzoic acid and benzaldehyde (Figure 10.27) [73].

10.11.10 Recycling of PE (HDPE, LDPE, LLDPE)

Mechanically recycled HDPE has been used for milk bottles in the UK; see, for example, http://www.dairyreporter.com/Markets/Nampak-to-hit-recycled-

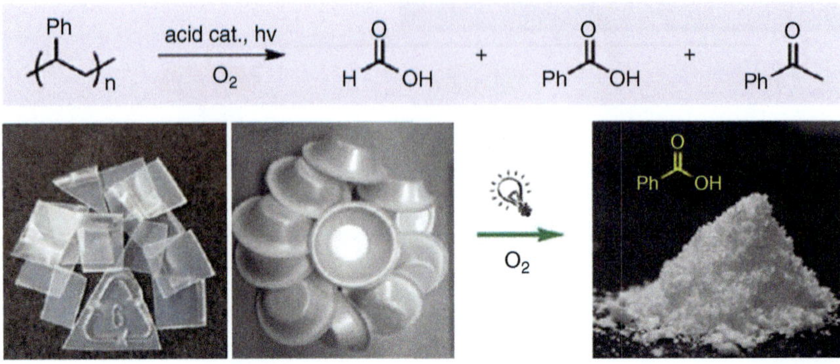

Figure 10.27 Conversion of PS into formic acid, benzoic acid, and acetophenone [73]. Image is reused from: ref. (75) / American chemical Society / CC BY 4.0.

content-target-for-milk-bottles-this-month. In the original uses of recycled HDPE in bottles, the recycled material was in an inner layer of a multilayer structure. This was the method, for example, that was employed in laundry detergent and fabric softener bottles.

In applications that are less sensitive to color and odor, such as black colored motor oil bottles, it was common to blend the recycled HDPE with virgin plastic, in a single-layer structure. Various uses of recycled HDPE bottles are shown in Figure 10.28.

10.11.11 Chemical Recycling of PE

PE can be chemically recycled to obtain different products such as oil, gas (including ethylene), waxes, and so forth. The challenge encountered with this approach is the low percentage of ethylene (that can then be used to produce PE) [74].

10.11.11.1 Catalytic pyrolysis

Catalysts have the potential to reduce pyrolysis temperatures and narrow the product distribution. High viscosity, low thermal conductivity, and relatively long molecular chains lead to a small catalyst/polymer contact area as well as catalyst poisoning (by nitrogen- and sulfur-bearing polymers).

Novel catalysts have been designed to address these issues. Some examples include Pt nanoparticles as a thin layer on $SrTiO_3$ nanocuboids; nanocrystalline HZSM-5. In addition, efforts have been undertaken to eliminate nylons from catalytic pyrolysis by eliminating nylons from multilayer plastics. Yields of various

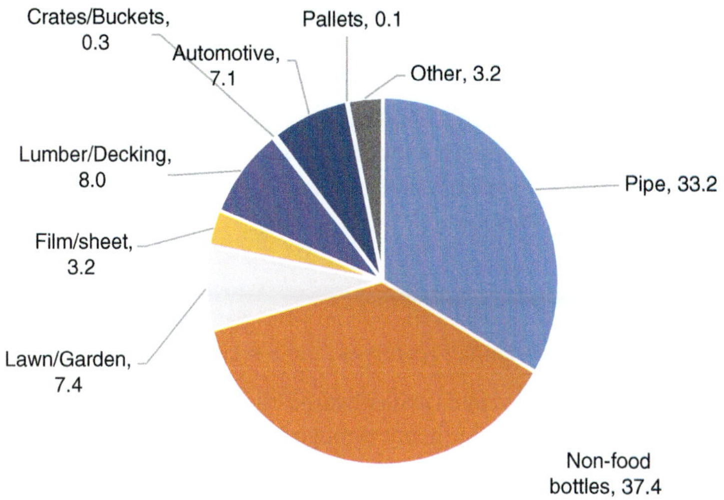

Figure 10.28 Uses of Recycled HDPE bottles, 2018.

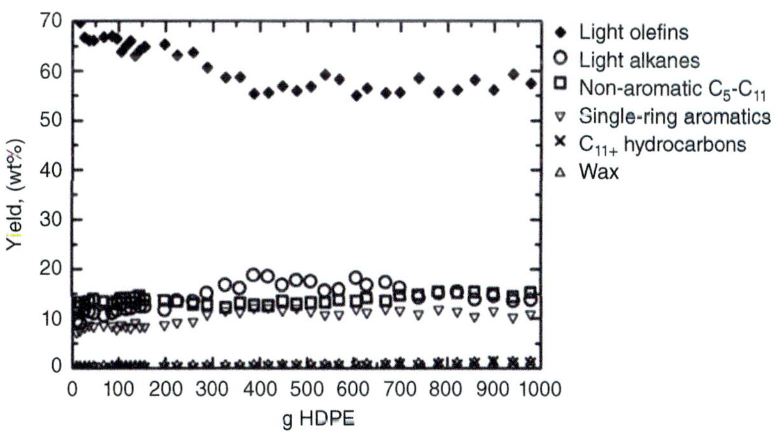

Figure 10.29 Yields of different products obtained over time using a HZSM-5 zeolite-based catalyst for the catalytic pyrolysis of HDPE [75]. Reproduced from [75] / with permission of Elsevier.

products obtained from the catalytic pyrolysis of HDPE using a HZSM-5 zeolite catalyst are shown in Figure 10.29.

Catalytic pyrolysis occurs via a combination of free-radical and carbocation mechanisms.

As the mechanism changes due to the catalyst addition, this also lowers the activation energy for the depolymerization/conversion to proceed so that it can occur at lower temperatures.

It should be noted that quick removal of the plastic degradation products during pyrolysis is important in order to prevent the formation of side products.

Stronger acid catalytic sites prefer chain-end scission, and thus generate high yields of shorter chain hydrocarbons.

10.11.12 Hydrocracking of Plastics

This process is termed thermochemical pyrolysis, and it serves a dual purpose:

- It breaks down large molecules into smaller hydrocarbons (Figure 10.30).
- It converts double bonds into single bonds through hydrogenation.

This transformation aids in the removal of impurities such as sulfur, nitrogen, and metals that are frequently present in heavy waste plastic. Among other benefits, this leads to reduced coke, reduced reaction temperatures and longer catalyst life. **How this process fits with circularity:** Saturated hydrocarbons undergo steam cracking, which then yields monomers that can be employed for PP and PE synthesis.

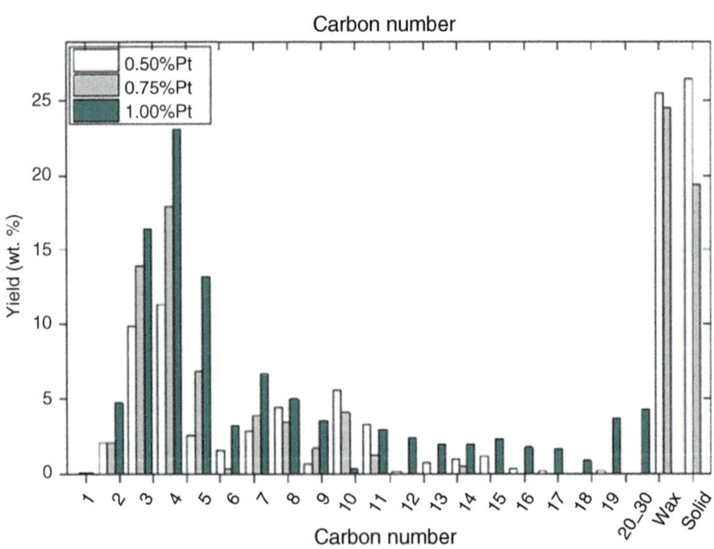

Figure 10.30 Effect of platinum loading (0.50, 0.75, and 1.00 wt%) at 330 °C [76]. Reproduced from [76] / with permission of American Chemical Society.

10.11.12.1 Ongoing Research

There is a strong interest in the development of catalysts that are economical, selective, stable (against poisoning, thermal decomposition), and active enough to allow for industrial implementation. A recently developed table salt-based catalytic approach enables rapid pyrolysis of PE/PP as well as metalized plastics into 86% oil, 14% gas, and 0% wax [77].

Other efforts in this area include the conversion of PE into diacids, surfactants, and other useful materials.

10.11.13 Polypropylene

As is the case with PE, PP can be reverted to its monomers, but it faces similar challenges such as low monomer yield. Currently, the recycling rate for PP is very low, at less than 0.5% in the United States. However, there is an ongoing effort to recycle more PP.

One breakthrough technology in this area is the PureCycle process, which was developed by Procter & Gamble, for polypropylene recycling after P&G increased its use of PP. This process utilizes supercritical alkane solvent mixtures in which only contaminants are extracted from PP. Filtration and extraction processes are followed to remove impurities and contaminants, including dyes and odors, from scrap PP.

Some characteristics of this process are outlined below:

- PVC is not tolerated as an impurity.
- The recommended purity of PP for this process is greater than 80%.
- This process successfully removes many contaminants.
- The recovery rate of the solvent is between 98% and 99%.

In the advanced PureCycle approach, solvent addition at the first stage of this process has the benefit (depending on the process) of dramatically reducing the viscosity of the melt, which allows economically attractive filtration of even the finest particles (Figure 10.31).

10.11.14 Gasification of Mixed Plastics to Syngas

Synthesis gas (also called syngas) is primarily composed of carbon monoxide (CO), hydrogen (H_2), and often some carbon dioxide (CO_2) and is a versatile intermediate that is used for the production of methanol, ammonia for fertilizer, and synthetic petroleum for diesel and other fuels [79].

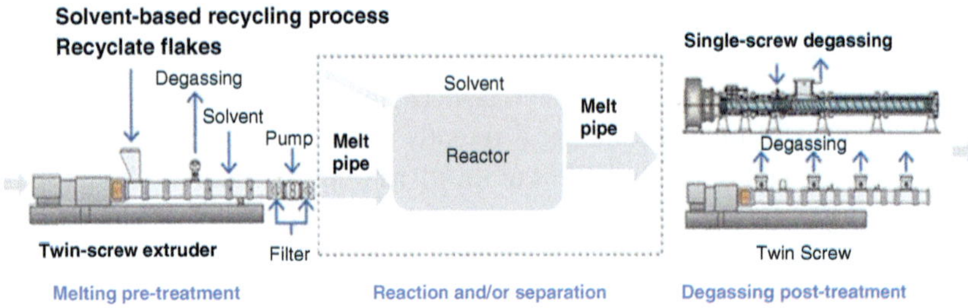

Figure 10.31 Flow scheme for the solvent-based recycling process for PP. Note that solvent addition is performed during the melting-pretreatment step. [78] / KrausMaffei Corporation.

Leading Technologies in chemical recycling of mixed plastics and/or Polyolefins

- Advanced Plasma Power Gasplasma process [80]
- Alter NRG plasma gasification process
- British Airways/Shell/Velocys Thermal Gasification Process
- Enerkem gasification process
- Fulcrum Bioenergy/BP thermal gasification
- OMV Refining Marketing GmbH (Austria) gasification
- Plastic Energy/SABIC anaerobic gasification
- Sierra Energy controlled-oxygen gasification
- Synpet thermal conversion
- Nexus Fuels, LLC pyrolysis
- Brightmark Energy and RES Polyflow pyrolysis process
- Plastic Advanced Recycling Corporation (PARC) catalytic pyrolysis
- Plastic2Oil (P2O) catalytic pyrolysis
- Recenso (Germany) developed a catalytic tribochemical conversion chemical recycling
- Enval microwave-induced pyrolysis
- Licella/Mura catalytic hydrothermal liquefaction
- Plastic waste conversion to downstream chemicals
- Eastman Carbon Renewal
- Green Mantra catalytic depolymerization

10.11.15 Use of Recycled Plastics in Food Packaging

Anyone considering the use of recycled plastics in food packaging, or other sensitive uses, must be concerned about the potential for migration of contaminants from the recycled plastic to the product. As is the case for paper, the processing conditions involved will generally involve sufficiently high temperatures to inactivate microorganisms, but will not destroy other types of contaminants. Under suitable conditions, these contaminants may migrate to the package contents.

There are four general approaches to the use of recycled plastics in such sensitive applications. The first is to use chemical recycling. In such systems, the recycled plastic is broken down into monomers, purified, and then repolymerized. Such materials are identical to virgin plastic. This approach is mostly suited to the condensation polymer family – in packaging applications, this means mostly PET. (It has also been applied to nylon, from materials such as carpet, as well as to other condensation polymers.) The main drawback of these systems is cost. For PET, the cost of chemical recycling, coupled with the cost of initial collection, means that these materials generally cost significantly more than virgin resin. However, since these systems can sometimes be applied to mixed and/or highly contaminated feedstocks, there is growing interest in chemical recycling. In early October, 2019, Coca Cola announced use of recovered and recycled ocean plastic in beverage bottles, relying on a chemical recycling process (https://www.beveragedaily.com/Article/2019/10/04/Coca-Cola-showcases-bottle-made-with-recycled-sea-plastic#).

The second option is to "sandwich" the recycled layer inside layers of virgin plastic. This "functional barrier" approach depends on the layer of virgin material to protect the contents from migrants originating in the recycled material. It is particularly suited to products with very short shelf lives, as it delays, rather than prevents, migration.

The third, and still most common, approach is to process the recycled material in a system that successfully removes most potential migrants. In some cases this is coupled with control over the source of the material, for instance, using deposit PET bottles rather than curbside bottles since the former is a "cleaner" stream. Increasingly, though, these systems can handle curbside material, relying on intensive cleaning of the recycled stream.

The fourth option is simply to restrict use of recycled plastics to less sensitive applications. In such cases, the "normal" recycling process may provide material that is suitable for use without additional clean-up. For example, one of the early uses of recycled PET was for egg cartons. Here, any potential for migration from the carton to the egg itself is minimal.

Potential users of recycled plastics in food packaging should consult the FDA document: "Guidance for Industry: Use of Recycled Plastics in Food Packaging: Chemistry Considerations" available at Ref. [81].

The FDA also presents a more general discussion of recycled plastics in food packaging at Ref. [82]. The agency maintains a list of submissions for which it has issued a letter of nonobjection (now referred to as a No Objection Letter, NOL) for food contact articles.

It is noteworthy that PET mechanical recycling has the highest number of No Objection Letters (NOL) whereas obtaining NOL for LDPE/LLDPE is challenging.

One reason for this greater hurdle is that LDPE/LLDPE are utilized in various applications that may not require food packaging standards. Additionally, LDPE and LLDPE are often collected together as flexible materials, and LLDPE is extruded with PFAS to prevent shark skin effects. Consequently, securing NOL approval for LDPE/LLDPE is very difficult.

Case Study: REMADE Institute

Developing sustainable recycling solutions for plastics is a daunting challenge, which often requires commitment from stakeholders in various disciplines. For example, although it is possibly to develop highly innovative technologies in academic settings such as university labs, it is often difficult for these new technologies to become widely adopted or viable for everyday use without cooperation from stakeholders in industry. The REMADE institute was established in 2017 as a public–private partnership which was partially funded by the U.S. Department of Energy's Office of Advanced Materials and Manufacturing with the aim of addressing this key challenge and facilitating the development of new technologies that will help the United States to transition to a circular economy. The REMADE institute includes in its membership almost 170 manufacturers, national laboratories, universities as well as trade organizations. Through these partnerships, the REMADE Institute supports applied research at its early stages that can generate jobs, provide a more robust supply of recycled materials, and combat climate change by reducing emissions and the embodied energy of materials [83].

REMADE Institute has a number of key goals as part of its mission. The first of these goals is to foster the development of technologies that can reduce energy emissions by lowering primary material consumption while promoting the use of more secondary feedstocks in energy-intensive sectors. The second goal is to foster the development of technologies so that secondary feedstocks will have "better than cost and energy parity." The third goal is to facilitate wide-ranging adoption of new technologies in a wide range of industries. Lastly, the fourth goal is to provide training, education, and development opportunities for the workforce, thereby facilitating the adoption of REMADE Institute-fostered technologies. REMADE Institute has various metrics to evaluate its success. These benchmarks include a 30% reduction of virgin or raw materials, along with a 30% increase of recycled materials being consumed. They also include a 20% reduction of greenhouse gas emissions, along with a 25% increase in embodied energy efficiency. Additional benchmarks include the promotion of cross industry reuse and achieving cost and energy parity for recycled materials.

Once developed and adopted, the technologies developed through REMADE Institute will have numerous key impacts, such as saving 1.2 quads of energy each year, enabling the creation and retention of hundreds of thousands of clean energy jobs, reducing greenhouse gas emissions each year by 67.2 MMT, and generating as much as $22 billion USD each year in new economic opportunities in the U.S. [84].

As of November 2023, REMADE Institute had launched or supported almost 100 projects valued at almost $100 million USD since it was established in 2017 [85]. The four classes of energy-intensive materials that REMADE Institute focuses on include metals, polymers (including plastics), fibers, and electronic scrap. Meanwhile, they focus on five key areas of technological development, including systems analysis and integration, recycling and recovery, manufacturing materials optimization, designing tools to improve material reuse and utilization at the EoL, and lastly remanufacturing and EoL reuse.

As mentioned earlier, plastics are one of the areas of interest for REMADE Institute, and this includes plastic recycling technologies and approaches. A few examples of plastic recycling focused research projects that are currently supported by REMADE Institute include "Recycling of PET in Sustainable Food Packaging Systems" which is focused on the development of a process through which multilayered materials can be replaced by a single mono-material packaging system that is manufactured from recycled poly(ethylene terephthalate) (PET); "Identification of mixed plastics and valuable electronics at the source" focusing on the development of sensing software that can facilitate sorting at recycling facilities; "Scalable High Shear Catalyzed Depolymerizations of Multilayer Plastic Packaging" which is investigating the catalytic depolymerization as a cost-effective strategy for the processing of these films into higher-valued products; "Novel, Transient, Thermal Barcode System for Highly Accurate, High-Speed, Automated Plastics Sorting" which is focusing on the development of thermal barcode technology for the efficient sorting of seven types of plastics using molecular fingerprinting and machine learning. Another such example is the project entitled "Pilot Plant Demonstration of Plastic Upcycling for the Production of Sustainable Petrochemical Alternatives" which is focusing on the development of plastic-derived waxes and lubricant base oils as petroleum alternatives using plastics from local waste streams as feedstocks. Yet another example is the project "Low-Cost, High-Value Aromatics From Upcycling of Polyolefins Through Microwave Catalytic Processing" focusing on the development of a one-step microwave recycling process for the preparation of high-value materials such as benzene, toluene, and xylene (BTX). The aforementioned examples are not a comprehensive list of projects supported by REMADE Institute, but they do help demonstrate how this institute is helping to facilitate the development and adoption of new sustainable technology in various areas, including plastics recycling.

HolyGrail Case Study

A major hurdle that has impeded recycling efforts is the sorting process, which tends to be time-consuming, costly, or can lead to a significant amount of materials being discarded rather than being recycled. The wide variety of materials as well as the wide range of shapes used for packaging creates a significant challenge during the sorting process. Sensors such as infrared sensors are often used to facilitate the sorting, but they can also have limitations. For example, a sleeve on a bottle made of a different type of plastic than the bottle itself may cause the bottle to be incorrectly sorted, as the sorting system (such as an infrared scanner) may misidentify the bottle based on the composition detected in the sleeve rather than the actual composition of the bottle itself. Meanwhile, near-infrared (NIR) sensors cannot detect black PET trays containing carbon black, which renders them invisible to the sensor. One way in which packaging sustainability is being greatly strengthened is through digital watermarking technology and the use of chemical tracers. The HolyGrail Project was launched in 2016 by around 30 companies led by Procter & Gamble and under the auspices of the Ellen Macarthur Foundation [86].

The motivation for the HolyGrail project was to develop an approach to address this key sorting challenge. As mentioned earlier, chemical tracers and digital watermarks are two ways to facilitate sorting. For example, a chemical tracer can have a distinct spectroscopic signal and can be incorporated into the plastic used to produce a packaging container, its label, or its sleeve. This tracer can be used to generate a binary code, based on whether or not the tracer is present in a particular material. More than one tracer can also be used, with each having their own unique spectroscopic signals, thereby generating a more complex code that could be read by a machine or sensor on a sorting line. Such codes could be used to identify whether or not the plastic is food-grade, or other criteria. Meanwhile, digital watermarks are optical codes that could be applied or embossed in a plastic container, applied onto packaging labels, or applied as a texture on the surface of plastic packaging. These digital watermarks would typically be invisible to human eyes but visible to cameras, sensors, or barcode scanners at sorting facilities. A diverse range of information could be provided through these codes, identifying information such as the type of plastic, manufacturer, the intended use of the packaging, or other relevant information.

Between chemical tracers and digital watermarks, the use of digital watermarks appears to be especially gathering momentum, and the HolyGrail project has progressed to HolyGrail 2.0, with even more organizations participating (130) in this endeavor [87]. It has recently undergone testing at a national scale in France. A validation study conducted early in 2023 using 5.6 tons of digitally watermarked PET bottles also demonstrated that this technology would be viable for use at scale [88]. In that study, 5.6 tons of digitally watermarked PET were manufactured and then mixed with postconsumer waste at a material recovery. This mixed waste containing 200,000 digitally watermarked bottles was then baled and sent to a recycling facility where sorting tests were performed with a digital detection system. It was found that the average detection and sorting efficiencies were 92.1% and 88.3%, respectively. These respective efficiencies increased further with two-pass sorting (that is standard practice in any case) to 95.9% and 95.1%.

Case Study: Solvent-Targeted Recovery and Precipitation (STRAP)

A promising example of an emerging technology that has recently been developed by George Huber, Reid Van Lehn, and their coworkers at University of Madison-Wisconsin is solvent-targeted recovery and precipitation (STRAP). This innovative approach that is progressing to pilot-scale through collaboration with researchers at Michigan Technological University, Green Bay Innovation Group, and Amcor is a new approach known as solvent-targeted recovery and precipitation (STRAP). This approach employs a selective solvent to specifically dissolve one of the polymers in a mixed plastic such as a multilayer film, and the solution of this dissolved polymer is then passed through a filter, and subsequently precipitated with the use of an anti-solvent (one in which the polymer is insoluble or has low solubility). This process of selectively dissolving and recovering a plastic can be repeated to recover the other plastics that are present in multilayer plastic packaging. In summary, the mixed plastic is "washed" several times by selective solvents with different plastics being recovered each time, and the plastic resin thus being recovered for subsequent reuse [89, 90]. Therefore, this technology serves to greatly expand the range of materials that can be effectively recovered. It is noteworthy that multilayer plastics have been particularly difficult to recycle due to their complex compositions with multiple plastics, and thus this approach could pave the way for the recycling of these challenging materials. This process enables the plastics to be restored in their initial state so that they can be reused or even upcycled into value-added applications. The selection of suitable solvents is facilitated by thermodynamic calculations about the solubility of each polymer, via an approach known as COnductor-like Screening MOdel for Realistic Solvents (COSMO-RS). This COSMO-RS model enables the researchers to predict the solubility of polymers in various solvent mixtures, at different temperatures. It is noteworthy that this approach is different than the PureCycle approach. One benefit of STRAP over the PureCycle approach is the operation for the STRAP at normal pressure, while PureCycle operates at high pressure. The recovery of PE from a multilayer film via STRAP is shown schematically in Figure 10.32.

While this technology was initially developed at laboratory scale, it is currently being scaled up for larger scale production, and a process development unit (PDU) has been developed for scalability studies. At a larger scale, the waste plastic is initially shredded into small particles so that

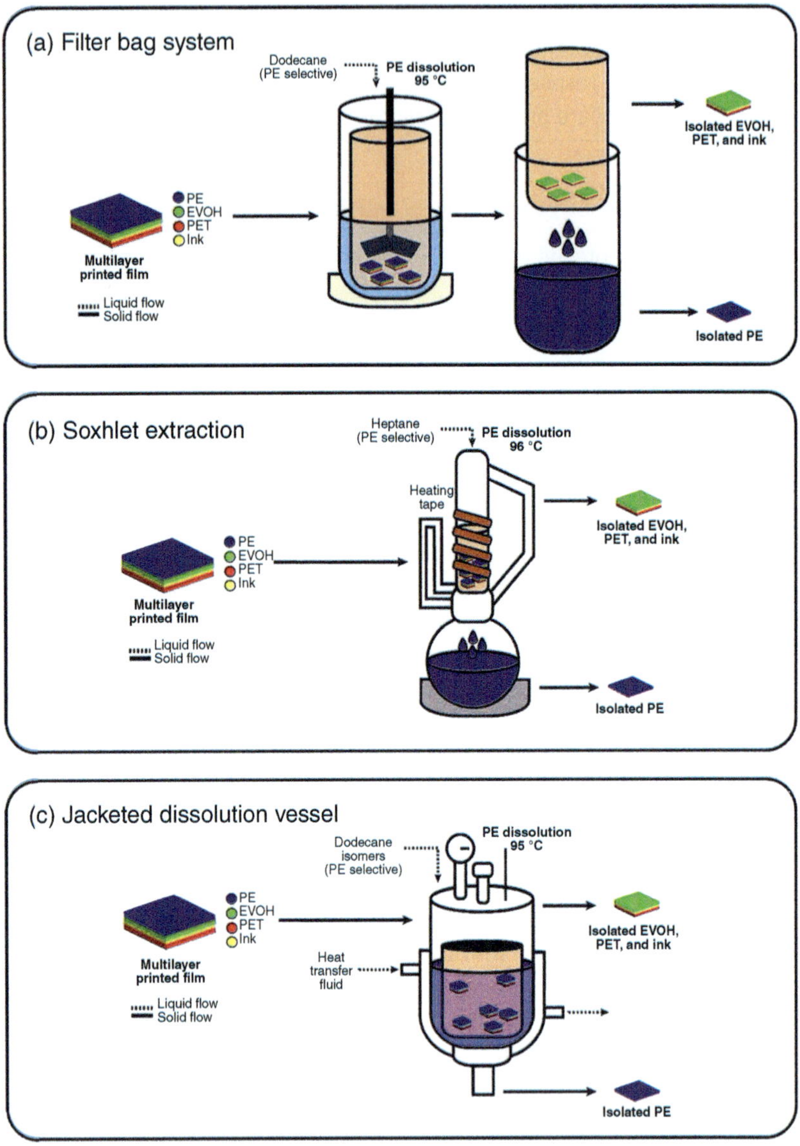

Figure 10.32 Schematic depiction showing the use of STRAP technology to recover PE from a multilayer film. Three different STRAP approaches are shown in this illustration, including (a) the use of a filter bag system, (b) Soxhlet extraction, and (c) the use of a jacketed dissolution vessel [91]. Reproduced from [91] / with permission of American Chemical Society.

they can readily flow through the reactor tanks of the PDU. The shredded plastic particles are firstly fed into a dissolution tank in which a selective solvent is used to dissolve one of the plastics present in the mixture. The dissolved plastic is then passed through a filter to separate it from the undissolved plastic. The undissolved plastic is then sent to a solvent recovery system. Meanwhile, the dissolved plastic is sent to a precipitator, where it is cooled so that it precipitates from the solution, and is thus recovered as a solid. The precipitated plastic is then sent to a solvent recovery system so that the solvent can be recovered for subsequent use. Lastly, the recovered plastic is extruded into pellet form, and it can then be sold to plastic converters.

It has been calculated that the recovery of polyethylene via the STRAP process can be achieved with 60–70% lower greenhouse gas emissions than would be produced for the production of virgin polyethylene [92]. This technology can also help divert difficult-to-recycle mixed plastics such as multilayer films from landfills or other undesirable EoL scenarios such as incineration. The reuse of these materials via STRAP can also help combat the proliferation of microplastics and ocean pollution.

10.12 Concluding Remarks

Given the widespread use of plastics as well as its desirable properties, such as its processability, its lightweight nature, and affordability, these materials will likely remain in widespread use for the foreseeable future. Therefore, plastics recycling can be anticipated to have a critical role in providing a circular economy for these materials. While mechanical recycling has been the traditional strategy for plastics recycling, chemical recycling strategies have made significant advances and are likely to have a growing role in the future.

Problem Set for Chapter 10

1 Which of the following answers is correct about open-loop and closed-loop recycling?
  - **A** Closed-loop recycling refers to recycling something into an identical or similar article, such as a new PET bottle out of PET bottle.
  - **B** Open-loop recycling refers to recycling something into a different type of product (and often of lower value), such as PET bottle to fiber.
  - **C** Both a and b are correct.
  - **D** None of the above.

2 What should be the path forward for effective plastic recycling?
  A Mechanical recycling only.
  B Chemical recycling only.
  C Both chemical and recycling are complementary and should be advanced.
  D None of the above.

3 Examples of chemical recycling include.
  A Hydrocracking of HDPE.
  B Pyrolysis of HDPE.
  C Melt-processing of recycled HDPE to another bottle.
  D a and b.

4 What is the major product during plastic gasification?
  A Syngas ($CO + H_2$) along with other minor impurities.
  B CH_4.
  C CO_2.
  D $CO_2 + H_2$.

5 Examples of Mechanical Recycling Include:
  A PET bottle into fiber by melt processing.
  B HDPE bottle into lumber by melt processing.
  C PET bottle into another PET bottle by melt processing.
  D All of the above.

6 Compare pyrolysis, catalytic pyrolysis, and hydrocracking. In your answer, describe the benefits and disadvantages of each process.

7 Which one is true about trends in mechanical recycling?
  A Digital watermarking.
  B Monomaterials instead of multilayer packaging.
  C Near IR sorting.
  D All of the above.

8 Key challenges in plastic recycling include:
  A Compositional fluctuations.
  B Contamination.
  C Lower price of virgin materials.
  D All of the above.

9 In terms of the CO_2 index, which plastic waste treatment method is the most climate-friendly (less CO_2 generation)?
  A Mechanical recycling.
  B Pyrolysis.
  C Incineration with energy.
  D Landfilling.

10 Which statement is True about the following chemical recycling processes:
  A Pyrolysis is the thermal decomposition of plastics without a catalyst.
  B Hydrocracking is a catalyzed thermal decomposition of plastics in the presence of hydrogen atmosphere.
  C Catalytic pyrolysis is a catalyzed thermal decomposition of plastic in the presence of a hydrogen atmosphere.
  D Both a and b.

11 If we want to use recycled PET in food packaging, what are the available options? Ignore the cost aspect.
  A Depolymerize the PET into its monomer, purify the monomer, and then repolymerize the monomer to virgin PET and then use this polymer for food packaging.
  B Use PET as an inner layer between virgin PET layers to avoid direct food contact.
  C Use the recycled PET directly without any cleaning efforts for food packaging.
  D Both a and b.

12 Why is the Dissolution-Precipitation approach for plastic recycling called chemical recycling?
  A Because in this process the polymer is decomposed into small molecules.
  B Because in this process the polymer is separated by using a chemical (solvent).
  C Both a and b.
  D None of the above.

13 Why is the chemical recycling of plastic waste a challenging task?
  A Costly processes.
  B Energy-intensive processes.
  C Not suitable for food packaging.
  D Both a and b.

References

1 Toto, D. (2019). APR study: recycled plastics lower energy consumption, greenhouse gas emissions. *Recycling Today* (29 January). https://www.recyclingtoday.com/news/apr-life-cycle-impact-.assessment-recycled-pet-hdpe-pp/ (accessed 30 January 2024).

2 Braun, S. (2023). Why most plastic can't be recycled. *Deutsche Welle (DW)* (17 March). https://www.dw.com/en/why-most-plastic-cant-be-recycled/a-64978847 (accessed 30 January 2024).

3 EPA (2018). Advancing Sustainable Materials Management: 2015 Tables and Figures, United States Environmental Protection Agency. https://www.epa.gov/sites/default/files/2018-07/documents/smm_2015_tables_and_figures_07252018_fnl_508_0.pdf (accessed 11 November 2023).

4 The Association of Plastics Recyclers (2021). 2021 U.S. Post-consumer Plastic Recycling Dashboard. https://circularityinaction.com/2021PlasticRecyclingData (accessed 30 January 2024).

5 National Association for PET Container Resources (NAPCOR) (2021). 2020 PET Recycling Report. https://napcor.com/wp-content/uploads/2023/12/NAPCOR_2020RateReport_FINAL.pdf (accessed 30 January 2024).

6 Association of Plastics Recyclers (APR) (2024). Informational Resources. https://plasticsrecycling.org/library (accessed 20 November 2024).

7 Gerlat, A. (2013). Rigid plastics recycling, rises in pounds, accessibility, *Waste 360*, (20 March). https://www.waste360.com/plastics/rigid-plastics-recycling-rises-pounds-accessibility (accessed 30 January 2024).

8 More Recycling (2019). 2017 National Post-Consumer Non-Bottle Rigid Plastic Recycling Report. Prepared by More Recycling for the American Chemistry Council. https://www.plasticsmarkets.org/jsfcontent/Non-BottleReport17_jsf_1.pdf (accessed 15 November 2023).

9 More Recycling (2019). 2017 National Post-Consumer Plastic Bag & Film Recycling Report. Prepared by More Recycling for the American Chemistry Council. https://www.americanchemistry.com/content/download/8115/file/2017-National-Post-Consumer-Plastic-Bag-and-Film-Recycling-Reports.pdf (accessed 15 November 2023).

10 More Recycling (2020). 2018 National Post-Consumer Plastic Bag & Film Recycling Report. Prepared by More Recycling for the American Chemistry Council. https://www.americanchemistry.com/content/download/8116/file/2018-National-Post-Consumer-Plastic-Bag-and-Film-Recycling-Reports.pdf (accessed 31 January 2024).

11 More Recycling (2019). 2017 National Post-Consumer Plastic Bag & Film Recycling Report. Prepared by More Recycling for the American Chemistry Council. https://www.americanchemistry.com/content/download/8115/file/2017-National-Post-Consumer-Plastic-Bag-and-Film-Recycling-Reports.pdf.

12 More Recycling (2019). 2017 National Post-Consumer Plastic Bag & Film Recycling Report. Prepared by More Recycling for the American Chemistry Council. https://www.americanchemistry.com/content/download/8115/file/2017-National-Post-Consumer-Plastic-Bag-and-Film-Recycling-Reports.pdf.

13 Stina Inc. (2021). 2019 U.S. Post-consumer Plastic Recycling Data Report [Internet].https://www.plasticsmarkets.org/jsfcontent/2019_Annual_Plastic_Recycling_Report_jsf_1.pdf (accessed 14 November 2023).

14 Stina Inc. (2024). 2022 U.S. Postconsumer Recycling Dashboard [Internet]. Association of Plastic Recyclers (Prepared by Stina Inc.). https://circularityinaction.com/2022plasticrecyclingdata/ (accessed 18 August 2024).

15 Kelechava, B. (2019). Resin Identification Codes (RICs), as Specified by ASTM D7611. American National Standards Institute. https://blog.ansi.org/2019/02/resin-identification-codes-rics-astm-d7611/#gref (accessed 15 November 2023).

16 Kelechava, B. (2019). Resin Identification Codes (RICs), as Specified by ASTM D7611. American National Standards Institute. https://blog.ansi.org/2019/02/resin-identification-codes-rics-astm-d7611/#gref (accessed 15 November 2023).

17 ASTM International (2022). Standard Practice for Coding Plastic Manufactured Articles for Resin Identification. ASTM International. https://www.astm.org/d7611_d7611m-21.html (accessed 31 January 2024).

18 Kelechava, B. (2019). Resin Identification Codes (RICs), as Specified by ASTM D7611. https://blog.ansi.org/2019/02/resin-identification-codes-rics-astm-d7611/#gref (accessed 31 January 2024).

19 Clauson, D (2013). Standardization News: Modernizing the Resin Identification Code. ASTM International. https://sn.astm.org/features/modernizing-resin-identification-code-ja13.html (accessed 15 November 2023).

20 Kelechava, B. (2019). Resin Identification Codes (RICs), as Specified by ASTM D7611. American National Standards Institute. https://blog.ansi.org/2019/02/resin-identification-codes-rics-astm-d7611/#gref (accessed 15 November 2023).

21 Paben J. (2023). US plastic exports fell at a faster rate in 2022. *Plastics Recycling Update* [Internet] (21 February). https://resource-recycling.com/plastics/2023/02/21/u-s-plastic-exports-fell-at-a-faster-rate-in-2022/ (accessed 30 January 2024)

22 Staub C. (2024). Scrap plastic exports drop to new low. *Plastics Recycling Update* [Internet] (14 February). https://resource-recycling.com/plastics/2024/02/14/scrap-plastic-exports-drop-to-new-low/ (accessed 18 August 2024)

23 Paben, J. (2023). U.S. fiber and plastic exports continued to fall in 2022. *Plastics Recycling Update* (21 February). Resource Recycling, Inc. https://resource-recycling.com/recycling/2023/02/21/u-s-fiber-and-plastic-exports-continued-to-fall-in-2022/ (accessed 31 January 2024).

24 Paben, J. (2022). US scrap plastic exports continue years-long decline. *Plastics Recycling Update* (2 March). Resource Recycling, Inc. https://resource-recycling.

com/plastics/2022/03/02/us-scrap-plastic-exports-continue-years-long-decline/ (accessed 31 January 2024).

25 EuRIC AISBL (2023). Plastic Recycling Factsheet. https://circulareconomy.europa.eu/platform/sites/default/files/euric_-_plastic_recycling_fact_sheet.pdf (accessed 16 November 2023).

26 Phys.org (2022). Only nine percent of plastic recycled worldwide: OECD. https://phys.org/news/2022-02-percent-plastic-recycled-worldwide-oecd.html (accessed 15 November 2023).

27 Vollmer, I., Jenks, M.J.F., Roelands, M.C.P. et al. (2020). Beyond mechanical recycling: giving new life to plastic waste. *Angew. Chem. Int. Ed.* 59: 15402–15423. https://doi.org/10.1002/anie.201915651.

28 Vollmer, I., Jenks, M.J.F., Roelands, M.C.P. et al. (2020). Beyond mechanical recycling: giving new life to plastic waste. *Angew. Chem. Int. Ed.* 59: 15402–15423. https://doi.org/10.1002/anie.201915651.

29 Di Mondo, D. (2023). In My Opinion: rethink how we evaluate molecular recycling. *Resource Recycling*. https://resource-recycling.com/recycling/2023/01/17/in-my-opinion-rethink-how-we-evaluate-molecular-recycling/ (accessed 15 November 2023).

30 Hartz, E. (2020). Tough questions — and honest answers — about molecular plastics recycling. *Plastics Today* (14 May). https://www.plasticstoday.com/advanced-recycling/tough-questions-and-honest-answers-about-molecular-plastics-recycling (accessed 15 November 2023).

31 Walker, T.W., Frelka, N., Shen, Z. et al. (2020). Recycling of multilayer plastic packaging materials by solvent-targeted recovery and precipitation. *Sci Adv.* 6 (47): eaba7599. https://doi.org/10.1126/sciadv.aba7599.

32 Devi, M., Rawat, S., and Sharma, S. (2021). A comprehensive review of the pyrolysis process: from carbon nanomaterial synthesis to waste treatment, Oxford Open. *Materials Science* 1: itab014. https://doi.org/10.1093/oxfmat/itab014.

33 Gonzalez-Aguilar, A.M., Pérez-García, V., and Riesco-Ávila, J.M. (2023). A thermo-catalytic pyrolysis of polystyrene waste review: a systematic, statistical, and bibliometric approach. *Polymers* 15 (6): 1582. https://doi.org/10.3390/polym15061582.

34 Biakhmetov, B., Dostiyarov, A., Ok, Y.S., and You, S. (2023). A review on catalytic pyrolysis of municipal plastic waste. *WIREs Energy and Environment* 12: e495. https://doi.org/10.1002/wene.495.

35 Pal, S., Kumar, A., Sharma, A.K. et al. (2022). Recent advances in catalytic pyrolysis of municipal plastic waste for the production of hydrocarbon fuels. *Processes* 10: 1497. https://doi.org/10.3390/pr10081497.

36 Maqsood, T., Dai, J., Zhang, Y. et al. (2021). Pyrolysis of plastic species: a review of resources and products. *J. Anal. Appl. Pyrolysis* 159: 105295. https://doi.org/10.1016/j.jaap.2021.105295.

37 PwC (2023). Plastic Packaging Tax. https://www.pwc.co.uk/services/tax/plastic-packaging-tax.html (accessed 16 November 2023).

38 Nightingale, S.D. (2004). New technologies for food traceability: package and product markers. *Food Safety Magazine* (1 August). https://www.food-safety.com/articles/4156-new-technologies-for-food-traceability-package-and-product-markers (accessed 16 November 2023).

39 Alliance to End Plastic Waste. (2023). Pioneering Digital Watermarks for Smart Packaging Recycling in the EU: Digital Watermarks Initiative HolyGrail 2.0. https://www.digitalwatermarks.eu/ (accessed 20 November 2024).

40 Lingle, R. (2020). Digital watermarks for recycling plastic packaging: who, what, why, and where?, *Packaging Digest* (9 November). https://www.packagingdigest.com/sustainability/digital-watermarks-recycling-plastic-packaging-who-what-why-and-where (accessed 16 November 2023).

41 Features (2022). From waste sorting to serialization: Exploring the potential of digital watermarks. *Packaging Europe* (1 December). https://packagingeurope.com/features/from-waste-sorting-to-serialization-exploring-the-potential-of-digital-watermarks/9151.article (accessed 16 November 2023).

42 Digimarc (2022). Digimarc Recycle: Helping Brands Achieve Circularity: Product Overview. Digimarc Corporation. https://dmrcmktdocs.blob.core.windows.net/documents/4E57NSXD3ZZP-373239220-22.pdf (accessed 16 November 2023).

43 The Association of Plastic Recyclers (2018). Near infrared (NIR) sorting in the plastics recycling process. https://plasticsrecycling.org/images/Design-Guidance-Tests/APR-RES-SORT-2-NIR-sorting-resource.pdf (accessed 16 November 2023).

44 Stone, M. (2023). Exxon's new 'advanced recycling' plant raises environmental concerns. *The Guardian* (10 April). https://www.theguardian.com/us-news/2023/apr/10/exxon-advanced-recycling-plastic-environment (accessed 16 November 2023).

45 Aramco (2023). Supporting the Energy Transition: crude oil to chemicals. https://www.aramco.com/en/sustainability/climate-change/supporting-the-energy-transition/crude-to-chemicals (accessed 16 November 2023).

46 Aramco (2023). TotalEnergies and SABIC complete MENA region's first processing of oil from plastic waste at scale to make certified circular polymers. Aramco. https://www.aramco.com/en/news-media/news/2023/aramco-totalenergies-and-sabic-complete-processing-of-oil-from-plastic-waste (accessed 16 November 2023).

47 Blas, J. (2017). Inside Saudi Arabia's big bet on plastics. *Bloomberg* (2 July). https://www.bloomberg.com/news/articles/2017-07-02/from-oil-gushers-to-golf-balls-saudi-aramco-bets-on-chemicals (accessed 17 November 2023).

48 Closed Loop Partners's Center for the Circular Economy (2021). Transitioning to a circular system for plastics: assessing molecular recycling technologies in the United States and Canada. https://www.closedlooppartners.com/research/

transitioning-to-a-circular-system-for-plastics-assessing-molecular-recycling-technologies-in-the-united-states-and-canada-2/ (accessed 16 November 2023).

49 Luu, Paula. (2023). Molecular Recycling: What is Chemical Recycling and Why Does it Have So Many Different Names, and Why Does It Matter? Closed Loop Partner's Center for the Circular Economy. (15 August) https://www.closedlooppartners.com/what-is-chemical-recycling-why-does-it-have-so-many-different-names-and-why-does-it-matter/?utm_content=260771139&utm_medium=social&utm_source=linkedin&hss_channel=lcp-3960972 (accessed 20 November 2024).

50 NAPCOR and The Association of Plastics Recyclers (2018). Report on Postconsumer PET Container Recycling Activity in 2017. https://napcor.com/wp-content/uploads/2018/11/NAPCOR_2017RateReport_FINAL_rev.pdf (accessed 17 November 2023).

51 NAPCOR (2021). 2020 PET Recycling Report [Internet]. Charlotte, North Carolina. https://napcor.com/wp-content/uploads/2023/12/NAPCOR_2020RateReport_FINAL.pdf (accessed 29 January 2024).

52 National Association for PET Container Resources (NAPCOR) and The Association of Plastics Recyclers (APR) (2018). Report on Postconsumer PET Container Recycling Activity in 2017, page 5. https://napcor.com/wp-content/uploads/2018/11/NAPCOR_2017RateReport_FINAL_rev.pdf (accessed 17 November 2023).

53 Chemical Processing (2019). Carpet Recycling Technology Yields Polymer For Chemical Industry. https://www.chemicalprocessing.com/environmental-health-safety/news/11307870/carpet-recycling-technology-yields-polymer-for-chemical-industry (accessed 17 November 2023).

54 Maile, K. (2019). Eastman to recycle discarded carpet with chemical recycling. *Recycling Today* (5 November). https://www.recyclingtoday.com/news/eastman-circular-polymers-partnership/ (accessed 17 November 2023).

55 DeLozier, R. and Thiess, R. (2023). First person perspective: carpet is the feedstock of the future. *Resource Recycling* (14 June). https://resource-recycling.com/recycling/2023/06/14/first-person-perspective-carpet-is-the-feedstock-of-the-future/ (accessed 17 November 2023).

56 Lang, W.T., Mehta, S.A., Thomas, M.M. et al. (2023). Chemical recycling of polyethylene terephthalate, an industrial and sustainable opportunity for Northwest of England. *Journal of Environmental Chemical Engineering* 11: 110585. https://doi.org/10.1016/j.jece.2023.110585.

57 Lang, W.T., Mehta, S.A., Thomas, M.M. et al. (2023). Chemical recycling of polyethylene terephthalate, an industrial and sustainable opportunity for Northwest of England. *Journal of Environmental Chemical Engineering* 11: 110585. https://doi.org/10.1016/j.jece.2023.110585.

58 Bergsma, G. and Broeren, M. (2019). Chemische recycling in het afvalbeleid (Chemical recycling in waste policy). *CE Delft*. https://cedelft.eu/wp-content/

uploads/sites/2/2021/03/CE_Delft_190156_Chemische_recycling_afvalbeleid_DEF-update-2020.pdf (accessed 20 November 2024).

59 Sustainable Plastics (2021). PET Advanced Recycling Startup Gr3n closes €6.3 million in Series B Funding. https://www.sustainableplastics.com/news/gr3n-now-ready-bring-plastic-recycling-technology-market (accessed 1 February 2024).

60 IBM (2019). IBM Researchers Develop Radical New Recycling Process to Transform Old Plastic. https://newsroom.ibm.com/2019-02-11-IBM-Researchers-Develop-Radical-New-Recycling-Process-to-Transform-Old-Plastic (accessed 1 February 2024).

61 Tullo, A.H. (2019). Unilever and Coca-Cola roll out recycling initiatives. *Chemical & Engineering News* (9 October). https://cen.acs.org/environment/recycling/Unilever-Coca-Cola-roll-recycling/97/i40 (accessed 17 November 2023).

62 JEPLAN, Inc. (2024). BRING Technology™. https://www.jeplan.co.jp/en/technology/ (accessed 1 February 2024).

63 Loop Industries (2024). Making a difference: Revolutionary Technology. https://www.loopindustries.com/en (accessed 1 February 2024).

64 Poseidon Plastics (2024). Bringing an End To Single Use Plastic. https://poseidonplastics.com/ (accessed 1 February 2024).

65 CuRe Technology. Coca-Cola European Partners (CCEP Ventures) funding for CuRe Technology. https://curetechnology.com/news/coca-cola-european-partners-ccep-ventures-funding-for-cure-technology/ (accessed 20 November 2024).

66 Carbios (2023). Enzymatic recycling: removing the constraints of current processes. https://www.carbios.com/en/enzymatic-recycling/ (accessed 17 November 2023).

67 Khan, A., Naveed, M., Aayanifard, Z., and Rabnawaz, M. (2022). Efficient chemical recycling of waste polyethylene terephthalate. *Resour. Conserv. Recycl.* 187: 106639. https://doi.org/10.1016/j.resconrec.2022.106639 (accessed 17 November 2023).

68 Petritz, H.-N. (2024). Mechanical Recycling vs. Chemical Recycling: Rivals or Partners?https://waste-management-world.com/resource-use/mechanical-chemical-recycling-rivals-or-partners/ (accessed 1 February 2024).

69 Zhao, X., Long, Y., Xu, S. et al. (2023). Recovery of epoxy thermosets and their composites. *Materials Today* 64: 72–97. https://doi.org/10.1016/j.mattod.2022.12.005.

70 Quirino, R.L., Monroe, K., Fleischer Iii, C.H. et al. (2021). Thermosetting polymers from renewable sources. *Polym. Int.* 70: 167–180. https://doi.org/10.1002/pi.6132.

71 Khan, A., Naveed, M., and Rabnawaz, M. (2021). Melt-reprocessing of mixed polyurethane thermosets. *Green Chem.* 23: 4771–4779. https://doi.org/10.1039/D1GC01232K.

72 Kumar, V., Khan, A., and Rabnawaz, M. (2022). Efficient depolymerization of polystyrene with table salt and oxidized copper. *ACS Sustainable Chem. Eng.* 10: 6493–6502. https://doi.org/10.1021/acssuschemeng.1c08400.

73 Huang, Z., Shanmugam, M., Liu, Z. et al. (2022). Chemical recycling of polystyrene to valuable chemicals via selective acid-catalyzed aerobic oxidation under visible light. *J. Am. Chem. Soc.* 144: 6532–6542. https://doi.org/10.1021/jacs.2c01410.

74 Orozco, S., Artetxe, M., Lopez, G. et al. (2021). Conversion of HDPE into value products by fast pyrolysis using FCC spent catalysts in a fountain confined conical spouted bed reactor. *ChemSusChem.* 14 (19): 4291–4300. https://doi.org/10.1002/cssc.202100889.

75 Elordi, G., Olazar, M., Lopez, G. et al. (2009). Catalytic pyrolysis of HDPE in continuous mode over zeolite catalysts in a conical spouted bed reactor. *J. Anal. Appl. Pyrolysis* 85: 345–351. https://doi.org/10.1016/j.jaap.2008.10.015.

76 Bin Jumah, A., Anbumuthu, V., Tedstone, A.A., and Garforth, A.A. (2019). Catalyzing the hydrocracking of low density polyethylene. *Ind. Eng. Chem. Res.* 58: 20601–20609. https://doi.org/10.1021/acs.iecr.9b04263.

77 Shaker, M., Kumar, V., Saffron, C.M., and Rabnawaz, M. (2023). Revolutionizing plastics chemical recycling with table salt. *Advanced Sustainable Systems* 8: 2300306. https://doi.org/10.1002/adsu.202300306.

78 Keilbach, F.-X. and Bagdahn, K.U. (2022). Solvent-based recycling: emerging technology for reclaiming & purifying polyolefins with twin-screw extruders. *Plastics Technology* (26 January). https://www.ptonline.com/articles/solvent-based-recycling-emerging-technology-for-reclaiming-purifying-polyolefins-with-twin-screw-extruders (accessed 18 November 2023).

79 Saebea, D., Ruengrit, P., Arpornwichanop, A., and Patcharavorachot, Y. (2020). Gasification of plastic waste for synthesis gas production. *Energy Reports* 6: 202–207. https://doi.org/10.1016/j.egyr.2019.08.043.

80 Ray, R., Taylor, R., and Chapman, C. (2012). The deployment of an advanced gasification technology in the treatment of household and other waste streams. *Process Safety and Environmental Protection* 90: 213–220. https://doi.org/10.1016/j.psep.2011.06.013.

81 U.S. Food & Drug Administration. (2021). Guidance for Industry: Use of Recycled Plastics in Food Packaging (Chemistry Considerations). https://www.fda.gov/regulatory-information/search-fda-guidance-documents/guidance-industry-use-recycled-plastics-food-packaging-chemistry-considerations (accessed 20 November 2024).

82 U.S. Food & Drug Administration. (2020). Recycled Plastics in Food Packaging. http://www.fda.gov/food/ingredientspackaginglabeling/packagingfcs/recycledplastics/default.htm (accessed 20 November 2024).

83 About REMADE, REMADE Institute, A Division of the Sustainable Manufacturing Innovation Alliance Corp. https://remadeinstitute.org/about/ (accessed 25 March 2024).

84 Impacts & Metrics, REMADE Institute, A Division of the Sustainable Manufacturing Innovation Alliance Corp. https://remadeinstitute.org/technology/our-impact/#tpms (accessed 25 March 2024).

85 Atkins, E. (2023). Remade Institute to fund 14 new projects. *Waste & Recycling* (3 November). https://wasterecyclingmag.ca/uncategorized/remade-institute-to-fund-14-new-projects/1003289768/ (accessed 25 March 2024).

86 Cooper, T.A. (2020). *Ten-Year Forecast of Disruptive Technologies for Recycled Plastic Packaging to 2030*. Smithers Information Ltd.

87 Cooper, T.A. (2020). *Ten-Year Forecast of Disruptive Technologies for Recycled Plastic Packaging to 2030*. Smithers Information Ltd.

88 Reynolds, Matt (2023). Column: is HolyGrail 2.0 proving viable at scale?*Packaging World* (27 June). https://www.packworld.com/design/package-design/article/22865876/holygrail-20-proving-viable-at-scale (accessed 19 March 2024).

89 Xu, Z., Sanchez-Rivera, K.L., Munguia-Lopez, A. et al. (2023). Recycling of plastic films through solvent targeted recovery and precipitation. *Chemical Upcycling of Waste Plastics* https://cuwp.org/wp-content/uploads/2023/10/1_STRAP_102023.pdf (accessed 15 March 2024).

90 Walker, T.W., Frelka, N., Shen, Z. et al. (2020). Recycling of multilayer plastic packaging materials by solvent-targeted recovery and precipitation. *Sci. Adv.* 6: eaba7599. https://doi.org/10.1126/sciadv.aba7599.

91 Sánchez-Rivera, K.L., Granger, C., Appiah, H. et al. (2024). Cast film production with polyethylene recycled from a post-industrial printed multilayer film by solvent-targeted recovery and precipitation. *ACS Mater Lett* 6: 4042–4050. https://pubs.acs.org/doi/abs/10.1021/acsmaterialslett.4c01048 (accessed 14 August 2024).

92 Xu, Z., Sanchez-Rivera, K.L., Munguia-Lopez, A. et al. (2023). Recycling of plastic films through solvent targeted recovery and precipitation. *Chemical Upcycling of Waste Plastics*. https://cuwp.org/wp-content/uploads/2023/10/1_STRAP_102023.pdf.

11

Legal, Regulatory, EPR, and Green Design

11.1 Introduction

We have seen thus far that there are many ways to reduce the environmental footprint of packaging by selecting the appropriate packaging material for a given application as well as through reuse and recycling efforts. In addition to these approaches, regulatory policies as well as green design are also important tools in our efforts to achieve sustainable packaging. While there is often a strong business case for producing and using sustainable packaging in its own right, given the public awareness of environmental issues, well-crafted policy can often provide further incentives for manufacturers to develop and adopt sustainable packaging. In addition, green design can help avoid many environmental issues and recycling challenges (such as costly and time consuming separation procedures, for example) at the end of a packaging product's service lifecycle. In this chapter, we will discuss how some of these approaches are and can be employed to achieve a sustainable future.

The sulfanilamide disaster in 1937 led to the creation of the United States Federal Food, Drug, and Cosmetic Act (FD&C Act) in 1938 [1]. The Act gives authority to the U.S. Food and Drug Administration (FDA) to oversee the safety of food, drugs, medical devices, and cosmetics.

Key provisions of the Act include:

- Prohibiting the sale of adulterated or misbranded food, drugs, devices, and cosmetics.
- Mandating pre-market approval for all new drugs, meaning drugs must be proven safe and effective before being sold.
- Requiring drug labels to provide adequate directions for safe use.
- Mandating that cosmetics be safe and properly labeled.

Degradation, Stabilization, and Recycling of Packaging Materials, First Edition.
Muhammad Rabnawaz, Susan E.M. Selke, and Ian Wyman.

- Establishing more robust regulations for food safety, including standards for the identity and quality of packaging containers.

The Act has been amended multiple times to adapt to the changing landscape, but the FD&C Act remains the foundational law of the FDA's mission to protect public health.

Food Packaging regulations:
The FD&C Act defines food additives as [2]:

Direct food additives are substances that are intentionally added to foods for a specific purpose, such as to improve taste, texture, preservation, or nutritional value. These are ingredients included in the recipe or formulation of the food product and are expected to be consumed as a part of the food.

Indirect food additives refer to substances that may migrate into food from the packaging, storage containers, or other materials that come in contact with food. These are not intentionally added to the food, but they can end up in it through processes such as leaching or transfer during manufacturing, processing, or storage.

11.2 EU Packaging Directives

The European Union (EU) is widely regarded as one of the leading jurisdictions in terms of its environmental regulations and its efforts to implement sustainable packaging policies. Therefore, it can be helpful to take a look at the EU packaging directives. EU packaging directives were introduced in 1994 and are referred to as **Directive 94/62/EC** [3]. The directives are the basis for the rules in the EU member states on managing packaging and packaging waste. Why do we need to understand EU packaging directives? There are several reasons for this:

1) Many major packaging corporations operate as multinationals; legal changes in one region can influence their practices in other parts of the world. A reason for this is that when a company adopts sustainable solutions in one region, implementing those new sustainable solutions in other parts of the global market often makes good business sense. For example, adopting one protocol for the global marketplace can offer a multinational corporation greater simplicity and help streamline its operations as opposed to having multiple protocols for different regions.
2) EU packaging laws are often used as a reference for making new laws in other parts of the world.
3) The EU has been a global leader in reducing packaging waste through strict regulations.

The **overarching goals of the EU** packaging directives include [4]:

- Protect the environment from packaging waste.
- Achieve standardized measures across all countries in the EU regarding packaging and package waste management.

The EU's packaging laws apply to all packaged products sold within the European Union, even if the packaging or packaged products are imported from outside the region. The amendments in 2018 focused on the circular packaging economy and mandated extended producer responsibility (EPR).

Key points of EU Packaging Laws [5]:

- Laws target packaging and packaging waste to protect the environment and human health.
- Promote more careful use of natural resources.
- Promote a circular economy in packaging and reduce greenhouse gases.
- Reduce the EU's dependency on imports.
- Increase the recycling of packaging waste to move to a circular economy.
- Work toward waste prevention by improving resource efficiency.
- Promote reuseable packaging and reduce reliance on nonrecyclable packaging.
- Offer financial contributions for reusable packaging under EPR schemes for packaging.
- Promote recyclable single-use packaging.
- Promote a sustainable bioeconomy by using renewable raw materials.
- Use bio-based recyclable packaging and compostable, biodegradable packaging where beneficial from a lifecycle perspective.
- Biodegradation of packaging waste to CO_2 and water is not counted toward recycling targets.
- The EPR rules devised in the waste framework Directive 2008/98/EC should be applied to packaging producers.

11.2.1 EU Directive Amendment (2018/852)

The EU 2018 Packaging and Packaging Waste amendment focuses on circular packaging by promoting reuse and recycling [6]. This amendment became law in the EU on 4 July 2020. Key points of the EU 2018 amendment include:

By 31 December 2025, the recycling targets include [7]:

Recycling targets for various packaging materials, measured by weight, are set as follows:

a) Aluminum: 50%
b) Wood: 25%
c) Plastic: 50%

d) Ferrous Metals: 70%
e) Glass: 70%
f) Paper and Cardboard: 75%

A minimum of 65% of all packaging waste is required to be recycled by weight.

By 31 December 2030, the minimum targets by weight for recycling packaging waste include:

Minimum recycling targets by weight for different packaging materials are set as follows:

a) Ferrous Metals: 80%
b) Paper and Cardboard: 85%
c) Glass: 75%
d) Plastic: 55%
e) Aluminum: 60%, and
f) Wood: 30%

Overall, at least 70% of all packaging waste is required to be recycled by weight.

The above recycling targets also emphasize that the recycled materials must not compromise safety and hygiene standards, and they must offer the required performance for the packaged goods. Quality standards should be set to promote high-quality and effective recycling of packaging waste. Under certain circumstances, the target's accomplishment can be delayed by up to five years.

11.2.2 How Are the Recycling Rates Calculated?

Generated packaging waste is quantified based on the amount (weight) of packaging placed by a country on the market (**input**) in a given calendar year. Meanwhile, the weight of packaging waste recycled is determined by measuring the **output** of any sorting operation. The amount of biodegradable packaging waste that undergoes aerobic or anaerobic treatment may be counted as recycled where that treatment generates compost, digestate, or other output with a similar quantity of recycled content in relation to input, which is to be used as a recycled product, material, or substance. Where the output is used on land, Member States may count it as recycled only if this use results in benefits to agriculture or ecological improvement. Waste-to-energy conversion, incineration, and landfilling are not counted toward the recycling targets. In addition, packaging waste that is collected in one country (an EU member) and sent to another EU member state for the purposes of recycling is counted toward the attainment of the targets where the packaging waste was collected and not where it was sorted or recycled.

By 31 December 2024, EU member countries will ensure that EPR schemes are established for all packaging in accordance with EU Directive 2008/98/EC.

The EPR provides financial support for packaging waste collection, disposal, and recycling.

Recently, a proposal for a new regulation in the EU on packaging and packaging waste by amending Regulation (EU) 2019/1020 has been proposed. These regulations recommend three areas of priority to reduce package waste [8, 9].

1) **Reduce packaging waste generation:** Here the focus is to reduce waste generation by ~19% by 2030. This can be achieved by reducing reliance on single-use packaging, avoiding packaging where possible, and by finding alternatives for multimaterial complex packaging structures.
2) **Promote a circular economy** for packaging by promoting packaging design features that are inherently recyclable and get rid of labels from packages that confuse consumers, to ensure that labels provide clear instructions on how to recycle that package.
3) **Promote the use of more recycled content** in packaging by minimizing the downcycling of recycled packaging materials.

How can the U.S. and other parts of the world benefit from the EU Packaging Directives?

While the European Union's geographic and cultural dynamics are distinct from other regions, the EU's model for handling packaging waste can offer valuable insights. Instead of starting from scratch, the United States could leverage the knowledge and effectiveness of the EU's waste legislation for handling packaging waste in the United States.

11.3 Extended Producer Responsibility (EPR)

EPR requires producers (polluters) to bear the financial responsibility for cleaning and managing their product's waste. The "Green Dot" system is one example of an EPR program. The **P** can also denote Product, and there are significant differences between these two meanings of EPR. To add to this confusion, the term "product stewardship" [10] is also used, sometimes meaning extended producer responsibility, sometimes extended product responsibility, and sometimes simply the broader concept of minimizing adverse impacts of products and their packaging.

In the context of packaging waste, producers are responsible for the recovery, recycling, and disposal of packaging waste, and these protocols are generally coupled with certain levels of recycling requirements. Until recently, calls to apply EPR policies toward packaging were mostly ignored because the US plastic scrap exporters were making money (and avoiding tipping fees) from mixed plastic bales by exporting them to numerous countries. This situation has changed in the last few years, as stricter regulations on waste plastic export, such as China's "National Sword" policy, have led to a drastic decline in U.S. plastic scrap exports. For example, China was the largest importer of U.S. plastic scrap (33% of all U.S. plastic scrap) prior to 2017, but now only accounts for 3–4%.

Who is a Producer? Maine's EPR bill defines a producer as an individual or company that:

1) Has **legal ownership of the brand** of a product sold, offered for sale, or distributed for sale in or into the state contained, protected, delivered, presented or distributed in or using packaging material [11]; **or**
2) Is the sole entity that **imports** a product into the state for sale, offer for sale, or distribution for sale in or into the state a product contained, protected, delivered, presented, or distributed in or using packaging material that is branded by a person.

EPR is an extension of the responsibility of producers and all entities involved in the product chain to reduce the cradle-to-cradle [12] impacts of a product and its packaging; the primary responsibility lies with the **producer or brand owner**, who makes design and marketing decisions.

For example, PepsiCo is the brand owner of Pepsi-Cola. Bottle suppliers for this product are known as packaging suppliers (such as Amcor and others), and bottling companies that are licensed to manufacture Pepsi products are called producers. Sometimes, producers and brand owners are the same entity. For example, the Coca-Cola Company owns both the brand and formula for its beverages and produces them.

11.3.1 Historical Background of EPR

In 1990, Thomas Lindqvist suggested that producers be responsible for handling their product waste at the end of pipe/end of life (EoL) [13]. Handling involves collection, recovery, recycling, and final disposal of the products at the EoL. In 1991, Germany adopted EPR for packaging by introducing a **"dual system"** for collection – by establishing a second collection system in parallel to the local municipality collections. The second collection system was sponsored by producers under the EPR program and was overseen by a third-party system. Companies participating in this parallel collection system put a "**Green Dot**" on their product that conveys to consumers that the companies are paying for the recovery of that package. Because of EPR and other relevant strategies, Germany has the highest recycling rates, reaching up to 66% of MSW [14].

By 2023, only Maine, Oregon, Colorado, and California have passed EPR regulations in the United States. Other states are expected to join this initiative in the near future.

11.3.2 What Are the Potential Benefits of EPR?

1) EPR revenues rather than taxpayers' money can be used to create an extensive collection and sorting infrastructure, thus diverting more packaging waste from landfills to recycling.
2) EPR, in principle, will shift recycling/disposal costs from U.S. taxpayers to producers.
3) EPR can improve the circular economy of packaging by incentivizing the adoption of sustainable and recyclable packaging.

An argument often put against EPR is that producers will transfer the EPR fee to consumers by raising the price. Producers also use this argument as a strategy to discourage the adoption of EPR or attempt to bend EPR laws in their favor. Carefully crafted EPR legislation will promote environmentally friendly packaging without adding extra tax burdens on the taxpayers.

A **well-crafted EPR** legislation means laws are aimed at **promoting the following outcomes:**

- Promote recyclable packaging.
- Minimize the use of nonrecyclable packaging.
- Promote the use of recycled content in packaging.
- Reduce packaging waste.
- EPR laws should ensure that required food safety and shelf stability are maintained.

Some recommended regulations that can make EPR more effective should include:

1) Asking for higher waste disposal charges on nonrecyclable packaging, thereby discouraging their production and encouraging the adoption of recyclable options.
2) Charging lower disposal costs on recyclable packaging.
3) Charging lower disposal costs on packaging that is biobased.
4) Higher taxes on plastic packaging compared to metal and paper to promote the usage of more eco-friendly packaging options.
5) Offering incentives for minimal material usage in packaging to spark competitive innovation for more efficient and lighter packaging designs.
6) Ensuring that new packaging still meets food preservation needs, as the environmental benefits of sustainable packaging can be negated by food waste if the packaging fails to adequately protect the contents.

7) Protecting small businesses by not overburdening them with regulatory costs. For example, in Maine, small-scale producers generating less than 15 tons of packaging can utilize a streamlined reporting system and are subject to a fixed fee of $500 or less per ton for their packaging waste [15].

The United States has not adopted any **general** producer responsibility systems for packaging materials. Information about U.S. EPR and product stewardship initiatives can be obtained from the Product Stewardship Institute, http://www.productstewardship.us/. **Pesticide containers** are currently the only packaging specifically listed on their map of EPR state laws (http://www.productstewardship.us/?State_EPR_Laws_Map) [16].

The State of California has adopted a policy on EPR that calls for placing "shared responsibility for EoL product management on the producers, and all entities involved in the product chain, instead of the general public; while encouraging product design changes that minimize a negative impact on human health and the environment at every stage of the product's lifecycle" [17]. This law has resulted in the passage of specific stewardship laws for carpet, paint, mercury thermostats, recalled products, auto refrigerants, and pesticide containers. In 2022, the Plastic Pollution Prevention and Packaging Producer Responsibility Act (SB 54) was signed by the governor of California and shifts responsibility for packaging waste from consumers to the plastics industry, with $5 billion USD to be raised from the plastics industry over a period of 10 years. By 2032, SB 54 requires that [18, 19]:

1) Single-use plastic packaging and food-ware be reduced by 25%;
2) Single-use packaging and food-ware be recycled at a rate of 65%; and
3) All (100%) single-use packaging and food-ware must be recyclable or compostable.

11.3.3 Recycled Content

Some states have laws requiring recycled content in packages (or occasionally other products). These likely grew out of the success of mandatory recycled content laws for newspapers. The most significant recycled content laws are in California and Oregon.

11.3.3.1 Oregon

Oregon's "Rigid Plastic Container Law" was passed in 1991, and came into effect in 1995 [20]. It requires that rigid plastic containers in sizes of 8 ounces to 5 gallons comply with the law through use of 25% postconsumer recycled content, through achieving a recycling rate of 25% (either statewide for all plastic containers or for particular types of containers), through being reusable or refillable five times or more (and it has to really happen; it can't just be theoretical), or through a source reduction option (which is good only for five years, and then more source reduction has to take place). It does not apply to food, cosmetics, medical, or certain other types of products (which obtain exemptions).

After passage of the law, the plastics industry invested heavily in efforts to increase plastics recycling in Oregon, funding a plastics recycling facility, and helping communities to include plastic bottles in curbside collection programs. It should also be noted that Oregon is a deposit state, so they started with a reasonably healthy base level of plastics recycling. As a result, by the time the law went into effect, the recycling rate in the state was over 25%, so all rigid plastic containers were automatically in compliance.

Oregon forecasted the recycling rate for compliance purposes by December 31 of each year, for the next year so that manufacturers would know what to expect. In December 2006, the Oregon DEQ (Department of Environmental Quality) announced that the 2004 rate was 24.0%, the 2005 rate was 24.3%, and for the first time issued a projected rate that was below 25%. However, the rate then began to increase sharply, and grew to about 30% in 2007. The Oregon DEQ attributed the increase to many curbside recycling programs beginning to collect recyclables in large wheeled carts rather than small bins, and accepting various other plastic containers in addition to bottles. It appears that the movement toward product stewardship in Oregon [21] may have resulted in decreased attention to plastics recycling, as no reports on rigid plastic recycling appear to have been issued since December 2008. However, there have been significant changes to Oregon's bottle bill, including a 2018 increase in the deposit to 10 cents per container (see https://www.oregon.gov/deq/recycling/Pages/Bottle-Bill.aspx) [22], which has had a significant effect on increasing redemption rates for covered containers.

Oregon also has a recycled content requirement for glass bottles. It requires that glass food, drink, or beverage containers that are manufactured in Oregon or within 750 miles of the state border and sold to packagers in Oregon contain at least 50% postconsumer recycled glass in the manufacturing process. There is an interesting provision in the law that provides that no enforcement action will be taken if the "combined amount of recycled glass generated in Oregon for secondary end uses" meets the 50% minimum percentage requirement. Exemptions are also allowed if glass cullet meeting specifications is not available (https://www.oregonlaws.org/ors/459A.550.html) [23].

11.3.3.2 California

California passed a law very similar to Oregon's law on recycled content for rigid plastic containers, at about the same time. In California, the 25% recycling rate for all plastic containers has not consistently been met. Furthermore, the state complicated the matter by determining the recycling rate after the fact, rather than forecasting the rate as Oregon does. For various reasons, California in 2004 passed a revision to the Rigid Plastic Packaging Container (RPPC) law that removed the option of complying through a statewide plastic recycling rate. Now, regulated containers must be made from at least 25% postconsumer resin, be source-reduced by 10%, be reused or refilled at least five times, or meet a brand-specific or container type-specific recycling rate of 45% [24]. The requirement can be met by "averaging" the postconsumer content, source reduction, or by refilling and reusing packaging across a company's products, or within sub categories. There are also rules on what must be included and excluded from the average. It should be noted that unlike Oregon's law, beverage containers are exempt in California. More information can be found at http://www.calrecycle.ca.gov/plastics/rppc/.

California, like Oregon, has mandatory recycled content requirements for glass containers. It applies to glass manufactured for sale in the state and requires 35% postconsumer recycled content. Originally, this was scheduled to go up to 65%, but that requirement was dropped. It also contains an interesting provision that if at least half of the postconsumer cullet used by a manufacturer is mixed color, the total requirement drops from 35% to 25%.

11.3.4 Plastic Bags

Plastic carry-out bags have been a hot topic in recent years. Some countries have banned them altogether; some have imposed fines or taxes on their use; some have restricted use to bags above a certain thickness, bags that are considered degradable, etc. A good discussion of U.S. plastic and paper bag legislation can be found at Ref. 25 There is a partial listing of such laws with links at Ref. 26 and there are also a number of other sources that provide similar information.

For information about other countries, sources include https://www.reusethisbag.com/reusable-baginfographics/plastic-bag-bans-world.php among others. There's also an interactive map about bag bans at http://inhabitat.com/track-the-single-use-plastic-bag-ban-movement-with-this-interactive-map/.

11.3.5 Single-Use Plastics

Bans on plastic bags are now spreading to more general restrictions on "single-use plastics," particularly driven by concerns about the proliferation of plastics in the ocean and microplastics in the environment [27, 28].

A discussion of the growing movement for bans of such materials can be found, among other sources, at https://theconversation.com/why-stop-at-plastic-bags-and-straws-the-case-for-a-global-treaty-banning-most-single-use-plastics-109857.

A story about the ban in Canada can be found at https://www.pbs.org/newshour/science/canada-wants-to-ban-single-use-plastics-would-it-make-a-difference [29]. It appears that there is not uniformity on the definitions used – in particular, if a certain plastic package has a high recycling rate, is it or is it not considered a single-use plastic and subject to such bans?

In response to such efforts, many companies have announced their own efforts to cut down on or eliminate "single-use plastics" for their products. There are examples discussed, if you wish to look at them, at https://www.rd.com/culture/companies-getting-rid-plastic/ [30], https://ourplnt.com/strategies-businesses-taking-reduce-single-use-plastic/#axzz63rbsgEMR [31], and many other sites.

11.4 Green Design

For decades, the purpose of packaging design has been to reduce costs and meet performance requirements. This was done through many approaches such as down-gauging (lightweight), multilayers and multimaterials, and so on. However, those days are in the past. **The time has come to decide where environmental sustainability** [31] **cannot only be brought into the equation but also be at the heart of future packaging design.** The Sustainable Packaging Coalition has provided criteria and attributes for sustainable packaging [32, 33]. Others have defined it differently. For example, according to Ref. 34, **"sustainable packaging"** is a package that does not create harmful products throughout its life cycle (before, during, and after use), is waste-free, recyclable/renewable/biodegradable, is consumer-safe, provides protection and shelf-stability to the products/goods being packaged, and is economically viable.

If a package is environmentally nonsustainable [35], it will likely be phased out, regardless of its economics and performance. It is strongly recommended that designing new packaging must ensure that it meets the **criteria in the order of priorities: environmental sustainability > performance > cost economics.** Performance and estimate cost are easy to determine, but for environmental sustainability, one needs to conduct life cycle assessment (LCA).

11.4.1 Problematic Materials and Their Alternatives

Certain packaging materials or formats pose recycling challenges, either due to their difficulty in recycling, interference with the recycling process or their toxicity. Examples include polyvinyl chloride (PVC), polyvinylidene chloride (PVDC), per- and polyfluoroalkyl substances (PFAS), and Foams. Alternatives are now emerging to address these challenges, as detailed in Ref. [36].

11.4.1.1 U.S. Plastics Pact

The U.S. Plastics Pact was established in 2020 as part of the Ellen MacArthur Foundation's global Plastics Pact collaborative initiative that is led by The Recycling Partnership and the World Wildlife Fund, which aims to work with stakeholders across the plastics sector to rethink the way plastics are designed, utilized, reused, and to help develop a circular plastics economy in the United States [37, 38].

11.4.1.2 Three Key Targets of the U.S. Plastics Pact

- Identify and eliminate all problematic plastics packaging materials by 2025.
- 100% of plastic packaging will be reusable, recyclable, or compostable by 2025 [38].
- Achieve an average of 30% recycled content or responsibly sourced biobased content in plastic packaging by 2025 [38].

11.4.1.3 The Do's of Recyclable and Compostable Packaging

The Association of Plastic Recyclers (APR) provides an informative set of guidelines around recyclable packages [39]. In addition, some guidelines on Circular Packaging are also reported by The Recycling Partnership [40]. For more guidelines, please refer to Ref. 34.

11.5 The Path Forward for Packaging Sustainability

- Better design and materials that enable lightweight packaging without compromising the package's essential function.
- Find alternatives to existing multilayer packaging, as it is challenging to recycle.
- Strengthen waste packaging collection.
- Utilize automated sorting processes at material recovery facilities (MRFs).
- Develop paper and metal packaging as replacements for plastics where feasible.
- Create new plastics that are biodegradable in soil, marine environments, and freshwater.

For more details, please refer to the "Decision table for green packaging design" discussed in Ref. 34.

11.5.1 Emerging Packaging Trends and Technology

The importance of sustainability has been gaining increasing recognition in the packaging sector in recent years, and there are a number of trends and new technologies that are currently emerging and/or are likely to emerge in the near future.

A shift toward new barrier materials is an important future direction, with materials such as nanoclays, cellulose nano- or microcrystals, and graphene oxides being promising alternatives to conventional barrier materials. The use of polymers such as polyethylene naphthalate (PEN) and polyglycolic acids as new barrier materials also have significant potential [34].

Chemical recycling [41] and other advanced recycling strategies hold significant promise in the packaging and recycling sectors. A key challenge for chemical recycling will be to find ways to minimize the energy costs associated with this process. Although mechanical recycling will maintain an important role as well, both chemical recycling (where the plastics are broken down into their monomers for use as feedstocks to regenerate polymers or to synthesize other materials) and mechanical recycling can have complementary roles.

The solvent-targeted recovery and precipitation (STRAP) approach [42] mentioned in one of our case studies which is being brought to larger-scale production also holds great promise as an emerging technology.

The simplification of packaging systems is also highly desirable, and efforts are likely to expand in this area. This could include the use of a single material to form a package, rather than different materials for the package itself (such as a bottle), cap or lid, and label. Using a single material can greatly facilitate sorting efforts during recycling processes. With regard to sorting, artificial intelligence (AI) is gaining an important role. While much of the AI-facilitated sorting will be occurring at MRFs, it can also be employed at waste receptables. One such example is "Oscar" sort stations, which have been developed by the company Intuitive AI and been introduced in 10 federal government buildings in Canada's National Capital region [43]. These systems look like typical waste collection systems with different bins to collect different types of waste, such as recyclable, compostable waste, and those destined for landfills, except that it has an interactive screen and a scanner which encourages people to put their waste into the correct bin. This is done by the individual holding the item up to the screen with the scanner, and it then indicates which bin should be used for that item. Due to its use of AI, these Oscar stations are able to learn more about which bins should be used the more it is used. It can also be programmed according to the recycling stream being used at a given location so that it can provide advice that is correct for that location. This feature is useful in Canada's National Capital region, part of which is in Ontario and part of which is in Quebec, as both provinces use different recycling streams with their own sorting protocols. Although not currently implemented in Canada's federal buildings, these stations were designed so that they can reward individuals who correctly dispose of their waste with a QR code that can offer discounts for products. These systems can help increase recycling and composting rates, while also helping to ease sorting requirements at municipal facilities.

The adoption of biobased packaging materials is another important trend, with a shift away from petroleum-based plastics. As was discussed in one of the case studies, one such example of this is the production of bio-based PE and bio-based ethylene vinyl acetate (EVA) by Braskem [44] as well as the production of bio-based PET bottles by Coca-Cola [45].

Another important trend is a shift toward lighter weight packaging materials for packaging, which can help minimize transportation costs as well as materials used for production. In some cases, this may involve a shift toward lighter materials such as aluminum or paper. For example, Aldi and Packamama have very recently introduced flat wine bottles that were produced entirely from recycled PET [46]. These bottles are significantly lighter than glass bottles, and their shape is designed so that twice as many bottles can fit into a standard wine case. In other cases, packaging weight loss is being achieved through efforts such as downgauging or other manufacturing approaches to develop lighter packaging formats, even with heavier materials such as glass [47].

Emerging trends in the use of paper as sustainable packaging materials are also hot favorites. Approximately, 32% of paper goes unrecycled each year, due to heavy contamination, wet-strength or laminated/coated features. The most unrecycled paper include paper that is coated with plastic or aluminum foil, paper used in cash registers, automated teller machines (ATMs), and fax machines, wax-coated paper, tissues and paper towels (due to contamination with grease, food, or personal waste), sticky notes, paper with food residue, and wet-strength papers. (Papers that have high resistance to rupture when wet, such as those used in some types of beverage cartons.) One solution to this issue is the use of enzymatic repulping to facilitate the removal of coatings and organic materials from paper. This occurs in two main ways: first, enzymes degrade the coating materials; second, enzymes break down the thin layer around the fiber, thereby freeing the coating from the fiber. Among other trends, the design of coatings for paper that do not hinder the recycling process is also gaining attention.

There are a number of fronts through which packaging sustainability is being advanced. In many cases, efforts toward these emerging technologies involve the cooperation between manufacturers, academia, government, policy makers. Not least of these drivers is the public, as pressure from society can lead toward the crafting of new policies, while public preferences toward sustainable products provide strong motivation for innovation to be pursued by manufacturers, food retailers, and other stakeholders in the packaging sector.

Case Study: Avery Dennison's BOPP Films

There are many things that need to be accounted for in the packaging sustainability equation, and one important consideration is the labels for packaged items. They have a critical role in conveying important information to consumers, while also providing a way for manufacturers and retailers to ensure that their products stand out from the competition. However, labels can also present their own sustainability challenges. The use of paper labels can impede the recycling of polyethylene terephthalate (PET) and high-density polyethylene (HDPE) rigid bottles, clamshell containers, and other types of plastic packaging. Directly printing onto packaging can be an option, but it can lead to issues such as contamination after recycling of the products by the inks used. In response to this issue, there has been a shift toward biaxially oriented polypropylene (BOPP) labels [48].

In 2020, Avery Dennison's clear BOPP films received certification with the APR Critical Guidance Test Protocol for HDPE recycling [49, 50]. This protocol is performed at a laboratory scale and is used to determine whether or not new packaging technologies are compatible with recycling systems. With this certification, these BOPP labels can remain on HDPE packaging without adversely affecting the recycling of these plastics. Consequently, the recycling of HDPE containers containing these labels becomes much simpler than would be the case if they had to be removed, and a wider range of HDPE packaging thus becomes available for recycling. Avery Dennison was the first to receive this certification for BOPP labels on HDPE plastics. Prior to receiving this certification, Avery Dennison had also received similar certification for their BOPP films on PET containers (Figure 11.1).

Figure 11.1 Photograph of a container with a BOPP label with CleanFlake technology. *Source:* Image used with permission from Avery Dennison.

11.6 Concluding Remarks

Significant advances have been made toward promoting the use and development of sustainable packaging. While some of this has been achieved through researchers and producers on their own, it is evident that policy has had an important role in promoting packaging sustainability. As awareness of environmental issues continues to grow, it is likely that the public will urge their local, state, and national governments to develop stronger and wider-reaching environmental regulations applying to packaging. As a result, environmental policies will continue to be seen as an important tool for governments and organizations to achieve sustainability goals. In many cases, regulations that are adopted in one jurisdiction eventually become adopted in other jurisdictions, as the public and policy-makers become aware of those policies. For producers, it is vital to be aware of upcoming policies so that they can adapt to a highly competitive landscape. In the case of multinational producers, packaging technologies that are mandated in one jurisdiction are sometimes also used in other jurisdictions where they are not required, due to production considerations. As a result, some policies can effectively have global reach, even if they are not universally adopted. In addition, employing the concept of green design helps to promote sustainability, as the packaging can be developed with its EoL scenarios in mind, thus facilitating its reuse and recycling, while also ensuring that the packaging is biodegradable. Addressing packaging waste issues can be a daunting task, but there are a number of effective tools and strategies to help meet sustainability targets.

Problem Set for Chapter 11

1 Which of the following is correct about Source Reduction?
 A Reduction of package weight.
 B Elimination of toxic materials from packaging.
 C Making thinner straws to reduce material used.
 D All of the above.

2 Which of the following is *not* a problematic material in packaging?
 A PVC.
 B PFAS.
 C Polystyrene foam.
 D Polypropylene.

3 Which of the following fits the design for recycling criteria?
  A Use of problematic plastic materials.
  B Use multilayer materials.
  C Use of niche materials.
  D None of the above.

4 Which of the following is *not* an example of Green Packaging design?
  A Reduce the weight of the package.
  B Use of monomaterials instead of multimaterial packages.
  C Use of recycled content in the package.
  D Design packaging that does not meet criteria for market performance.

5 Which of the following is true about the U.S. Plastics Pact?
  A All plastic packages will be recyclable by 2025.
  B Undertake steps to ensure 50% of plastic packages are recycled by 2025.
  C Eliminate all problematic plastic packaging by 2025.
  D All of the above.

6 What is/are the objective(s) of the Maine EPR?
  A It bounds producers to pay for the cost of recycling their wasteful packaging.
  B It bounds consumers to pay for the recycling.
  C It bounds municipalities to pay for the recycling.
  D All of the above.

7 Define EPR and explain how it functions.

8 What is the U.S. Plastics Pact, and what are its primary goals?

References

1 U.S. Food & Drug Administration. (2023). The 1938 Food, Drug, and Cosmetic Act. https://www.fda.gov/about-fda/histories-product-regulation/1938-food-drug-and-cosmetic-act (accessed 23 November 2023).
2 U.S. Food & Drug Administration. (2023). Food ingredient & packaging terms. https://www.fda.gov/food/food-ingredients-packaging/food-ingredient-packaging-terms#:~:text=Food%20Additive%20%2D%20A%20food%20additive,substance%20intended%20for%20use%20in (accessed 23 November 2023).

3 European Environment Agency. (2023). Directive 94/62/EC on packaging and packaging waste. https://www.eea.europa.eu/policy-documents/directive-94-62-ec-on (accessed 23 November 2023).
4 European Commission. (2023). Packaging waste. https://environment.ec.europa.eu/topics/waste-and-recycling/packaging-waste_en (accessed 23 November 2023).
5 Directive (EU) 2018/852 of the European Parliament and of the Council of 30 May 2018 amending Directive 94/62/EC on packaging and packaging waste (Text with EEA relevance), EUR-Lex, Publications Office of the EU. https://eur-lex.europa.eu/eli/dir/2018/852/oj (accessed 2 February 2024).
6 The Compliance People. (2023). Directive (EU) 2018/852 Amending Directive 94/62/EC on Packaging and Packaging Waste. https://thecompliancepeople.co.uk/updates/legal/directive-eu-2018-852-amending-directive-94-62-ec-on-packaging-and-packaging-waste/ (accessed 24 November 2023).
7 Packaging and packaging waste, EUR-LEX, Publications Office of the EU. https://eur-lex.europa.eu/EN/legal-content/summary/packaging-and-packaging-waste.html (accessed 2 February 2024).
8 EXECUTIVE SUMMARY OF THE IMPACT ASSESSMENT REPORT accompanying the document Proposal for a Regulation of the European Parliament and the Council on packaging and packaging waste, amending Regulation (EU) 2019/1020, and repealing Directive 94/62/EC, European Commission, Brussels. (2022). https://eur-lex.europa.eu/legal-content/EN/TXT/PDF/?uri=CELEX:52022SC0385 (accessed 27 November 2023).
9 Buckby, D. (2023). Will the EU's Flagship Packaging Legislation Be the Largest Driver of Plastic Waste Reduction? Forbes. https://www.forbes.com/sites/woodmackenzie/2023/07/14/will-the-eus-flagship-packaging-legislation-be-the-largest-driver-of-plastic-waste-reduction/?sh=321c8a5a6818 (accessed 14 July 2023).
10 Northwest Product Stewardship Council. (2023). What is product stewardship? https://productstewardship.net/about/what-product-stewardship (accessed 27 November 2023).
11 Title 38: WATERS AND NAVIGATION, Chapter 24: SOLID WASTE MANAGEMENT AND RECYCLING, Subchapter 3: WASTE REDUCTION AND RECYCLING, Maine Legislature, Maine Revised Statutes. https://legislature.maine.gov/statutes/38/title38sec2146.html (accessed 27 November 2023).
12 Cradle-to-grave is the full life cycle assessment from resource extraction ('cradle') to the use phase and disposal phase ('grave'). Cradle-to-cradle is a specific kind of cradle-to-grave assessment, where the end-of-life disposal step for the product is a recycling process. Cradle-to-gate is an assessment of a partial product life cycle from resource extraction (cradle) to the factory gate (i.e., before it is transported to the consumer).

13 Lindquist, T. (2023). The inventor of EPR reflects on 33 years of failure and future prospects for recovery. Packaging Insights. https://www.packaginginsights.com/news/thomas-lindquist-the-inventor-of-epr-reflects-on-33-years-of-failure-and-future-prospects-for-recovery.html (accessed 27 November 2023).

14 Botham, D. (2022). Which country recycled the most in 2021? RecyclingBins.co.uk. https://www.recyclingbins.co.uk/blog/which-country-recycled-the-most-in-2021/ (accessed 27 November 2023).

15 Maine Department of Environmental Protection. (2023). Extended producer responsibility program for packaging. https://www.maine.gov/dep/waste/recycle/epr.html (accessed 27 November 2023).

16 Product Stewardship Institute. (2023). PSI is a policy advocate and consulting nonprofit that powers the emerging circular economy. https://productstewardship.us/?State_EPR_Laws_Map (accessed 27 November 2023).

17 State of California. (2023). Extended producer responsibility (EPR). http://calrecycle.ca.gov/epr/ (accessed 27 November 2023).

18 State of California. (2023). Plastic Pollution Prevention and Packaging Producer Responsibility Act SB 54. https://calrecycle.ca.gov/packaging/packaging-epr/ (accessed 27 November 2023).

19 Gartner, R. and Kaufmann, O. (2023). A Brief Overview of California's Plastic Pollution Prevention and Packaging Producer Responsibility Act SB 54, McDermott Will & Emery. https://www.mwe.com/insights/a-brief-overview-of-californias-sb-54/ (accessed 27 November 2023).

20 Oregon Department of Environmental Quality. (2023). Rigid plastic container recycling in Oregon. https://www.oregon.gov/deq/recycling/pages/rigid-plastic-containers.aspx (accessed 27 November 2023).

21 Oregon Department of Environmental Quality. (2023). Product stewardship https://www.oregon.gov/deq/mm/production/Pages/Product-Stewardship.aspx (accessed 27 November 2023).

22 Oregon Department of Environmental Quality. (2023). Oregon's evolving bottle bill. https://www.oregon.gov/deq/recycling/Pages/Bottle-Bill.aspx (accessed 3 February 2024).

23 ORS 459A.550 Report on use of new and recycled glass. OregonLaws. https://oregon.public.law/statutes/ors_459A.550.html (accessed 3 February 2024).

24 CalRecycle(2024). Rigid Plastic Container (RPPC) Compliance Options. State of California, https://calrecycle.ca.gov/Plastics/RPPC/Enforcement/Compliance/ (accessed 20 November 2024).

25 Wang, Q., Tweedy, A., and Wang, H.G. (2022). Reducing plastic waste through legislative interventions in the United States: development, obstacles, potentials, and challenges. *Sustain. Horizons* 2: 100013. https://doi.org/10.1016/j.horiz.2022.100013.

26 PlasticBagLaws.Org. California Local Ordinances & Documents for Reference. https://www.plasticbaglaws.org/legislation/state-laws/ (accessed 27 November 2023).

27 UN Environment Programme. (2023). Microplastics: the long legacy left behind by plastic pollution. https://www.unep.org/news-and-stories/story/microplastics-long-legacy-left-behind-plastic-pollution (accessed 27 November 2023).

28 da Costa, J.P., Mouneyrac, C., Costa, M. et al. (2020). The role of legislation, regulatory initiatives and guidelines on the control of plastic pollution. *Front. Environ. Sci.* 8: https://doi.org/10.3389/fenvs.2020.00104.

29 Government of Canada. (2022). Canada's ban on certain harmful single-use plastics starts to take effect this month. https://www.canada.ca/en/environment-climate-change/news/2022/12/change-is-here-canadas-ban-on-certain-harmful-single-use-plastics-starts-to-take-effect-this-month.html (accessed 27 November 2023).

30 West-Rosenthal, L.B. (2022). 22 big companies that are getting rid of plastic for good. Reader's Digest. https://www.rd.com/article/companies-getting-rid-plastic/ (accessed 3 February 2024).

31 An environmentally sustainable package does not create harmful products throughout its life cycle (before, during, and after use), is waste-free, and is recyclable/renewable/biodegradable.

32 Definition of sustainable packaging. Sustainable Packaging Coalition, A Project of GreenBlue. (2011). https://kidv.nl/media/rapportages/definition_for_sustainable_packaging.pdf?1.2.1 (accessed 3 February 2024).

33 Kachook, O. (2023). Definition of sustainable packaging: 5 principles of sustainable packaging. Sustainable Packaging Coalition, A Project of GreenBlue. https://sustainablepackaging.org/wp-content/uploads/2024/01/SPC_Definition-of-Sust-Packaging_Landscape.pdf (accessed 3 February 2024).

34 Rabnawaz, M. (2023). *Sustainable Packaging Materials: Winning Solutions*. ASIN: B0BY7VYFX5.

35 U.S. Plastics Pact. (2020). U.S. Plastics Pact launches to ignite change toward circular economy for plastic., https://usplasticspact.org/launch-august2020/ (accessed 3 February 2024).

36 Fogg, J. (2020). Exploring new options in more sustainable packaging formulations, paper, film & foil converter. https://www.pffc-online.com/news/16146-exploring-new-options-in-more-sustainable-packaging-formulations (accessed 22 March 2024).

37 Ellen MacArthur Foundation. (2024). The U.S. Plastics Pact. https://www.ellenmacarthurfoundation.org/us-plastics-pact (accessed 3 February 2024).

38 U.S. Plastics Pact. (2023). U.S. Pact 2021 Annual Report. https://usplasticspact.org/2021-report/ (accessed 28 November 2023).

39 The Association of Plastic Recyclers. (2023). APR Design® Guide for Plastics Recyclability. https://plasticsrecycling.org/apr-design-guide (accessed 28 November 2023).

40 The Recycling Partnership. (2023). Take action to improve recyclability. https://recyclingpartnership.org/circular-packaging-assessment/ (accessed 28 November 2023).

41 Shaker, M., Kumar, V., Saffron, C.M., and Rabnawaz, M. (2023). Revolutionizing plastics chemical recycling with table salt. *Adv. Sustain. Syst.* 8: 2300306. https://doi.org/10.1002/adsu.202300306.

42 Walker, T.W., Frelka, N., Shen, Z. et al. (2020). Recycling of multilayer plastic packaging materials by solvent-targeted recovery and precipitation. *Sci. Adv.* 6 (47): eaba7599. https://doi.org/10.1126/sciadv.aba7599.

43 Thompson, E. (2024). Meet Oscar, the Canadian-built, AI-driven recycling system helping governments go green. CBC. https://www.cbc.ca/news/politics/artificial-intelligence-recycling-government-1.7069063 (accessed 27 March 2024).

44 Braskem to invest US$61 million to increase biopolymer production. Bioplastics Magazine. (2021). https://www.bioplasticsmagazine.com/en/news/meldungen/20210225-Braskem-invests US 61 million to increase-biopolymer-production.php (accessed 25 March 2024).

45 Packaging World. (2022). Coca-Cola, suntory cross the 100% plant-based bottle finish line. https://www.packworld.com/sustainable-packaging/article/22171823/cocacola-suntory-debut-100-plantbased-bottles (accessed 25 March 2024).

46 Packaging Europe. (2024). Flat, lightweight rPET wine bottles revealed by Aldi and Packamama. https://packagingeurope.com/news/flat-lightweight-rpet-wine-bottles-revealed-by-aldi-and-packamama/11110.article (accessed 25 March 2024).

47 SGD Pharma. (2024). SGD Pharma unveils NOVA lightweight glass bottle packaging innovation, addressing the commitment to environmental responsibility from cosmetic and beauty brands Paris. https://www.sgd-pharma.com/sgd-pharma-unveils-nova-lightweight-glass-bottle-packaging-innovation-addressing-commitment (accessed 26 March 2024).

48 Cooper, T.A. (2020). *Ten-Year Forecast of Disruptive Technologies for Recycled Plastic Packaging to 2030*. Smithers Information Ltd.

49 Plastics Today. (2020). A packaging first: BOPP film labels are HDPE recyclable. https://www.plasticstoday.com/packaging/a-packaging-first-bopp-film-labels-are-hdpe-recyclable (accessed 23 March 2024).

50 Avery Dennison Corporation. (2020). BOPP films recognized for HDPE recycling. https://cwp-test-origin.averydennison.com/content/dam/averydennison/lpm-responsive/na/doc/product-overviews/product/Film/rigid-film/po-bopp-hdpe-rigid-films.pdf (accessed 30 May 2024).

Further Reading

American Forest & Paper Association. (2024). What is Extended Producer Responsibility. https://www.afandpa.org/news/2024/what-extended-producer-responsibility (accessed 15 January 2025).

Rabnawaz, M., Sustainable Packaging Materials: Winning Solutions, 2023, ISBN: 9781569901625

Read EPA's description of extended product responsibility (product stewardship). http://archive.epa.gov/wastes/conserve/tools/stewardship/web/pdf/eprbrochure.pdf (an archived document).

Read the Product Stewardship Institute's statement about definitions and principles of product stewardship and EPR. http://www.productstewardship.us/?page=Definitions and http://www.productstewardship.us/?page=Principles_of_EPR.

Appendix 1

Solutions to Problem Sets

A.1 Problem Set for Chapter 2

1 Why do recycled plastics have weaker mechanical properties compared to virgin plastics?
Answer: Recycled plastics have lower mechanical properties compared to virgin plastics due to chain scission or degradation, which results in a decrease in the molecular weight (Mw). For mixed recycled plastics, the poor compatibility between different polymers also contributes to diminished mechanical properties.

2 How can we determine the thermal degradation of plastics?
A X-ray studies.
B Thermogravimetric analysis.
C Microscopic methods (e.g., SEM).
D None of the above.
Answer: b

3 Which polymer will thermally oxidize faster among the following?
A Linear PE.
B Lightly branched PE.
C Densely branched PE.
D All degrade at the same rate.
Answer: c

Degradation, Stabilization, and Recycling of Packaging Materials, First Edition.
Muhammad Rabnawaz, Susan E.M. Selke, and Ian Wyman.

4 What are the two potential causes of environmental stress cracking (ESC) in plastics?

Answer: Selective absorption of the agent in localized regions, such as in the amorphous regions of semicrystalline polymers, results in plasticization, therefore lowering the glass transition temperature (T_g). As a result of this, softening and weakening occur in these regions, consequently leading to localized failure. Another reason is the selective solvation of the polymer by a packaged product in localized areas, such as low molecular weight fractions and oligomers, leading to a weak point on the surface, perhaps amplified by stress-concentrating surface defects (for example, scratches), leading to initial failure.

5 Based on figure below, what is the impact of CB-D on polymer stability?

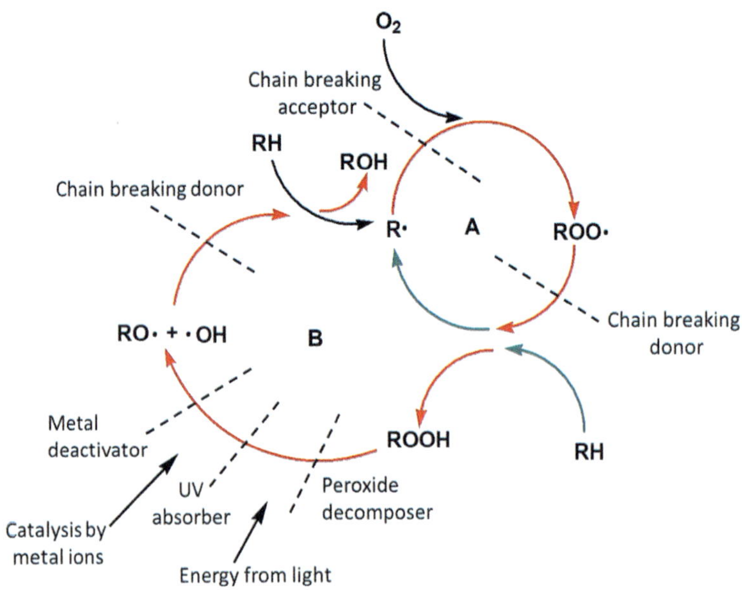

Answer: In the absence of CB-D, one reactive radical and one precursor for reactive radical (ROOH) form. CB-D provides a hydrogen atom to the ROO radical, inhibiting the ROO radical from abstracting hydrogen from another polymer chain. Meanwhile, CB-D, after proton loss, forms a less reactive radical (stable radical) that cannot abstract hydrogen from the polymer to make a new radical. Thus, in the presence of CB-D, only one ROOH forms per ROO radical.

A.2 Problem Set for Chapter 3

1 In the case of wood, what are the factors that affect weathering?
A Moisture.
B Temperature.
C Light.
D All of the above.
Answer: d

A.3 Problem Set for Chapter 4

1 You have two fillers available: A and B. Filler B is mildly basic while A is acidic. Which filler will you use for the paper to obtain a more stable paper, considering all other functions are the same?
A Filler A.
B Filler B.
C Either one.
Answer: b (Filler B)

2 What strategies are used to prevent biological degradation in paper?
A Slimicide.
B Copper salt.
C Chlorinated phenols.
D All of the above.
Answer: d

3 How can the wet strength of paper be improved?
Answer: The wet strength of paper can be improved by a variety of approaches. For example, a cross-linked resin network at the fiber interface reduces fiber swelling. Another approach is using a water-insoluble adhesive between the fibers. A higher degree of interatomic attraction (hydrogen bonding) can also help improve wet strength.

4 What are the advantages and disadvantages of using a polylactic acid (PLA) laminated/coated paper cup compared to a low-density polyethylene (LDPE) laminated/coated cup?
Answer: PLA-coated paper is compostable and biobased compared to LDPE-coated paper. However, both PLA- and LDPA-coated paper are difficult to recycle.

5 If the dry tensile strength of paper is 200 lbs/in and it decreases to 10 lbs/in when it becomes fully saturated with water, would this paper be suitable for wrapping wet materials like those used in butcher wrap? Why or why not?
Answer: This paper has a wet-strength of 5%, which is too low to be used for any wet-strength applications.

6 What is the advantage of polyamide-polyamine-epichlorohydrin (PPE) resin over urea-formaldehyde resin in paper wet strength applications?
- **A** Use of an acidic medium for curing, thus providing better stability in an acidic medium.
- **B** Use of neutral or alkaline conditions for curing, thus providing better stability in an acidic medium.
- **C** Both a and b.
- **D** None of the above.

Answer: b

7 Why do some paper substrates turn yellow faster than others?
- **A** Because they contain cellulose.
- **B** Because they contain lignin.
- **C** Because they contain hemicellulose.
- **D** Because they contain some inorganic salts.

Answer: b

8 Discuss the end-of-life (EoL) solutions for a wet-strength paper.
Answer: Wet-strength paper is difficult to recycle, and they thus end up in landfills. However, wet-strength paper can be designed to be recycled by enzymatic or chemical methods.

A.4 Problem Set for Chapter 5

1 Which of the following packaging materials cannot be further oxidized?
- **A** Glass.
- **B** Metal.
- **C** Plastics.
- **D** None of the above.

Answer: a

2 Which glass is more stable against chemical corrosion?
- **A** Soda lime.
- **B** Borosilicate.
- **C** Potash glass.
- **D** None of the above.

Answer. b

3 What is the major function of using **flux** in glass?
  - **A** To lower the melting point of the silica used to make glass and thus minimize the energy requirements.
  - **B** To increase the melting point of the silica in making the glass.
  - **C** To improve glass water resistance and durability.
  - **D** All of the above.

  Answer: a

4 Describe two methods that are used to improve the strength of glass.
  Answer: Methods such as SO_2 treatment and strengthening via the Fused Salt Mixture Approach.

A.5 Problem Set for Chapter 6

1 What is the purpose of coating tin on steel?
  - **A** To minimize/prevent steel dissolution.
  - **B** To minimize/prevent tin dissolution.
  - **C** Tin plating has no real function.
  - **D** None of the above.

  Answer: a

2 What is the role of water in corrosion?
  - **A** Carrier for corrosive species such as dissolved oxygen or hydrogen ions.
  - **B** Acts as a corrosive species/electrolyte (source of H^+ and OH^- ions and free radicals).
  - **C** It generates nitrogen gas.
  - **D** Both a and b.

  Answer: d

3 Why does aluminum (Al) corrode slower than steel despite Al reacts faster with oxygen than is the case with the steel?
  - **A** The formation of a protective Al_2O_3 layer.
  - **B** The formation of a protective Fe_2O_3 layer.
  - **C** The formation of a protective FeS layer.
  - **D** All of the above.

  Answer: a

4 A polymer upon pyrolysis gives 90% of its parent monomer; which of the following answers are correct about this polymer?
  - **A** It degrades primarily by random scission.
  - **B** It degrades primarily by depolymerization (unzipping).
  - **C** It degrades equally by both random scission and depolymerization.
  - **D** None of the above.

  Answer: b

5 If you need to pack a product with aggressive components, such as nitrate or sulfur, into a steel can, how would you approach this problem?
Answer: Applying a tin coating on the top of steel, followed by chromium oxide, and then a thick lacquer coating would help.

A.6 Problem Set for Chapter 7

1 What different categories does the EPA use to classify municipal solid waste (MSW)?
  A Durable goods, nondurable goods, containers and packaging, and other wastes.
  B Durable goods, nondurable goods, containers and packaging, and electronics.
  C MSW, hazardous waste, and non-hazardous industrial waste.
  D None of the above.
Answer: a

2 In the United States, what percent of the MSW was recycled, composted, incinerated for energy recovery, and sent to landfills/other disposals in 2015?
  A Recycled 20%, composted 8.9%, incinerated 12.8%, and landfill/other disposals 62.5%.
  B Recycled 10%, composted 8.9%, incinerated 22.8%, and landfill/other disposals 63%.
  C Recycled 40%, composted 5%, incinerated 5%, and landfill/other disposals 52.5%.
  D Recycled 25.8%, composted 8.9%, incinerated 12.8%, and landfill/other disposals 52.5%.
Answer: d

3 Why does the United States utilize landfills more extensively than the European Union for waste management?
  A The United States has more land for landfilling.
  B European Union Packaging directives discourage landfilling.
  C European Union has better infrastructure for recycling and recovery than the United States.
  D All of the above.
Answer: d

4 Which of the following answers denotes the Solid Waste Management Hierarchy **from the most preferred (left) to the least preferred (right)?**

A Source reduction and reuse, recycling and composting, energy recovery, treatment and disposal.

B Source reduction and reuse, energy recovery, recycling and composting, treatment and disposal.

C Energy recovery, source reduction and reuse, recycling and composting, treatment and disposal.

D None of the above.

Answer: a

5 The EPA divides solid waste solid wastes into:

A Three categories

B Two categories

C Five Categories

D Four Categories

Answer: a

6 Why is waste incineration common in the European Union while landfilling is more common in the US?

A More landfill space available in the United States.

B European Union regulations facilitate incineration as it classifies incineration as recovery.

C Landfilling is cheaper than incineration.

D All of the above.

Answer: d

A.7 Problem Set for Chapter 8

1 Why is recycling aluminum more environmentally friendly than producing new aluminum from its ore?

A Because more energy is required to extract aluminum from its ore.

B Because $CaCO_3$ is used in the aluminum extraction process that becomes converted into CO_2.

C There is no significant difference between obtaining aluminum from ore or by recycling.

D None of the above.

Answer: a

2 Discuss why using recycled glass is more carbon dioxide efficient compared to producing it from raw materials.

Answer: Recycled broken glass (cullet) has a lower melting temp and thus requires less energy to reprocess. Also, recycled glass does not undergo chemical reactions during melting, which is in contrast to the glass produced from raw materials that releases CO_2 from the chemical reaction (e.g., $CaCO_3$ is converted into CO_2 and CaO).

3 In the United States, in 2015, the most recycled packaging materials **by percentage** among the following was:

A Glass.
B Wood.
C Steel.
D Aluminum.

Answer: c

4 Which of the following answers is true about material recovery facilities (MRFs)?

A In MRFs, mixed recyclables are separated from each other.
B The existing MRFs have no problem with the separation of plastic bags from other recyclables.
C Dirty MRFs are used to separate single streams recyclable from each other.
D a and c.

Answer: d

5 Which of the following metals is not banned by Model Toxics Laws in Packaging?

A Nickel.
B Cadmium.
C Mercury.
D Lead.

Answer: a

6 Which of the following answers is true about closed-loop recycling?

A Closed-loop recycling refers to recycling something into an identical or similar article.
B Making new aluminum beverage cans out of old aluminum beverage cans is one example.
C Making fiber for carpets from recycled PET bottles is an example of closed-loop recycling.
D a and b.

Answer: d

A.8 Problem Set for Chapter 9

1 Which of the following packaging materials is compostable?
- **A** Glass
- **B** Metal
- **C** Polyethylene
- **D** Uncoated paper (and some coated paper).

Answer: d

2 Which of the following answers is true about the recycling of paper and paperboard?
- **A** Paper and paperboard are recycled via closed-loop only.
- **B** Paper and paperboard are recycled via open-loop only.
- **C** Both closed- and open-loop recycling is used for paper and paperboard recycling.
- **D** None of the above.

Answer: c

3 What is the difference between repulpable and recyclable paper?
- **A** Repulpable refers to the ability of paper to be broken down to its pulp under standard repulping conditions without ensuring its suitability for producing high-quality recycled paper.
- **B** Recyclable paper can be broken down to its pulp under standard repulping conditions, with the resulting pulp being suitable for creating high-quality recycled paper.
- **C** Both a and b are correct.
- **D** None of the above.

Answer: a

4 How is de-inking of recycled paper performed?
- **A** Flotation
- **B** Surfactant treatment
- **C** Both a and b
- **D** None of the above

Answer: c

A.9 Problem Set for Chapter 10

1 Which of the following answers is correct about open-loop and closed-loop recycling?

A Closed-loop recycling refers to recycling something into an identical or similar article, such as a new PET bottle out of the PET bottle.
B Open-loop recycling refers to recycling something into a different type of product (and often of lower value), such as PET bottle to fiber.
C Both a and b are correct.
D None of the above.

Answer: c

2 What should be the path forward for effective plastic recycling?
A Mechanical recycling only.
B Chemical recycling only.
C Both chemical and recycling are complementary and should be advanced.
D None of the above.

Answer: c

3 Examples of chemical recycling include.
A Hydrocracking of HDPE.
B Pyrolysis of HDPE.
C Melt-processing of recycled HDPE to another bottle.
D a and b.

Answer: d

4 What is the major product during plastic gasification?
A Syngas ($CO + H_2$) along with other minor impurities.
B CH_4.
C CO_2.
D $CO_2 + H_2$.

Answer: a

5 Examples of mechanical recycling include:
A PET bottle into fiber by melt-processing.
B HDPE bottle into lumber by melt-processing.
C PET bottle into another PET bottle by melt-processing.
D All of the above.

Answer: d

6 Compare pyrolysis, catalytic pyrolysis, and hydrocracking. In your answer, describe the benefits and disadvantages of each process.
Answer: Pyrolysis – plastic is heated in the absence of oxygen at high temperatures. A large number of products are formed. Energy-intensive process. **Catalytic Pyrolysis-** plastic is heated in the absence of oxygen at milder

temperatures in the presence of a catalyst. More selective products are formed via this approach. However, the catalyst is expensive and catalytic poisoning can also happen. **Hydrocracking** – This is a thermochemical conversion where plastic is converted into saturated products in the presence of hydrogen gas and catalyst. Here, catalyst poisoning, and char formation is minimized.

7 Which one is true about trends in mechanical recycling?
A Digital watermarking.
B Monomaterials instead of multilayer packaging.
C Near IR sorting.
D All of the above.
Answer: d

8 Key challenges in plastic recycling include:
A Compositional fluctuations.
B Contamination.
C Lower price of virgin materials.
D All of the above.
Answer: d

9 In terms of the CO_2 index, which plastic waste treatment method is the most climate-friendly (less CO_2 generation)?
A Mechanical recycling.
B Pyrolysis.
C Incineration with energy.
D Landfilling.
Answer: a

10 Which statement is True about the following chemical recycling processes:
A Pyrolysis is the thermal decomposition of plastics without a catalyst.
B Hydrocracking is a catalyzed thermal decomposition of plastics in the presence of a hydrogen atmosphere.
C Catalytic pyrolysis is a catalyzed thermal decomposition of plastic in the presence of a hydrogen atmosphere.
D Both a and b.
Answer: d

11 If we want to use recycled PET in food packaging, what are the available options? Ignore the cost aspect.
A Depolymerize the PET into its monomer, purify the monomer, and then repolymerize the monomer to virgin PET and then use this polymer for food packaging.

B Use PET as an inner layer between virgin PET layers to avoid direct food contact.
C Use the recycled PET directly without any cleaning efforts for food packaging.
D Both a and b.
Answer: d

12 Why is the Dissolution–Precipitation approach for plastic recycling called chemical recycling?
A Because in this process, the polymer is decomposed into small molecules.
B Because in this process, the polymer is separated by using a chemical (solvent).
C Both a and b.
D None of the above.
Answer: b

13 Why is the chemical recycling of plastic waste a challenging task?
A Costly processes.
B Energy-intensive processes.
C Not suitable for food packaging.
D Both a and b.
Answer: d

A.10 Problem Set for Chapter 11

1 Which of the following is correct about Source Reduction?
A Reduction of package weight.
B Elimination of toxic materials from packaging.
C Making thinner straws to reduce the material used.
D All of the above.
Answer: d

2 Which of the following is *not* a problematic material in packaging?
A PVC.
B PFAS.
C Polystyrene foam.
D Polypropylene.
Answer: d

3 Which of the following fits the design for recycling criteria?
A Use of problematic plastic materials.
B Use multilayer materials.
C Use of niche materials.
D None of the above.
Answer: d

4 Which of the following is *not* an example of Green Packaging design?
A Reduce the weight of the package.
B Use of monomaterials instead of multimaterial packages.
C Use of recycled content in the package.
D Design packaging that does not meet criteria for market performance.
Answer: d

5 Which of the following is true about the US Plastics Pact?
A All plastic packages will be recyclable by 2025.
B Undertake steps to ensure 50% of plastic packages are recycled by 2025.
C Eliminate all problematic plastic packaging by 2025.
D All of the above.
Answer: d

6 What is/are the objective(s) of the Maine Extended Producer Responsibility (EPR)?
A It bounds producers to pay for the cost of recycling their wasteful packaging.
B It bounds consumers to pay for the recycling.
C It bounds municipalities to pay for the recycling.
D All of the above.
Answer: a

7 Define EPR and explain how it functions.
Answer: EPR requires producers (polluters) to bear the financial responsibility for cleaning and managing their product's waste. The "Green Dot" system is one example of an EPR program. Producers pay a certain amount per ton of their waste to recover/recycle/dispose.

8 What is the US Plastics Pact, and what are its primary goals?
Answer: The US Plastics Pact brings together businesses, not-for-profit organizations, government agencies, and research institutions that work together toward a common vision of a circular economy for plastics, as outlined by the Ellen MacArthur Foundation's *New Plastics Economy Initiative*. This vision aims to ensure that plastics never become waste by eliminating the plastics we don't need, innovating to ensure that the plastics we do need are reusable, recyclable, or compostable, and circulating all the plastic items we use to keep them in the economy and out of the environment.

Index

Degradation, Stabilization, and Recycling of Packaging Materials, First Edition.
Muhammad Rabnawaz, Susan E.M. Selke, and Ian Wyman.

b

c

d

u

v

w

x

y

z

Printed and bound by CPI Group (UK) Ltd, Croydon, CR0 4YY

17/09/2026

14964330-0003